W9-BGV-465

# CONQUEST,
# TRIBUTE,
## *AND*
# TRADE

HOWARD J. ERLICHMAN

# CONQUEST, TRIBUTE, *AND* TRADE

## THE QUEST FOR PRECIOUS METALS *and the* BIRTH OF GLOBALIZATION

**Ⓟ Prometheus Books**

59 John Glenn Drive
Amherst, New York 14228–2119

Published 2010 by Prometheus Books

Cover illustration *Defeat of the Spanish Armada* by Philippe-Jacques de Loutherbourg, 1796.

Inquiries should be addressed to
Prometheus Books
59 John Glenn Drive
Amherst, New York 14228–2119
VOICE: 716–691–0133
FAX: 716–691–0137
WWW.PROMETHEUSBOOKS.COM

14  13  12  11  10      5  4  3  2  1

Library of Congress Cataloging-in-Publication Data

Erlichman, Howard J., 1953–
   Conquest, tribute, and trade : the quest for precious metals and the birth of globalization / by Howard J. Erlichman.
       p. cm.
   Includes bibliographical references and index.
   ISBN 978–1–61614–211–7 (cloth : alk. paper)
   1. Economic history—16th century. 2. Commerce—History—16th century. 3. Precious metals—History—16th century. 4. Imperialism—History—16th century. 5. Portugal—History—16th century. 6. Spain—History—16th century. 7. Netherlands—History—16th century. 8. Gold mines and mining—History—16th century. 9. Silver mines and mining—History—16th century. 10. Copper mines and mining—History—16th century. I. Title.

HC51.5.E75  2010
333.8'54209031—dc22

3 9547 00346 9140

2010020039

Printed in the United States of America on acid-free paper

# CONTENTS

# MAPS

# INTRODUCTION

Ten years into a new millennium, Western civilization is being challenged on a variety of fronts. Attacks against Western culture, free trade, and globalization are regular features of the daily news. Ancient religious animosities will not go away. Most third world nations show little sign of breaking out of their economic malaise. Americans are puzzled by the unpopularity of their armed forces, investors, and even humanitarian organizations abroad. Meanwhile, the centuries-long economic ascendancy of the Western world is being threatened increasingly by China, Japan, and India and a petroleum cartel centered in the Middle East. The Eastern powers have induced the US economy into a series of balance-of-payment deficits, the squander of capital (and human lives) in places like Korea, Vietnam, Afghanistan, and Iraq, and investment policies tilted more toward speculative securities than to "bricks and mortar." American taxpayers were stuck with a massive bailout tab in 2008–2009 to prevent an out-of-control financial sector from bringing down the entire system.

While most of these pressures are reversible, the underlying issues are related and all can be linked historically to the birth of globalization in the sixteenth century. The following work is intended to explain how and why Europe's spectacular geographical conquests in this one epoch, the turbulent sixteenth century, laid a foundation of Western dominance (and animosities) that has lasted for roughly five hundred years, included patterns of economic development that are being repeated in the twenty-first century, and generated a host of lessons that should be studied by anyone interested in the processes through which commercial empires are won and lost. I believe that readers will find some intriguing parallels, for example, between the Spanish state bankruptcies of 1557–1560, 1575–1578, and 1596–1597 and the massive government bailouts of 2008–2009.

More specifically, I have attempted to answer three interrelated economic questions that have been ignored or underanalyzed by conventional humanities-oriented historians. First, how and why the discovery, extraction, and distribution of precious metals (gold, copper, and, above all, silver) enabled three insignificant but closely related states—Portugal, Spain, and the fledgling Dutch Republic—to check the powerful Ottoman Empire, supersede the great Italian city-states, and overturn centuries of Muslim commercial domination in Africa and Asia. Second, how and why the metallic wealth of Africa, the Americas, and even Japan was squandered, destroying indigenous societies across the globe and leaving a wake of colonial subjugation that has lasted for five centuries. And third, how and why the control of precious metals allowed the Dutch Republic to supersede the Iberians and serve as a "bridge" to the later triumphs of England (and the United States).

This book is intended to serve as a very modern tale of business and economic history that just happens to be set in the sixteenth century. Having worked professionally with a host of multinational corporations, start-up ventures, and models of economic development, I essentially stumbled into the subject matter via research completed for my first book, *Camino del Norte: How a Series of Watering Holes, Fords, and Dirt Trails Evolved into Interstate 35 in Texas.* My research explained how and why a succession of Mexican silver strikes had driven the development of royal highways (*caminos reales*) north to Texas and New Mexico, but left me with an unanswered, double-barreled question—whatever happened to the metallic wealth of New Spain, and why had Mexico remained a second-tier economy for most of the last five hundred years?

There was no simple answer to this double-barreled question. In fact, further research suggested that no one had adequately explained the economic experience of colonial Mexico and no one had adequately linked this experience, and that of other unfortunate subject states, with the global trading patterns that emerged during the sixteenth century. I discovered that the Portuguese, Habsburg, and Ottoman empires all loomed heavy in the economic activities taking place in New Spain and in a host of other sixteenth-century locales. When concurrent developments in South America, Africa, South Asia, and the Far East were added to the global economic mix, I was presented with an excellent opportunity to write my second book.

The following chapters are focused on the birth of globalization. They detail the central, driving, multidimensional (and almost unbelievable) role played by a handful of mines in Africa, Central Europe, the Americas, and Japan; the Muslim spice monopoly that was shattered by new supplies of precious metals; the larger-than-life characters who discovered, financed, or exploited these resources and created the world's first multinational corporations; the boomtowns that sprang up seemingly out of nowhere; and the spectacular squander, in financial and human terms, that resulted from a seemingly endless series of military campaigns and shifted the balance of global power to (and from) a succession of would-be empires.

Surprisingly, not one of the principal sixteenth-century powers—Portugal, Spain, Ottoman Anatolia, the Dutch Republic, and China—possessed significant amounts of precious metals in their homeland. Like the ancient Phoenicians, the sixteenth-century powers captured vast quantities of precious metals through exploration, conquest, tribute, and trade. They also established state-backed multinational enterprises to manage the process. Unfortunately for them, the volume of captured metals was never sufficient to maintain the would-be empire's far-flung military obligations. In fact, the century's last-arriving global power, the Dutch Republic, served mainly as a bridge between the Iberians and the subsequent triumphs of England and the United States. The Anglo-Americans formalized east-west/north-south trading patterns (and animosities) that had already been established by earlier Europeans.

One prominent economist, John Maynard Keynes, even suggested that the control of precious metals was the single most important factor in the development of premodern empires. In *A Treatise on Money* (1930), Keynes offered the following:

It would be a fascinating task to re-write Economic History [from] its remote beginnings [to] conjecture whether the civilisations of Sumeria and Egypt drew their stimulus from the gold of Arabia and the copper of Africa ... in what degree the greatness of Athens depended on the silver mines of Laurium ... how far the dispersal by Alexander of the bank reserves of Persia [was] responsible for the outburst of economic progress in the Mediterranean basin, of which Carthage attempted and Rome ultimately succeeded to reap the fruits ... [and] if the long stagnation of the Middle Ages may not have been more surely and inevitably caused by Europe's meagre supply of monetary metals than by monasticism or Gothic frenzy.[1]

Keynes never bothered to assume his "fascinating task," and few (if any) historians have attempted to integrate his suggested course with Adam Smith's somewhat hyperbolic declaration in *The Wealth of Nations* (1776) that Columbus's discovery of America and da Gama's passage to India were the two most important events in recorded history. However, it is of great interest that the epic voyages of Columbus (1492) and da Gama (1498) occurred within a single decade, were commissioned by Iberian royalty (Manoel of Portugal, Isabella and Ferdinand of Castile), and were driven by a desire to break the Muslim-dominated spice markets of Asia—not to find new sources of precious metals. When massive deposits of minerals were inadvertently uncovered or catalyzed, the discoveries created two accidental Iberian empires and set off a chain of events that would shape the course of global history.

Portugal declined to invest in mines and smelters, but her near-control of African gold and Hungarian copper provided the "venture capital" through which age-old Muslim trading monopolies in Africa and Asia were broken. In similar fashion, Caribbean gold (1500–1519) provided Castile (and Habsburg Spain) with the "venture capital" through which the American mainlands were explored, invaded, and exploited in stepping-stone fashion. The plunder of ceremonial treasures in Mexico (1520) and Peru (1533–1534) was followed by the discovery of huge silver lodes at Cerro de Potosí,

Bolivia (1545), and Zacatecas, Mexico (1546), and then alluvial gold in upper Colombia (1550). Even larger fortunes were realized from the adoption of a new smelting process (amalgamation) at Zacatecas (1550s) and Potosí (1570s), developments that placed silver at center stage during the final decades of the sixteenth century.

While a portion of America's metallic wealth stayed in the Western hemisphere, most of the treasure was squandered on a never-ending series of Habsburg military campaigns against the Ottoman Empire, France, German Protestantism, and rebellious provinces of the Spanish Netherlands. American silver helped to separate the Protestant Dutch Republic from the Catholic Spanish Netherlands, and Japanese silver helped to catalyze Dutch interest in the increasingly overstretched Portuguese Empire in the Far East. When the smoke had cleared, the Dutch had triumphed, the state treasuries of Portugal, Spain, France, and Istanbul were nearly bankrupt, Japan was shifting to Chinese-style isolationism, Protestantism was flourishing, and Native American populations were almost completely destroyed.

However, a century of feverish exploration activity had essentially completed a map of the world, and a single nation, carved out of seven insignificant Dutch provinces, had somehow come to dominate global economic affairs. When this book concludes in 1609, with the arrival of a wayward Dutch fleet in New York harbor, Amsterdam represented a position of global economic power that could easily have been assumed by Lisbon, Madrid, Istanbul, or even Beijing. The Dutch triumph was as short-lived as that of the Iberians, but it demonstrated how a dedicated commercial enterprise could attack the established (Iberian) economic order, co-opt a century's worth of (Iberian) infrastructural investments, and exploit fresh opportunities in an increasingly interdependent global marketplace.

The story unfolds in eleven chapters. In chapter 1, "The Portuguese Head East," Jacob Fugger diversifies the family textile business into precious metals thanks to a Central European mining boom, shrewd investments, and a series of Habsburgian loan defaults. Fugger copper and silver are central to the Portuguese commercial empire launched by da Gama's second voyage to India. The fledgling empire is facilitated by the exchange of Hungarian copper and African gold for Asian pepper and spices. The subsequent conquests of Hormuz, Goa, Malacca, and the Spice Islands create a Lisbon-based multinational enterprise and a Golden Age under Manoel I.

In chapter 2, "The Spanish Head West," Isabella and Ferdinand hire Columbus to find a westerly route to "Asia" to check Portugal's pending invasion of the Indian Ocean. Modest gold strikes made on Hispaniola and Cuba launch a feverish treasure hunt in the uncharted Caribbean Rim. Depleting mineral deposits cause Genoese investors to focus on sugar and prompt an unauthorized Hernán Cortés to invade Mexico in search of new sources of gold. Cortés circumvents Diego Velásquez's political authority, plunders Moctezuma's treasures, and establishes New Spain.

In chapter 3, "Convergence with the House of Habsburg," Jacob Fugger & Nephews expand their dealings with the House of Habsburg, consolidate their business activities in Antwerp, and "elect" Charles I of Spain as Holy Roman Emperor. Huge profits are generated from European metals, Portuguese spices, and papal indulgences. Fugger capital, supplemented by American plunder and a ransom windfall from France, finance the emperor's military campaigns against France, the Ottomans, and German Protestants (and also the Magellan expedition). The imperial finances are managed skillfully by Francisco de los Cobos.

In chapter 4, "The Great American Treasure Hunts," the emperor needs new sources of American treasure to fund his next campaigns. The Pizarros find and plunder the spectacular treasures of Cajamarca and Cuzco in Inca Peru and unleash a wave of treasure hunts throughout the Americas. Charles V squanders the Inca plunder in Europe and schemes to beat the Pizarros to the geological source of the Inca treasures. Scores of would-be conquistadors penetrate the interior reaches of Paraguay, Bolivia, Colombia, and Venezuela. Coronado is hired (over Cortés's protestations) to lead an expedition into northern Mexico.

In chapter 5, "An Unlikely Bankruptcy," accidental bonanzas struck at Potosí (Bolivia), Zacatecas (Mexico), and upper Colombia are squandered in Habsburgian campaigns against France, Istanbul, and German Protestantism. Charles V confiscates the Pizarro fortune and shifts his fundraising schemes to Antwerp and the undertaxed Habsburg Netherlands. Infantry-intensive siege campaigns like Metz are wildly expensive. Mounting debts and a new revolt by German Protestants force Charles to abdicate, leaving Felipe II (and the gnomes of Antwerp) to work through the Spanish state bankruptcy of 1557–1560.

In chapter 6, "Mining Revolution in Spanish America," American mining revenues are disappointing after the surface-level deposits are excavated. Bartolomé de Medina demonstrates a revolutionary separation technique at Pachuca (Mexico) that can treat deep, low-quality silver ores for the first time. Medina's amalgamation process is adopted immediately at Zacatecas and eventually finds its way to Cerro de Potosí. Viceroy Francisco de Toledo's massive investment program at Potosí raises productivity significantly. Genuine mining industries are established in New Spain, Bolivia, and Colombia (gold) after a variety of logistical problems are resolved.

In chapter 7, "Cash Flow Squeeze in the Spanish Netherlands," Felipe is forced to deal with the Ottoman navy in the Mediterranean and a revolt by Calvinist "Sea Beggars" in the Netherlands. The Dutch rebels are outraged by Spanish tax policies and Felipe's anti-heresy laws. The Duke of Alba crushes the rebellion with his Army of Flanders, but Felipe is undermined by huge war debts, a temporary falloff in Bolivian silver receipts, and another state bankruptcy in 1575–1578. The face-off between Felipe and his Genoese creditors inadvertently allows the breakaway northern (Dutch) provinces to consolidate their gains.

In chapter 8, "Convergence in the Far East," downward pressure on spice profits in the Indian Ocean force Portuguese entrepreneurs (and Jesuits) to investigate opportunities in the Far East. The Portuguese handle the exchange of Japanese silver for Chinese gold and silks after a series of silver strikes are made in Japan. Profits from the Macao-Nagasaki trade fuel the unification of Japan under warlords Nobunaga and Hideyoshi. New Spain counters by colonizing the Philippines to trade American silver for Chinese goods. The union of the Iberian crowns supports Manila, Macao, and Nagasaki and attracts some curious visitors.

In chapter 9, "Unimaginable Wealth and Squander," record silver receipts from Bolivia are allocated to the Army of Flanders and the Invincible Armada. The Duke of Parma's reconquest of the southern Netherlands (including Antwerp) is checked by a Dutch-English alliance and the armada's defeat off Gravelines. The naval defeat is a minor blow to Spanish power but a major blow to Spanish prestige and finances. Record volumes of American bullion are overwhelmed by the costs to rebuild the armada, resume the offensive in the Netherlands, and deal with France. Another state bankruptcy is declared in 1596.

In chapter 10, "The Dutch Advance," insignificant Amsterdam becomes a commercial dynamo after the fall of Antwerp. A massive infusion of capital, talent, and business contacts is supplemented by revolutionary ship designs; a dominant position in the carriage of grain, timber, and salt; and the timely lifting of the Spanish trade embargo. Ex-Antwerpers lead Amsterdam's diversification into the "rich trades" (spices, sugar, and metals), financial services, and overseas business activities. The Dutch East India Company (VOC) is organized to wage a full-scale assault against the vulnerable Portuguese.

In chapter 11, "Capital of the World," the Dutch Republic fuses the two Iberian commercial empires into one in the midst of the Eighty Years' War with Spain. Amsterdam becomes an unrivaled commercial colossus, fueled by Spain's balance of payment deficits, huge stocks of precious metals, a dominant position in grain and other commodities, and a diversified economy. Portuguese Asia is attacked by the VOC, Dutch trading posts are established throughout the world, and Dutch freighters dominate the transatlantic trade. Henry Hudson sails into New York harbor on behalf of the VOC.

One of the many ironies of the Dutch triumph was not that it was facilitated by the Spanish state bankruptcies of 1557–1560, 1575–1578, and 1596–1597, which it was, but that the bankruptcies themselves were induced by unprecedented mineral wealth. No matter that the centerpiece of this wealth, Cerro de Potosí, was located in the remote southern altiplano of Bolivia and was connected only indirectly to the fledgling Dutch system. The shape of world history hinged on the timing and scale of Potosí's boom-bust-boom cycles between 1545 and 1600. The ebbs and flows of Potosí silver catalyzed military campaigns, induced Spanish state bankruptcies, and created global trading relationships that would not have occurred otherwise. Potosí silver also reduced the viability of the Ottoman mining industry in the Balkans, helped to split the Spanish Netherlands in two, persuaded Felipe II to invade England in 1588, and raised the scale of silver-for-silk trading activities in the Far East. These developments matter because they undermined three great empires (Portuguese, Spanish, and Ottoman) and created an unexpected fourth—the Dutch. It was the ascendant Dutch who fused the global economic achievements of the Iberians and paved the way for the subsequent (and more lasting) triumph of England (and the United States).

This book is not focused on Cerro de Potosí, but it is focused on the remarkable chronology of events that converted remote, inaccessible territories like Bolivia and the Spice Islands into critical components of the globalization process. The fact that Bolivia contained the largest silver bonanza in American history reverberated across the Atlantic and Pacific Oceans in the second half of the sixteenth century. Few cared that remote Potosí was an accidental discovery that emerged from the feverish global hunt for precious metals and trading revenues. Unexpected surprises like Potosí had become expected after Columbus and da Gama had discovered new routes to the globe's western and eastern fringes, after the general "bullion famine" of the fifteenth century had been reversed, and after insignificant, mineral-poor kingdoms like Portugal and Spain had become the first truly global powers.

# CHAPTER ONE

# THE PORTUGUESE HEAD EAST

The seeds of the Portuguese Empire were sown by a Bavarian entrepreneur in 1367. In that year, a back-country weaver named Hans Fugger traveled down the Lech River from Graben to pursue trading opportunities in Augsburg. The ancient city had been founded in 15 BCE as Augusta Vindelicorum, along a tributary of the Danube River, and was becoming an important textile center. Few Augsburgers, including the newly arrived Hans Fugger, could have predicted in a post–Black Death year like 1367 that the river-trading town would achieve her greatest fame in association with a Central European mining boom in the late fifteenth century and that the descendants of Hans Fugger would amass the single-largest private fortune of the age.[1]

In the meantime, the continuing ravages of the Black Death caused tremendous hardships and occasional opportunities during the second half of the fourteenth century. In bleak 1367, it was a rare opportunity in the weaving industry that induced Hans Fugger to journey north to Augsburg. He prospered by supplementing his own modest line of fustian, a coarse cotton cloth, with an array of textiles manufactured by other Augsburgers. The expanded line was marketed to the trade fairs at Nuremberg, Ulm, Donauworth, and Frankfurt. Hans's business skills were apparently superior to those of his competitors. He eventually gained possession of fifty looms, a few wagons, and three homes in Augsburg to complement the family homestead in Graben.[2]

At his death in 1408, Hans Fugger left an estate valued at three thousand florins and had married off his son Jacob to a daughter of the master of the Augsburg Mint. While the medieval-styled mar-riage would prove to be fortuitous for three of Jacob Fugger's eleven children, Jacob was content with his position as a master of the Augsburg Weavers Guild. He focused all of his energies on trading local fustian for cotton, silks, textiles, and spices. Most of these items had to be imported from Venice—the commercial colossus of medieval Europe but only a ten-day journey from Augsburg. Jacob Fugger created a major textile and trading business prior to his death in 1469.[3]

Despite a family connection to the Augsburg Mint, Jacob Fugger had declined to participate in the recovery that was taking hold in the Central European mining industry. A group of local rulers, metals-needy Venetians, and Bavarian entrepreneurs were attempting to reverse the so-called bullion famine that had spread through Europe during the Black Death century of depression (1350–1450) and had even worsened during the subsequent Ottoman onslaught. Annual production from the once-prolific silver mines of Central Europe had fallen to less than ten thousand marks (2.5 tons) in 1453 when Ottoman sultan Mehmed II conquered Constantinople, the east-west trading capital of Byzantium, and the Balkan silver mines of Serbia, Kosovo, and Bosnia. Mehmed's seizure of Novo Brdo (Kosovo) in 1455 deprived the Italian city-states of roughly nine tons of annual silver output that had flowed their way. Further dislocations caused by a series of Ottoman cam-paigns against Venice (by sea) and Hungary (by land) would shortly push the bullion famine to epic proportions in Western Europe. The sixty-five thousand pounds sterling minted in England in 1474 was just half of that issued in 1350. The mint output of the Netherlands was down to one-third. When currencies were tight, business activ-ities grounded to a halt.[4]

While the Ottoman advance was depressing to Christian Europe, the Venetians, and to a lesser extent the Florentines and Genoese, had the most to lose. The great Italian city-states had dominated Europe's east-west trade for nearly two centuries and needed a fresh supply of precious metals to maintain their cross-Mediterranean exchanges with the Mamluk spice merchants of Egypt and Syria. No matter that the Italians had helped to drain Europe of her metal supplies during this same period. The more encouraging news for the Italians, however, was that the European economy was finally on the mend. The general population had started to grow after the Black Death century had run its course, the textile industries of northwestern Europe were flourishing, and the upstart Portuguese were importing modest quantities of West African gold by sea. If the Portuguese were inevitably depriving the Italians (and Mamluks) of caravan gold that had sustained their economies for centuries, their efforts were at least making a contribution to Western Europe's metal supply. Precious metals were essential because European trade goods were lowly regarded in the Muslim world and Christian rulers needed money to recruit armies to meet the Ottoman threat—especially after the traumatic fall of Constantinople.

As economic conditions improved, a number of Venetian merchants decided to take action. They would revisit the depleted mining districts of Saxony, Tyrolia, and Bohemia. The same Central European mines that had supported the rise of the Italian city-states in the twelfth century had not benefited from the metallurgical innovations that Saxon emigrants had applied to the more recent strikes in Hungary, Serbia, Kosovo, and Bosnia. The opportunity had been apparent to local investors as well. In 1451, the Duke of Saxony recruited teams of Saxon engineers to extend the depths of existing mines and to improve smelting and drainage processes that had remained unchanged for centuries. The duke paid special attention to the ancient cupellation technique for separating silver from copper and lead. The current separation methods were cumbersome and too expensive to process deep-seated ores economically. If Saxon emigrants could revitalize the Balkan mines in earlier centuries, a current generation of Saxon engineers could rescue mining industries closer to home.

The initiatives sponsored by the Duke of Saxony, the Venetians, and a small group of Bavarian capitalists paid off spectacularly. Ludwig Meuting of Augsburg jump-started the process in 1456 when he advanced a loan of thirty-five thousand florins to Archduke Siegmund of Tyrolia. The key point was that the loan was secured by the right to purchase a discounted share of silver from the Schwaz mines (near Innsbruck). When the loan went unpaid, as most Habsburg loans did, Meuting was in the silver mining business. By 1479, Augsburg's Imhof, Wolff, and Welser families had gained interests in the archduke's Schneeberg mines simply by following Meuting's lead. The investments paid huge dividends. Annual silver production from Schneeberg (Saxony) and Schwaz (Tyrolia) averaged around ten tons during the 1480s and allowed Tyrolia and Saxony to mint their first silver coins in over a century. The collective efforts moved Augsburg, with a population of twenty thousand, to the center of Bavarian merchant banking.[5]

Jacob Fugger had stuck to textiles and making money. It was left to one of his sons, Jacob Junior, to make money and history. Born in Augsburg on March 6, 1459, the youngest of Jacob's eleven children was slated for a career in the Catholic Church had his father not passed away in 1469. Young Jacob was pulled into the family business. During an apprenticeship at the Fondaco dei Tedeschi, a German merchants association that had operated in Venice for nearly two hundred years, Jacob learned Venetian business practices and the art of double-entry bookkeeping. He returned to Augsburg in 1479 in time to assist his older brothers Ulrich and George diversify into a new (and very controversial) line of business— transferring papal taxes between Scandinavia and the papal treasury in Rome. The following year, Ulrich raised the family's profile still higher by securing a contract to furnish textiles and silks for the pending marriage of Maximilian, son of Holy Roman Emperor Frederick III, and the daughter (Maria) of Duke Charles of Burgundy. The textile contract was the first of the Fuggers' many dealings with the House of Habsburg, and the marriage itself would have important implications.[6]

Backed by a relationship with the House of Habsburg and his father's access to the Augsburg Mint, Jacob Fugger finally entered the mining fray in 1487. He was so anxious to work with a sophisticated Genoese partner, Antonio de Cavalli, that he even dissolved a partnership with a small-time Nuremberg-based merchant (Hans Kramer) when a greater opportunity arose. Fugger and Cavalli arranged a twenty-four-thousand-florin loan to Archduke Siegmund that was secured by unpledged mining revenues from Schwaz. When Siegmund defaulted, true to form, the Fuggers were in the mining business as well. The perpetually cash-strapped archduke continued the process by requesting a personal advance of as much as one hundred thousand florins, many times the value of the Fuggers' net worth, secured by the right to purchase the archduke's entire silver production from Schwaz at a very low price. The default clauses on the second loan would give the Fuggers a large slice of the Tyrolian silver mining industry.[7]

The emerging Fugger mining empire was bolstered by the transfer of Siegmund's Tyrolian domains to his cousin Maximilian I in 1490. The hapless son of Holy Roman Emperor Frederick III would soon be regarded as the most inept ruler in Habsburg history. The change in regimes was beneficial to the extent that Maximilian's regular defaults triggered the transfer of an increasingly large portion of Habsburg mining properties to the Fuggers and other Augsburg firms. These and other merchant bankers were building sizable fortunes from Habsburg mismanagement and the earlier investments in mining and metallurgy. A genuine silver boom was taking shape in Tyrolia and Saxony. Average annual silver production had soared to over forty-four thousand marks (over eleven tons) and much more was on the way.[8]

The increasingly sophisticated Fuggers established a formal business partnership under the name of Ulrich Fugger & Brothers in 1494. Capitalized at a modest fifty-four thousand florins, the partnership was actually misnamed. It was the twenty-five-year-old Jacob Fugger who had assumed virtual control over family affairs. While Jacob's extensive business experience in Venice, Tyrolia, and Hungary had shifted the direction of the family business from textiles to mining, the junior partner did not neglect the family's traditional enterprises. He even came up with a way to circumvent the church prohibitions against lending at interest by promoting the theses of Johannes Ech—the German theologian who argued that an interest rate of 5 percent (as a starting point) was ethically acceptable. Otherwise, interest was typically embedded within the principal of a medieval loan.[9]

Since Central Europe also contained huge deposits of copper, the Fuggers were prepared to diversify their mineral holdings. No matter that copper had always been less highly prized than silver and gold. Europe's economic recovery and Portuguese trading activities in West Africa had raised demand for the metal significantly. The Fuggers had already acquired some copper properties in Carinthia (Austria) when Jacob formed a business partnership with Johann Thurzo of Cracow. The Fugger-Thurzo partnership struck in Venice in 1494 was intended to develop underexploited copper deposits held by the Kingdom of Hungary, near Neusohl (present-day Banska Bystrica, Slovakia), as well as those controlled by Maximilian in Tyrolia. In 1496, the Fuggers managed to exchange 160,000 florins in unpaid Habsburgian debts, accumulated from the military campaigns in Italy, for a large share of Maximilian's copper deposits. By then, demand for copper, bronze, and brass was expanding on all fronts.[10]

The copper partnership with Johann Thurzo was critical to the Fuggers' spectacular success. Thurzo was regarded as the leading mining engineer of his day, but he had insufficient funds to revise and commercialize his latest separation technique (the Saiger process) and implement some of his hydraulic (drainage-related) innovations. As early as 1469, his father Johann Thurzo the Elder had developed a rudimentary version of the Saiger process to treat the "black" (silver and lead) argentiferous copper deposits that predominated in the Carpathian Mountains of upper Hungary. The elder Thurzo learned to separate the silver component of the "black copper" by mixing in lead in specially designed hearths (*Saigerhutten*) and heating the copper-lead-silver mixture to a desired temperature. At a certain point, the silver separated from the copper, combined with the lead, and became extractable by the conventional cupellation method. The copper-lead alloy, in turn, was oxidized to separate a copper residual that could be resmelted into pure copper. Jacob Fugger was also intrigued by the nine ounces of silver that could be separated from every hundredweight (*quintal*) of ore.[11]

Unfortunately for the Thurzos, the Saiger process was not a sure thing in 1494 and the ever-present risk of an Ottoman invasion clouded the investment climate in Hungary. Merchant bankers had been scarred by the archduke's earlier machinations in Tyrolia and refused to invest in high-risk Hungary until some of his outstanding debts had been repaid. Metal production in Hungary had peaked in the late fourteenth century, and recent discoveries in Kremnica (near Neusohl) and Baia Mare (northern Romania) were contributing no more than three tons of production in 1494. But Fugger capital, analytical skill, and access to Thurzo technology changed everything in 1495. Thurzo tested and tweaked the Saiger process at Neusohl and added a new smelting furnace and rolling mill. He also introduced an animal-powered hydraulic system to drain the deep, frequently flooded mines that were common to the region. Copper production in Hungary soared and soared further after Vasco da Gama identified a huge, untapped Indian market for copper of any kind.[12]

Copper would be king during the first two decades of the sixteenth century. As the Fugger-Thurzo copper inventory at Antwerp rose to a hard-to-believe 34,202 quintals (worth over 200,000 florins) in 1497, Jacob Fugger schemed to build the world's first vertically integrated mining enterprise. This enterprise would generate nearly 1.4 million florins in profits during the 1494–1525 period by controlling the mining, smelting, distribution, and pricing of copper sheet and plate. The logistics were daunting. Most of the Fugger copper was sent north (via Cracow) to the family's branch offices in the Baltic Sea—Danzig, Breslau, Stettin, Copenhagen, and Lubeck—and then shipped south in Hanseatic ships to Antwerp. Some of this copper was shipped to Portugal to support the West African trade while other quantities were reserved for the bronze- and brass-making centers at Brunswig, Nuremberg, Aachen, and Milan. Neusohl was so successful that the Thurzo process was transferred shortly to the Fugger smelting works at Hohenkirchen (Thuringia), Mansfeld (Thuringia), and Fuggerau (Tyrolia). The pure copper mined at Fuggerau (near Villach) was well suited for the bronze and brassware prized in West Africa and the bronze cannon, naval equipment, and hardware prized in Europe. Antwerp's leading merchant, Erasmus Schetz, made a small fortune in copper and brass before being overwhelmed by Jacob Fugger.[13]

The Fuggers' modest branch office in Antwerp, established in 1494, presided over an emerging copper monopoly of epic proportions. There would be much more to come. Jacob Fugger gradually redeployed his metal profits into bills of exchange; the forward-contract buying of huge quantities of metals, spices, and other commodities; foreign exchange arbitrage; and even an intelligence-gathering "news service." The House of Fugger revolutionized European commerce. The publication of the first manual on double-entry bookkeeping in 1494 had made their business practices available to anyone, but few were in a position to follow Jacob Fugger's pathway. By 1499, Maximilian had become so outraged by a copper monopoly of his own making that he realigned himself with a displaced Venetian copper cartel. The archduke was too late. He discovered painfully that he needed Fugger capital to finance his next military campaigns in Switzerland and Italy. Maximilian's promotion of Antwerp (over Bruges) even ensured that the House of Fugger would have a convenient trading center with which to serve the House of Habsburg and the Portuguese Crown—Manoel I placed a royal agent in Antwerp in 1499. Jacob Fugger was building a princely palace (Fuggerhaus) in Augsburg along the old Roman Road, but Antwerp would be his primary business headquarters.[14]

Jacob Fugger's timing was excellent. The family's copper monopoly and silver holdings were assembled just in time to cash in on one of the greatest business opportunities of all time—South Asia. Vasco da Gama's maiden voyage to India in 1498–1499, following over fifty years of Portuguese trading activity in West Africa, achieved a commercial objective that had eluded the great Italian city-states for over two centuries. The Portuguese were now trading directly with Asians. There were no Muslim middlemen with which to deal. No matter that European products had been found to be worthless in the Indian Ocean and Asian merchants would accept only precious metals in trade. But Central Europe was in the midst of a mining boom, and the Portuguese were already importing sizable quantities of gold from West Africa. Asian demand for silver, copper, bronze, and brass was likely to soar.

The convergence of trading profits from the Central European mining industry and Portuguese Asia at Antwerp would catapult the House of Fugger to a position of staggering wealth in an age in which large concentrations of capital were extremely rare. If one combined the wealth and power of the Rockefellers (oil) with that of J. P. Morgan (capital) in the early 1900s, and then placed this concentration of economic power in a single city (Antwerp), one would have a rough idea about the scale, physical location, and economic terror of the Fuggers's achievements. That the family name was initially registered in Augsburg as Fucker is another interesting aspect of their legacy. If that were not enough, the Fuggers were even presented with a third card to play after Amerigo Vespucci (and Goncalo Coelho) had traced most of the Brazilian coastline in 1501–1502. The tracing proved that Columbus's recent island discoveries were not an eastern part of a Japanese archipelago or a western extension of the Azores. A huge mass of land, possibly rivaling in length the recently discovered western coastline of Africa, was waiting to be exploited by men like Jacob Fugger. If a tiny island like Hispaniola contained intriguing amounts of gold, the vast, uncharted mainland territories were bound to contain a whole lot more.[15]

Of course, less than ten years earlier, Jacob Fugger had not met Johann Thurzo, the Indian Ocean was a Muslim lake, and the entire Western Hemisphere did not even exist in European minds. The global economic system was dominated by a powerful chain of Muslim empires stretching between the eastern Mediterranean and the Malay Peninsula. China and Japan were exotic lands that were closed to foreign visitors (including each other). The kingdoms of Portugal and Castile were minor factors in global affairs. Central

Europe was a patchwork of kingdoms, duchies, and princedoms. All of this was shattered and rearranged by Columbus's discovery of Hispaniola in 1492, da Gama's voyage to India in 1498, and the remarkable series of geopolitical developments that followed. This new geography was created almost entirely from prospecting, excavating, trading, and distributing precious metals.

While the Central European mining boom benefited all of Western Europe, the Venetian economy had been revived greatly. The Venetians expanded their business dealings with the Mamluk Empire of Egypt and Syria to 660,000 ducats annually, and recaptured much of the Eastern trade that had been lost during the Ottoman advance. The Florentines and Genoese, outflanked by the Venetians and the Ottomans, were left to explore business opportunities with the upstart Portuguese. Venice's looming problem, however, was that her Mamluk business partners were incurring increasingly large trading deficits with the spice merchants of Calicut (India) and Malacca (Malaysia). The growing availability of European silver had raised the relative value of Eastern spices and forced Egyptians, Syrians, Persians, and Gujaratis to compete more aggressively for business. The intensified competition, in turn, was eroding profit margins in the West and filling up the coffers of Eastern merchants, rajas, and sultans. As the bulk of the spice profits flowed in an easterly direction, centuries-old trading patterns in the Indian Ocean appeared to be in flux.[16]

The Venetians' second problem was that the otherwise insignificant Kingdom of Portugal was scheming to end-run both Venice and the great Muslim emporiums at Tunis, Alexandria, Aleppo, and Aden. While the Venetians (and Bavarians) were revitalizing the mining industry of Central Europe, in an attempt to relieve the Ottoman-enhanced bullion famine, the cash-strapped Portuguese Crown was doing them two better. Portugal was importing large quantities of West African gold, a metal that carried eleven times the value of closer-to-home German silver, and Portuguese sailors were tracing the entire western coastline of Africa. If a new easterly route to the Indian Ocean could be discovered and exploited, the Portuguese stood to gain access to the lucrative spice markets of South Asia and achieve a business objective that had eluded the Italians for over two centuries—circumvent the Muslim middlemen and formalize a direct route to China.

The Portuguese expansion was fostered, in part, by recent advances in geographic thinking. Europeans had been shocked by the arrival of the Mongol horde in 1241 and the subsequent establishment of a trading-oriented dynasty in China under Kublai Khan. The Mongol invasion path suggested that China was a lot closer to Europe than previously believed. It also added credence to the minority belief that the world was round. While Ptolemy had made the discovery of the globe in his *Geography* in the early 100s, the unexpected arrival of the Mongols prompted an Englishman, John Holywood (Sacrobosco), to prepare an Arab-to-Latin translation of *Geography* in the mid-1200s. Sacrobosco's *De Sphaera Mundi* demonstrated that the globe really was a globe and encouraged the papacy and a handful of Italian merchants to organize a series of expeditions to the Far East. Marco Polo's remarkable chronicles of Cathay (China), Cipangu (Japan), Sumatra, and the Indian Ocean between 1271 and 1291 helped to clarify a region that had always been a mystery.[17]

Unfortunately for the Italians, the commercial exploitation of Marco Polo's market research was halted by the Black Death (1348), the subsequent economic depression in Europe, the Ottoman advance, and the Ming dynasty's overthrow of the westward-leaning Mongols in 1368. If Europe's learned men were confident that the world was round, there was little to be done about it. Research activities continued. The *Catalan Atlas* was completed in 1375 by a largely Jewish school of cartographers in Majorca. A Greek-to-Latin translation of Ptolemy's *Geography* was prepared in 1406. These handwritten works informed a new generation of scholars that the globe could be sliced into 360 degrees of latitude and longitude. This, in turn, suggested opportunities to fix precise location (and mapping) coordinates and to explore distant lands more scientifically. The incentive to explore was enhanced by a mathematical error that had remained in place since the 100s. The error, still uncorrected when a printed edition of *Geography* appeared in Italy in 1475, was a huge one—Ptolemy and his successors had inadvertently assigned the full 360 degrees of longitude to the "known world." Since the "known world" stretched only between the Canary Islands and China, round-earth professionals in Italy, Portugal, and Spain were left to believe that Asia was much closer to Europe than it really was.[18]

## THE AFRICAN COASTLINE

One interested party, Infante Dom Henrique, the third son of King João I of Portugal, decided to expand his geographic horizons in 1415. Barely familiar with the western limits of the "known world," Prince Henrique focused his attention on something closer to home—West Africa. For many centuries, Portuguese merchants had traded with the Berber caravan ports that dotted the Muslim coastline of North Africa. The camel caravans delivered gold, ivory, and other exotica and dominated the trade routes into the African interior. No matter that the African interior itself was somewhat vague. Genoese sailors had reached no further south than the Canary Islands and generations of Berber, Arab, Italian, and Jewish traders, try as they might, failed to discover the whereabouts of the fabled "Wangara" gold deposits of West Africa. The Berbers had always relied on local middlemen to access deposits that were located somewhere between the upper Senegal and upper Volta rivers. In short, Europe's knowledge of the western coastline of Africa essentially stopped in southern Morocco.[19]

Motivated by a number of factors—the acute shortage of precious metals in Europe, Muslim control of Morocco, the Ottoman advance in the Balkans, and the "unfinished" legacy of the Crusades—Dom Henrique, Prince Henry the Navigator, jump-started Portugal's expansion program in 1415. In August of that year, he captured Ceuta (opposite Gibraltar) and reestablished the Muslim gold depot as a Portuguese staging area. The prince may have speculated that if Muslims controlled only the caravan routes, not the actual gold supplies, a geopolitical opportunity was presented to convert heathen West Africans to Christianity and then jointly evict the Muslims from the entire African continent. The ancient Ethiopian Christian community at Axum might even be persuaded to join the crusade. While these ambitious plans could only be realized by sea, Portugal had a proud and extremely capable maritime tradition.[20]

The first wave of southern voyages, starting in 1419, went badly. As many as fifteen attempts were made to round Cape Bojador, a landmark that the *Catalan Atlas* of 1375 had depicted as the end of the known world, before a Portuguese vessel under Gil Eanes finally accomplished the feat in 1434. Eanes obtained small quantities of gold along the coast of the Western Sahara but not much else. The management of these early expeditions alternated between Prince Henrique, Afonso V (1432–1481), and syndicates of nobles, government officials, and private merchants. Revenues were generated by occasional fishing expeditions, slave trading, and, to a lesser extent, gold. While a modest string of coastal *feitorias* ("factories") were established as the Portuguese moved southward, few of them were significant.[21]

Prince Henrique hedged his bets by commissioning a series of expeditions into the Atlantic Ocean. Driven by fishing excursions and the introduction of a lateen-rigged caravel, a triangular-sailed ship that sailed closer to the wind than a northern cog, Portuguese sailors visited the Madeiras, the Azores, and the otherwise unknown Cape Verdes between 1419 and 1444. Portuguese attempts to colonize the seven Canary Islands were thwarted by Castile, however, after Castilians had defeated the native Guanches, begun colonization activities, and signed the Treaty of Alcacovas with Portugal in 1479. While the treaty left Portugal with the Madeiras, the Azores, and the Cape Verdes, the Madeiras were the crown jewel. A slave-based sugar industry was launched there by Prince Henrique and other Portuguese nobles under a feudal-styled captaincy system borrowed from the Italians.[22]

Madeira served as the staging area for Portugal's next round of voyages. Virtually all of these expeditions were ad hoc and unplanned, authorized by Prince Henrique but left to private enterprise to generate the results. In addition to the West African coast, Portuguese sailors fanned out in all directions. It is possible, if not conclusive, that Portuguese fishing vessels penetrated as far west as Greenland, Newfoundland, northeastern Brazil, and even the Caribbean. Atlantic fishermen claimed occasionally to have sighted eastward-moving birds and driftwood from seemingly out of the nowhere. If these possible sightings (or landings) were achieved decades before Columbus, who was certainly in a position to have heard about them, it helps to explain Columbus's confidence in his voyage of 1492.[23]

Supported by improvements to the caravel, which evolved into a two-masted, lateen-rigged vessel of fifty-sixty tons, Portugal's southern expeditions continued with Nino Tristão's rounding of Cape Blanco (Mauretania) in 1442 and the establishment of a feitoria on Arguin Island in 1445. Located at the mouth of the Senegal River, Arguin served as a model for future feitorias. It was intended to trade for slaves, ivory, and gold and to provide another staging area for the next round of expeditions. The general pattern was that Portuguese traders arrived only after suitable ports (and feitoria sites) had been investigated by the pilots. The Portuguese rarely attempted to penetrate the African interior. Malaria and yellow fever were pervasive and local tsetse flies posed a regular threat to horses, mules, and oxen. Trade goods were limited to lightweight items like gold and ivory that could be carried by human porters. Since the region lacked natural harbors other than at Dakar and Lagos, shallow-drafting caravels were required to enter and exit local ports. If these issues weren't enough, the commercial viability of the new trading posts hinged on access to gold, which wasn't clear, and the willingness of local African chiefs and Muslim merchants to sell native captives to Portuguese traders. During the 1450s, Arguin sent no more than fifty-five pounds of gold and no more than nine hundred slaves to Lisbon on an annual basis. Of course, these were just small fractions of the future scale.[24]

Starting with the feitoria on Arguin Island, the Portuguese learned about local political conditions as they went along. None of the coastal tribes had assembled anything approaching the status of a Jenne, Gao, or Timbuktu—inland Niger River city-states that had been organized into caravan trading centers—and only limited attempts were made to convert native tribes to Christianity. Gold, slaves, and ivory were obtained from coastal chiefs, native merchants, and mulatto traders on an ad hoc basis, hauled by human porters to the fortified Portuguese feitorias, and transferred onto Portuguese caravels. West Africa was so fragmented politically that organized resistance was minimal. Market strategies varied by region. The lower Senegal River trade was oriented more to slaves than to gold, but the inland port of Kantora, located 450 miles up the Gambia River, was one of the few interior markets that could be reached by caravels. This capability helped to drive a lucrative salt-for-gold exchange during the late 1400s.[25]

The Italians made half-hearted attempts to capture some of the sub-Saharan caravan trade. In addition to the Portuguese advance, they were pushed into action by the prolonged bullion famine in Europe and the diversion of caravan gold to the eastern Mediterranean. The Ottoman conquests in the Balkans had forced the Mamluk sultans of Egypt and Syria onto a gold standard to fill their escalating trading deficits with the spice merchants of Calicut and Malacca. Since the sub-Saharan caravans had delivered most of Europe's gold supply during the preceding five hundred years, the Italians schemed to cut out some of the middlemen between North Africa and the Niger Bend depots. The Genoese Centuriones arranged a partnership with the Jewish and Berber traders of Tuat (Algeria) in 1447, followed by the arrival of Florentine traders at Timbuktu (Mali). These and other ventures failed miserably. Tuat collapsed (or was destroyed) and the Italians were left to deal with the Portuguese.[26]

At the time of Prince Henrique's death in 1460, Afonso V had pushed Portuguese interests as far south as present-day Sierra Leone and had even gained legitimacy for his kingdom's commercial conquests. A series of papal bulls issued between 1452 and 1456 essentially gave Afonso carte blanche to continue his nation's aggressive assault against the heathen coast of West Africa. The quid pro quo was that Afonso and his successors were expected to reserve a share of the royal fifth (*quinto*) for the papal treasury in Rome. The same formula had been applied to confiscated Muslim plunder during the Iberian Reconquest. Except for a temporary contract with Fernão Gomes between 1468 and 1474, the royal trading monopoly was strictly enforced along the coastline. In the interior, the absence of Portuguese businesses (or colonies) allowed more than a few daring entrepreneurs to cut their own deals.[27]

Portugal's most profitable feitoria was established at São Jorge da Mina, near present-day Accra (Ghana), following the accession of João II in 1481. The fortress was constructed in 1482 by as many as five hundred troops, possibly including Christopher Columbus, and provided Portugal with sizable volumes of gold, slaves, ivory, and pepper for the next 150 years. Always a trading and administrative center, São Jorge da Mina was manned by no more than sixty Portuguese and was strong enough to secure a Madeira-styled captaincy, offices of the later Mina Company (1509), and the regional gold trade. The actual source of da Mina's gold supplies is unclear. Intermediaries obtained gold from the fabled Wangara deposits of the upper Senegal River or from tributaries of the Volta River in Ghana. Their location remained a mystery, even to the intermediaries, and were well defended by hostile natives, unnavigable rivers, and malaria. But the evidence was clear—the entire region was rich in gold. Local chiefs eventually established their own state (Ashanti) and ensured that the "Gold Coast" would live up to its name.[28]

The local chiefs received a variety of European trade goods in exchange—copper, brassware, utensils, horses, carpets, and textiles. The Berber caravans had carried copper to the bronze centers of Ife and Benin since the eleventh century and knew that the metal was especially prized in Nigeria, Cameroon, and the lower Congo River. Shaving bowls, urinals, and chamber pots were also in demand. Preferences varied by region, but at least twenty tons of copper, brass, and metalware were delivered annually to the Portuguese factor at São Jorge da Mina during the early 1500s. As unsettling as it may be, a human being could be purchased for around ten pounds of copper in Benin or for eight to ten copper bracelets in Cameroon—ten times the value of an ivory-bearing elephant. While

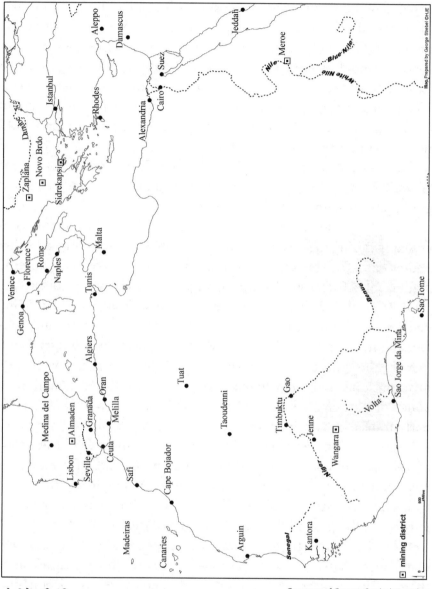

## MAP 1.
## PORTUGUESE
## AFRICA,
## 1415–1485

Courtesy of George Strebel, © author.

rising demand would raise the cost of a Benin-based slave to nearly forty pounds of copper by 1517, the slave trade was a small fraction of the one in gold. Portuguese West Africa was Europe's primary source of gold. The Genoese traders of Safi, a gold depot located west of Marrakech, secured much of the north-bound metal as collateral through which Genoese (and Florentine) merchants could invest in the next round of Portuguese ventures. Lisbon's gold receipts supported the Portuguese *cruzado*. Introduced in 1457 to replace the Muslim-Castilian *dobla*, the 3.58-gram, 23.75-carat cruzado was slightly superior to the Venetian ducat and remained gold-pure through 1536.[29]

Lisbon would receive an average of 170,000 cruzados of gold annually during the 1490s, delivered mainly from São Jorge da Mina and/or Safi. The Italians and Mamluks had feared correctly that the loading of African gold onto Portuguese vessels would divert the trade from the traditional Muslim caravans. Gold was now being delivered by twelve or so caravels annually (not camels) and converted into cruzados at the Royal Mint (not ducats or *ashrafis*). From Lisbon, the cruzados were distributed to the European merchants, investors, and creditors who had either advanced the outbound trade goods or participated directly in the African ventures. These multiparty transactions helped to move the otherwise isolated Portuguese economy into the European mainstream. Expanding gold imports, and the risk of contraband activities, had necessitated the establishment of a royal Casa da Mina (House of the Mine) during the 1460s to manage the African trade. The Casa da Mina was conveniently located on the ground floor of the royal palace.[30]

The West African trade was also dependent on the shipping companies of the Hanseatic League. It was the Hansa cargo ships that carried southbound copper and metalware to Lisbon and northbound African (and Portuguese) goods to England, the Netherlands, and the Baltic states. In addition to metals, Lisbon received textiles, grain, and other commodities from a string of feitorias in London, Bruges, Danzig, and other northern ports. These feitorias were no different than those operating at Arguin and

São Jorge da Mina and were all modeled on the Venetian trading compounds that had operated for centuries in the eastern Mediterranean. The Portuguese Crown admired the Venetians so much that a Venetian-styled *armazém* (arsenal) and *ribeira das naus* (dockyard) were eventually added to Lisbon's waterfront.[31]

The impact of these developments on the home country, an erstwhile producer of fish, salt, cork, wine, and fruits, was positive but not striking. Portugal's population had started to expand after 1450, reaching somewhere in excess of one million, but the vast majority of her citizens were peasants. The prime beneficiary was Lisbon. The city expanded to nearly sixty thousand inhabitants during the early 1500s, including five thousand or so slaves, making it one of the largest towns in Europe. But other than shipbuilding and, of all things, biscuit making, Lisbon's industries were very limited. The city was a trading town, and most of the import trade was controlled by a strict royal monopoly that benefited the Crown, a small group of nobles, and licensed merchants. Portugal's global expansion was managed in the same feudal character as the domestic economy. More profits were reinvested in land, palaces, and luxuries than in business enterprises. The middle class was tiny and would grow even tinier with the pending expulsion of the kingdom's Jews and Muslims. Prior to their eviction in 1497, Lisbon's Jewish community was confined to three separate quarters (*judarias*) that were walled and gated. The city's Muslim population was forced to live entirely outside of town.[32]

Lisbon's improving position, supported by authorizations in the Treaty of Alcacovas, prompted João II to commission a new round of expeditions in 1483. He recruited professional navigators to lead them and contracted with private shipbuilders to construct a new fleet of shallow-drafting caravels. The assumptions were that Africa's southern tip, wherever it was, was impassable and Portuguese pilots should concentrate on finding a cross-continental river route to the Indian Ocean. Both assumptions turned out to be wrong. In the course of tracing 1,450 miles of West African coastline between 1483 and 1486, Diogo Cão discovered that the leading candidate, the Congo River, was not an inland river route to the Indian Ocean. It was blocked one hundred miles upriver by Ielala Falls. Cão reached as far south as Walvis Bay (Namibia) until he was halted by increasingly less-favorable winds. His successor, Bartholeméu Dias, would pay close attention.[33]

In his earth-shattering voyage of 1487, Dias scrapped the inland river route idea and heeded Cão's warnings about unfavorable winds. He veered his three-ship fleet (two caravels and a supply vessel) in a southwesterly direction into the center of the South Atlantic, hoping to access the prevailing westerlies before shifting back to the coastline. The strategy worked. When Dias finally sighted land, he was near Mossel Bay on the eastern side of the Cape of Good Hope. He proceeded another six hundred miles up the coast to the Great Fish River before halting his advance then and there. Possibly confused by the extensive South African coastline, Dias did not sight (and name) the cape (Cabo da Boa Esperanca) until he had passed the landmark on his return trip to São Jorge da Mina. When the first man to round the Cape of Good Hope arrived in Lisbon in December 1488, the achievement was shocking. Less than six years after Portugal had established a major feitoria at da Mina, a Portuguese sailor had entered the Indian Ocean—the gateway to the spice markets of Asia.[34]

If João II was elated by Dias's triumph, a follow-up voyage was delayed until the unknown territories of the Indian Ocean could be better understood. In fact, a more conventional eastern expedition was already in progress under Pero da Covilha. Dispatched from Lisbon in May 1487, Covilha followed the traditional spice route via Egypt and Yemen (Aden) to Calicut on the southwestern (Malabar) coast of India. After returning to Cairo in late 1490, he probably (but not certainly) dispatched an intelligence report to João II by secret courier. The report would have described regional trade routes, spice supplies, and possible invasion points for the Portuguese fleet. Unfortunately for Covilha, his market research was clouded by his capture (or deliberate resettlement) in Ethiopia during a follow-up journey in 1493. Since da Gama's famous voyage did not depart until 1497, reports by Dias and (or associated with) Covilha probably influenced João's decision to reject Columbus's business proposals in 1484 and 1488. The cape route was a lot more promising.[35]

As the Portuguese prepared to invade the Indian Ocean, the shock waves from Dias's achievement were setting in. Not only was Portugal diverting West African gold from the Muslim caravans, the North African trading depots, and the Italian city-states, she had a papally authorized platform to strike at the very heart of the Mamluk-Venetian spice monopoly. This platform would be reconfirmed by the Treaty of Tordesillas with Castile in June 1494. But one problem remained for João II (and his successor Manoel I) in his quest to trade directly with Calicut and the other pepper ports of Malabar. As the Italians had known for centuries, European trade goods were next to worthless in the Indian Ocean. Ample supplies of precious metals would be needed to effect trade with the Muslim spice merchants who dotted the coastline. Since the Fugger-Thurzo copper venture was not established until 1494 and expanding supplies of Central European silver were pledged to Maximilian's many creditors, Portugal's modest supplies of copper, metalware, and gold would need to be supplemented.

Portugal was set to begin a Golden Age under Manoel I (1495–1521), but it would not be easy. While Castile would have the straightforward (if arduous) task of conquering and colonizing primitive Caribbean peoples, Portugal was forced to challenge the military and commercial might of sophisticated Muslim empires. The invasion of the Indian Ocean was inexplicably halted after Dias rounded the Cape of Good Hope in 1487. The Crown apparently needed time to digest Covilha's intelligence reports, weigh the impact of Columbus's western discoveries, and negotiate a binding authorization from a less-than-accommodating Spanish pope, Alexander VI. If that were not enough, João was forced to deal with his own succession problems. João's only son Afonso had died in 1491, and Manoel, a mere brother-in-law, probably needed time to consolidate his throne and assess the dynamic global environment. Following Columbus's bombshell of 1492, the Atlantic situation was clouded by two more discoveries—Pero de Barcelor and João Fernandes Lavrador reached Greenland (for Portugal) in 1495, and John Caboto (Cabot) arrived on Newfoundland (for England) in 1497.[36]

## DA GAMA INVADES INDIA

For whatever reason, it was not until July 8, 1497, that a four-ship fleet under Vasco da Gama was dispatched to the Indian Ocean with trade goods, 170 men, and twenty guns. The twenty-nine-year-old da Gama improvised as he went along. First, he sailed so far to the west of Dias's South Atlantic route of 1487 that he nearly landed in uncharted Brazil. Then, after struggling against the same unfavorable winds that had plagued both Cão and Dias, he eventually rounded the cape in late November only to lose his supply ship en route to the Great Fish River. Since this had been Dias's furthest point east, da Gama was now on his own. He sailed north to the mouth of the Zambezi River (Mozambique) in late January and began to notice that the entire East African coastline was controlled by unfriendly Muslims. Other than Malindi (Kenya), an iron-trading port that was less Muslim-influenced than the others and usually at war with Mombasa, the northerly moving Portuguese were unwelcome guests at virtually every stop. Da Gama seized local hostages before dispatching any of his landing parties and had to be persuaded to hire a Gujarati pilot in friendly Malindi.[37]

Few Europeans understood that East Africa had a very different history than West Africa. The Indian Ocean trade had launched a series of powerful entrepôts between 900 and 1200 CE, and most of these ports were still in business in 1500. Mogadishu (Somalia) and Kilwa (Tanzania) controlled the prime gold-trading routes into the upper Zambezi region. A Great Mosque had been constructed on Zanzibar Island (Tanzania) in 1107. Each of these ports traded with the Niger Bend gold depots in Mali as well as with a string of thirty other East African ports—all Muslim, all with their own ruler, and nearly all Swahili-speaking. Until recently, Kilwa had been the dominant economic power in East Africa. She and her vassal port of Sofala (Mozambique) had expanded their exchanges of Indian and Chinese goods for African gold, copper, and ivory. Kilwa's Sultan Hasan ibn Talut grew so wealthy from these transactions during the late thirteenth century that he minted his own gold coins and built a Great Mosque that was compared with the Great Mosque of Córdoba.[38]

Da Gama and his sponsors were also ignorant of the sophisticated Muslim system that lay to the east. This world was not necessarily a happy place. Ottomans were battling against Persians and European Christians. Mamluk Egyptians and Syrians were in conflict with each other, as well as with rival Persians and Ottomans. India was fragmented into scores of warring Muslim, Mongol, and Hindu kingdoms. Mongols and Turkoman tribes threatened everyone. An ancient pillar of the east-west trade, China had turned isolationist in 1436 and had come to rely on Muslim traders. However, the global chaos did not appear to threaten the commercial status quo—it was essentially a Muslim economic system. Muslims controlled the entire coastlines of North and East Africa, the Persian Gulf, Gujarat in northwestern India, and the pepper ports of Calicut and Cochin on India's Malabar coast. On the eastern side of the Indian Ocean, Muslims controlled the Bay of Bengal, the Straits of Malacca (between Sumatra and the Malay Peninsula) and the entire spice trade between India and China. Muslims had also gained footholds in Sumatra, Java, and the fabled Spice Islands (the Moluccas). Chinese politics and the control of tin, pepper, cloves, gold, and timber had transformed the instant city of Malacca into one of the world's greatest (and most exotic) marketplaces. The challenge for Portugal was to grab a share of this wealth for herself.[39]

Da Gama had observed a mere slice of this Muslim system in East Africa. So when a Gujarati pilot was hired to guide the fleet across the Indian Ocean in late April, the excitement was high—the Portuguese were finally on their way to India. Following a wind-aided, twenty-three-day sail, da Gama cruised the Malabar coastline and anchored in Calicut harbor on May 21, 1498. This was the first

European ship to visit India since Roman times. The Portuguese remained at Calicut until late August in anticipation of a meeting with the ruling Samudri (Zamorin) Raja. The meeting never took place. The Zamorin's Muslim intermediaries laughed off the Portuguese presents—cloth, coats, hats, coral, sugar, butter, and honey—and refused to present the insulting gifts to the great Zamorin. Da Gama declined to offer any gold, and a subsequent proposal to trade Portuguese cloth and hardware for pepper and spices was also rebuffed. In fact, the insulting Portuguese were asked to leave Calicut immediately. Most of the local Muslim merchants had paid hefty license fees to operate in Malabar, and European trade goods were regarded as nearly worthless. Muslims controlled virtually all of India's pepper and spice trade and refused to accept anything other than precious metals (and horses).[40]

An otherwise humiliating affair was overwhelmed by the magnitude of da Gama's achievement. The Portuguese returned to Lisbon in 1499 with astonishing firsthand information about a network of Muslim trading relationships stretching between East Africa and the western (Malabar) coast of India. Manoel was presented with a string of gold, pepper, and spice ports to exploit, possible access to the exotic markets of the eastern Indian Ocean, fabled Malacca, and maybe even China. If hostility had greeted the Portuguese at virtually every stop, it failed to prevent Manoel from organizing a syndicate of Florentine and other investors in early 1500. The thirty-three-year-old Pedro Alvarez Cabral was handed a massive fleet of thirteen ships and over one thousand men for a second expedition to India. Unfortunately for Cabral and his investors, only five of the ships reached their final destinations—four went down in a South Atlantic storm, three were either lost at sea or sent back to Lisbon, and the eighth was abandoned at Malindi on the return trip. The fact that Cabral and the surviving ships managed to discover Brazil accidentally on April 22, 1500, landing at a point north of Porto Seguro, was of small consolation. Brazil was viewed as an insignificant island, especially after a wayward Spanish expedition of 1499–1500 had found nothing of value in neighboring Guiana.[41]

Cabral and his investors were focused on India, not America (or even East Africa). The Crown had instructed Cabral to establish a factory at unfriendly Calicut and, if possible, block Muslim spice traffic to and from the Mamluk-controlled Red Sea. When Cabral's fractured armada finally reached Calicut in mid-September, he discovered that the new Zamorin was at least willing to talk—especially if the Portuguese were willing to join Calicut's military campaign against the rival city-state of Cochin. Relations with the Zamorin deteriorated in mid-December, however, when the unprovoked Portuguese decided to seize a spice-laden vessel en route to the Red Sea. Calicut's Muslim merchants responded by destroying the new Portuguese factory and killing over fifty of Cabral's men. Cabral, in turn, captured ten Muslim trading ships in Calicut harbor and then bombarded Calicut and neighboring ports. As many as five hundred Indians were killed in the action.[42]

Cochin was suddenly looking good to the Portuguese. Cochin's ruler, Unni Goda Varna, was delighted to enlist Cabral's forces against his northern rival. He permitted the Portuguese to purchase a large quantity of pepper in January and sent word to his allies at Cannanore (further north of Calicut) to welcome his newfound friends. Despite these successes, the accidental discovery of Brazil, and the provision of some useful advice to Amerigo Vespucci and Goncalo Coelho during a layover in Senegal, Cabral was in deep trouble when he returned to Lisbon in July 1501. He had lost eight of his thirteen ships and had failed to establish trading relations with Calicut. Cochin and Cannanore were regarded as second-tier and Brazil was dismissed as another mineral-poor Atlantic island—until proven otherwise.[43]

With Cabral still at sea, Manoel had hedged his bets with a more modest, Florentine-financed, four-ship affair under João da Nova in early March 1501. Inconsequential stops at virtually all of the East African trading posts persuaded Nova that European trade goods were in fact worthless to sophisticated Muslim merchants. However, Nova was allowed to purchase pepper, cinnamon, and ginger at friendly Cannanore and Cochin in September and managed to plunder defenseless trading vessels in and around Calicut. The Zamorin of Calicut was outraged. He dispatched another Muslim fleet to challenge the Portuguese and was again defeated. Unfortunately for the Zamorin, his defiance of Portuguese authority had made him a marked man. Even before Nova had returned to Lisbon with record spice profits in September 1502, a massive fleet under the recently ennobled Dom Vasco da Gama, Admiral of the Indies, was on its way east.[44]

King Manoel, enriched by spice cargoes and attentive to the unrealized ambitions of Prince Henrique, was beginning to think big. Not only was he preparing to break the Mamluk spice monopoly in the Indian Ocean, he was scheming of ways to evict the vulnerable Mamluks from the Holy Land and to restore Christianity to Jerusalem. Manoel may have assumed that the Ottoman Empire was planning the very same thing (without the Christianity) and needed to be stopped. If Manoel's instructions to block Muslim shipping to and from the Red Sea was an impossible task, it was viewed as a necessary step in the larger strategy. Another Christian state, Venice, had arrived at the opposite conclusion. It was bad enough that Venetian shipping and trading activities in the eastern Mediterranean were being checked by the refurbished Ottoman navy. Far worse was the emerging (Florentine-financed) Portuguese threat to Venice's age-old spice trading alliance with the Mamluks. If the Red Sea trading route was circumvented or even blocked, the Venetian economy would be crippled.

Venice's vaunted spy network monitored da Gama's second voyage to India in 1502. Focused on revenge and profits, da Gama's twenty-ship armada rounded the cape with intentions to alter Portugal's Indian Ocean strategy from "requests" to "demands." When the heavily armed fleet reached Sofala (Mozambique) in mid-June, local officials permitted the Portuguese to build a fort in the vicinity. At Kilwa (Tanzania), a port that had controlled the gold-trading routes into the upper Zambezi for centuries, da Gama had little time for diplomacy. After a proposed agreement was rebuffed, he sacked the town and forced her rulers to accept a protection-for-tribute deal. Kilwa's spectacular run as an independent Muslim kingdom was ended in mid-July. Follow-up Portuguese raids along the East African coastline were only partly successful. Zanzibar (Tanzania) eventually submitted, but stubborn Mombasa (Kenya) withstood a combined Portuguese-Malindi assault.[45]

Da Gama left East Africa in July 1502 to avenge the Zamorin of Calicut's treatment of Cabral and Nova and to assert the authority of the Portuguese Crown. The Portuguese fleet was well armed for its day. Each of da Gama's caravels carried four heavy guns (*camelos*) capable of firing twelve to eighteen pounds of stone shot, six four-pounders (*falcons*), and ten breech-loaded swivel guns (*bercos*). After arriving in Cannanore in mid-August, da Gama demonstrated the revised form of Portuguese diplomacy almost immediately. A Calicut-bound vessel, the *Miri*, was seized, shelled, and destroyed upon its return from a trading and pilgrimage trip to Jeddah and Mecca. A large number defenseless passengers—wealthy Muslim merchants, women, and children—were killed during the carnage.[46]

Da Gama's brutal retaliation persuaded the Zamorin to offer compensation for Portuguese losses incurred during the Cabral affair. However, da Gama demanded nothing less than the expulsion of the entire Muslim trading community from Calicut—a demand that the Zamorin could not possibly meet. The Portuguese fleet responded by blockading the harbor and shelling Calicut repeatedly in early November. The largest Portuguese vessel carried no more than twenty guns, but in those days, twenty guns was a frightening capability. Da Gama then headed south to friendly Cochin, remaining there through January buying pepper and spices for the Crown (and his own personal account) and establishing relations with centuries-old Nestorian Christian pepper traders. Da Gama discovered that cloves, cinnamon, incense, and other spices could often be had for Asian textiles (and later brazilwood), but pepper, the king of spices, could only be purchased with precious metals. No matter that the Portuguese had declined (to date) to carry any metals around the cape.[47]

Few Europeans knew much about pepper, a product that had

been prized by Chinese emperors, Muslim caliphs, and Venetian traders for centuries. The humble plant grew in a vine-like fashion and delivered fruit for a period of ten to twelve years. Since pepper thrived in warm, frost-free climates that received at least one hundred inches of rain per year, the wet forests of Malabar, Kanara, and the foothills of the Western Ghats were ideal. The pepper fruit itself grew in clusters of small green berries, comprising a soft core encased in pulp that reddened and became transparent during ripening. Black pepper was produced after the whole fruit was crushed, dried, and blackened by the sun. White pepper was derived from the dried core after the skin and pulp were removed. The annual crop was typically sold in advance to middlemen—Hindus, Muslims, and Nestorian (and Saint Thomas) Christians—and delivered to the principal pepper ports of Malabar and Kanara via oxcarts and riverboats.[48]

Pepper was on everyone's mind when da Gama decided to make a return visit to Calicut in January 1503. Greeted by a nighttime ambush of over seventy Muslim ships, da Gama and his crew were saved only because of the timely arrival of a Portuguese splinter fleet from Cannanore. Two of the Zamorin's men were taken as hostages, attached to the upper masts of da Gama's flagship, and then paraded around Calicut harbor. The Zamorin was undeterred. In February, he organized his largest fleet to date—thirty-two large ships, scores of smaller vessels, and nearly sixteen thousand Muslim troops who had sworn an oath of ritual suicide. Most of the ships and recruits were extorted from Calicut's Muslim merchant community. But once again, the Zamorin was disappointed. His local navies were designed to board and plunder enemy ships. They were no match for Portuguese cannon.[49]

Having had enough of Calicut, da Gama established a factory at Cannanore and then sailed directly across the Indian Ocean to Mozambique—arriving there in mid-April. The thirteen ships that returned safely to Lisbon by late October rewarded their sponsors with a staggering 32,500 quintals of pepper and spices. A modest quantity of East African gold was presented courtesy of Kilwa.

Everyone was thrilled. Da Gama's personal share, valued at nearly forty thousand ducats, must have intrigued subordinates like Vicente Sodre and his brother Bras. Left behind to defend the new Portuguese factories at Cannanore and Cochin, the independent-minded Sodre brothers, with or without da Gama's tacit approval, decided to leave their posts for a series of raiding expeditions off the Horn of East Africa. The raids were successful but the brothers were not—Vicente was killed in a storm, Bras apparently died in a ship-wreck, and the wily Zamorin of Calicut moved to invade Cochin while the Sodres were out at sea.[50]

Da Gama's success and the Sodres' failures prompted the Por-tuguese Crown to accelerate its timetable. By mid-September 1503, fleets under the command of Francisco and Afonso de Albuquerque and Nicolau Coelho arrived in Cochin to fortify the abandoned fac-tory and restore order. The Albuquerques even attempted to arrange a commercial treaty with the defiant Zamorin of Calicut. The answer was still no. So a fourth captain, Lopos Soares de Alber-garia, was dispatched from Lisbon in April 1504 with eleven vessels and the largest outbound shipment to date—162 tons of Fugger copper, twenty-nine tons of lead, seventeen tons of cinnabar, and nearly two tons of silver. Manoel's simple instructions were to bom-bard and blockade Calicut, leave a fleet of four ships behind to replace the Sodres' botch-up, exchange the outbound cargo for pepper and spices, and sail home. Albergaria completed his mission successfully. He returned to Lisbon in July 1505 with enough pepper and spices to deliver a whopping 175 percent return to Manoel and an Italian-Bavarian investment syndicate.[51]

Albergaria's success ensured that the India trade would be dom-inated by the exchange of copper for pepper and spices. Over two hundred tons of Fugger copper were carried by the outbound fleet of 1505–1506. The Casa da India, the agency responsible for man-aging Portuguese "India," worked with most of the major merchant bankers but instituted a strict royal monopoly on the outbound metals and inbound spices. A piece of almost every deal was reserved for royal officials, the papacy, bureaucrats, and even

crewmembers. This left Florentine merchant bankers like Bartholeméo Marchione, Girolano Frescobaldi, and Filippo Gualterotti
to handle the front-end cash and the back-end distribution. Florentines had led Europe's financial revolution in the fourteenth century,
and the sons of Giovanni de Medici, Cosimo and Lorenzo, had sex-
tupled the family fortune to 560,000 ducats by 1469. While the
Medicis remained the chief bankers to Rome, they left the spice
business to a new generation of Florentine bankers. They knew
what they were doing. The spices provided the collateral through
which the outbound ships, equipment, and trade goods were
financed. This meant that few of the inbound spice cargoes
remained in Lisbon for long. Most were redirected immediately to
the Crown's Antwerp office either to repay the Florentines or to be
warehoused for European distribution.[52]

The Florentines were always happy to take business away from
the Venetians. They also recognized that the Genoese had shifted their
attention (and capital) to the Caribbean sugar industry. In the early
years, the Florentines usually provided three-fourths of the cost of the
outbound Portuguese fleet either as cash or in trade goods. Capital
required to purchase spices in Malabar was a separate transaction. In
return for a profit-sharing arrangement with the Crown, Marchione
and his partners had outfitted one of Cabral's ships in 1500, two of
da Gama's in 1502, and four of Albergaria's in 1504. They were also
pressed to lend 182,812 cruzados to facilitate Manoel's personal
acquisition of 393 tons of spices (mainly pepper) in the Malabar venture of 1505. Marchione's difficulties in getting repaid—half of the
personal loan was still due in 1514—opened the door to the Affaitatis
of Cremona. The Gualterottis and Frescobaldis had also participated
in the metals-for-spices venture of 1505, but they and other Florentines played second fiddle to the Affaitatis in subsequent years. The
Affaitatis had influence. They had been permitted to send one of their
factors with da Gama's fleet of 1502 and had assembled a formidable
network of spice factories in Europe. The family would play a critical
role in arranging the purchase of Fugger copper for the expanding
pepper and spice trade with Malabar.[53]

The Affaitatis, Florentines, and Venetians viewed the spice trade as an extension of their earlier investments in the European mining industry. Spices and minerals were both viewed as an attractive source of collateral, and cash advances could be repaid with discounted shares of production. The Italians pioneered the purchase of entire Portuguese spice cargoes in port and made their greatest profits by splintering off these cargoes to scores of European merchants. The concept worked extremely well when demand outstripped supply. The prices per pound received in Antwerp in 1505 —pepper (three florins), cloves (ten), cinnamon (five), and nutmeg (five)—reflected huge markups from ship to port despite the fact that the quality of the finest Asian spices could be reduced by prolonged exposure to salt water. Venetian products obtained from the traditional overland routes were usually preferred.[54]

The metals-centric Bavarians became aware of the similarity between metals and spices as well. Otherwise, they continued to rely on Venice for most of their Eastern trade goods and left the distribution end of the spice business to the experienced Italians. But one firm, the House of Welser, had a change of heart in 1503. In February of that year, the Welsers gained Portuguese trading privileges for themselves (and other Bavarian firms) plus a royal permit to open a metals-and-spices factory in Lisbon. The Welsers were a distant second to the Fuggers, but they were always more entrepreneurial. Anton Welser had established a mining-oriented merchant bank in Augsburg long before the Fuggers and built a European trading network that was nearly as extensive as the Affaitatis. The family's creativity was rewarded in 1516 when an unexpected bonanza was uncovered at Joachimstahl on the Bohemian side of the Erzgebirge range. Joachimstahl would be Europe's single largest source of silver during the first half of the sixteenth century. In the meantime, it was the Welsers' professional skills and trading network that appealed to the Portuguese Crown. A grateful Manoel even organized riots against possible competitors—Lisbon's Jewish community had already been expelled in 1497, but the city's Converso businessmen were subjected to robbery, violence, and even killings between 1504 and 1506.[55]

If Manoel thought that the Bavarians and Italians would be willing to switch their trading patterns to Lisbon, he was badly mistaken. Antwerp, located near the mouth of the deep-water Scheldt River, was the primary center for European spices, metals, and capital and it was dominated by the House of Fugger. After Albergaria's spice fleet of 1504–1505 had returned a profit of 175 percent on a 58,000-florin investment—30,000 from the Italians, 20,000 from the Welsers, and a modest 4,000 each from the Fuggers and Hochstetters—Manoel decided to modify the role of the middlemen. He issued a royal decree that restricted Indian Ocean shipping to the Portuguese Crown alone. The merchant bankers would continue to be "permitted" (sometimes begged) to extend credit for the royal spice fleets, typically collateralized by the forward sale of spice cargoes, and to supply the outbound copper, silver, and trade goods to the royal factor in Antwerp. They would also continue to handle the distribution of the inbound spices (from Antwerp) throughout Europe. But they were no longer allowed to invest directly in the spice side of the transaction. Unlicensed merchant bankers, whether they were Italians or Bavarians, were now prevented from trading directly with the Portuguese spice monopoly in South Asia.[56]

Manoel's shift in strategy was a blatant quest for profits. Guided by the series of low-budget enterprises that had categorized the West African trade, the Crown had been cautious during the initial years of the Indian effort. With an uncertain profit opportunity and the ever-present risk of an attack or shipwreck, Manoel preferred to control risk by arranging investment syndicates with skilled professionals. This prudent course surrendered most of the profits to the Italians, but it minimized the issuance of debt and preserved the royal balance sheet. Some Italians guessed that this cautious strategy had limited the Crown's annual spice profits to a mere 50,000 ducats during the 1500–1506 period. The assumption was that annual spice revenues (350,000) were eroded by the total costs to outfit the fleets, stock the outbound trade goods, purchase the spices in India, and manage the perilous, fifteen-month round-trip voyage around the Cape of Good Hope (300,000). Whether or not these figures were accurate, it was clear that the India trade required huge amounts of capital. The cost to outfit a single outbound ship was around 10,000 ducats, and another 10,000 in cash (precious metals) was needed to purchase pepper and spices in Malabar. Manoel's decision to handle these heavy expenses on his own was risky, but it would prove to be very successful.[57]

The copper-for-spice fleets of 1505 and 1506 reflected the change in royal strategy. Although the Portuguese had gained control over most of Malabar (except Calicut) and most of East Africa (except Mombasa), the incoming governor of the newly created Estado da India, Francisco de Almeida, was taking no chances in 1505. As an ally of the entrepreneurial da Gama faction, Almeida had no patience for time-consuming negotiations. Backed by a twenty-ship armada and 1,500 men, he sacked Kilwa for the second time and constructed a local fortress to safeguard the royal administrators. Almeida proceeded north to Mombasa only to discover that the sultan was waiting with 1,500 archers and a few cannon salvaged from an abandoned Portuguese warship. This was enough to withstand the Portuguese assault for all of two days. Almeida had Mombasa sacked and plundered. When Mombasa dared to revolt again in 1528, the Portuguese burned the city to the ground and installed a nephew of the reviled Sultan of Malindi as its ruler.[58]

Mozambique and Sofala were established shortly thereafter as captaincies, based on the models in Madeira, São Jorge da Mina, and São Tome. However, the economic value of these and other East African conquests turned out to be disappointing. In 1505, the Portuguese factor at Sofala was receiving large amounts of ivory, but only modest quantities of gold from the upper Zambezi River. These gold receipts were well below those that had built Kilwa's fortune and suggested that Muslim traders had diverted Zambezi gold to other outlets. In addition, malaria took its toll on Sofala's Portuguese population and checked any notions to explore the interior. While the royal trade monopoly was even relaxed in Kilwa to encourage business activity, the Portuguese were too late. The population of this once-powerful sultanate had plummeted from a peak of around thirty thousand to less than four thousand. Kilwa's wealthiest merchants had relocated to Mombasa. The Crown eventually replaced local sultans with governors after a series of puppet rulers were murdered.[59]

Annual gold shipments from Portuguese Africa, East and West, were still valued at 200,000 cruzados. Nearly one ton of African gold was sent either to Lisbon or to Portuguese India on an annual basis between 1500 and 1520. While this volume only matched the one ton or so of Caribbean gold that was being shipped annually to Seville during this period, most of the Caribbean gold would run out by 1520, and Lisbon had more attractive reinvestment opportunities in the forms of pepper, spices, and textiles. Over half of Portuguese African gold shipments was contributed by São Jorge da Mina until Quelimane, an ivory-trading post located near the mouth of the Zambezi, began to handle modest amounts of gold in the 1530s. As increasingly more African gold was shipped to Portugal's eastern empire, da Mina shifted to the more lucrative, if relatively small-scale, slave trade. An average of 750 West African slaves had been shipped annually to Lisbon, Madeira, or the Cape Verdes between 1450 and 1500, but the volume accelerated with the expanding sugar industries of West Africa, the Caribbean, and Brazil.[60]

India was always valued more highly than Africa. With large quantities of pepper and spices to be had, the stakes were much greater and the competitive environment appeared to be manageable—Portugal had powerful armed fleets and the Muslims did not. Everyone in Malabar knew what gold was worth, but copper and spices varied widely in price. Portugal's challenge was to exchange marked-up supplies of Fugger copper for low-cost supplies of pepper and spices that could be marked up back home. The goal was to generate profit margins in excess of 100 percent. Backed by well-armed Portuguese fleets, there would be no need to compete fairly with Muslim merchants or to negotiate rights and privileges with local sultans and rajas. The Portuguese had only to blast their way into the eastern markets in the same way that the East African coastline had been subdued. Indian textiles, foodstuffs, and other products were left to the traditional marketplace, but the Portuguese were ruthless when it came to spices. They struck one-sided supply contracts with the overmatched rulers of Cannanore, Chaul, Cochin, and Quilon and introduced a system of regulated *cartazes* (passes) to separate friends from foes (mainly Calicut). Any ship found without an authorized cartaz, signed by a royal factor, could be attacked at will.[61]

The Mamluk Empire, possibly at the urging of their frantic allies in Venice and Calicut, finally woke up to the Portuguese threat in 1507. Their spice monopoly was being threatened in the same way that they had destroyed an earlier Cairo-based (Karimi) monopoly in the fifteenth century. Amir Husain Mushrif al-Kurdi, ruler of Jeddah and guardian of a chain of Mamluk ports between Cairo and Aden, even managed to organize a naval alliance among Mamluk Egypt, Venice, Gujarat, and Calicut. With much to gain, the Venetians sent two war galleys, shipwrights, and artillery to the amir's naval base at Suez. The ambitious objective was to remove the Portuguese from the Indian Ocean and restore the Muslim status quo.[62]

Unfortunately for the amir and his allies, a fleet of lightly armed galleys, feluccas, and dhows were simply no match for heavily armed Portuguese carracks. The amir also made the mistake of winning a naval battle against a Portuguese squadron in early 1508. In March of that year, an allied Mamluk-Gujarat-Calicut armada of fifty-two ships and over 1,500 sailors caught up with a smallish Portuguese fleet off Chaul, south of present-day Mumbai, and crushed the Europeans completely. Viceroy Almeida's son Lourenco was among the hundreds of Portuguese killed in the action. If the victory was thrilling to Cairo, Calicut, and Venice, it shocked Lisbon and prompted the always-cautious Gujaratis to reconsider their position—the victorious Mamluks now posed a greater threat to their textile trade than the spice-centric Portuguese. Fearing either the Mamluks or the inevitable Portuguese response, the Gujaratis backed out of the Muslim alliance. It was a good move. An enraged Almeida left Cannanore in December 1508 with a powerful fleet of nineteen ships and 1,200 men and destroyed the Mamluk navy off Diu on February 3, 1509.[63]

## ALBUQUERQUE'S BOLD PLAN

The naval triumph at Diu was so complete that there would be no further opposition to Portuguese authority in the western Indian Ocean. The textile-centric Gujaratis were prepared to remain neutral and leave the high seas to the heavily armed Portuguese. No matter that the governorship of India was transferred from the entrepreneurial Almeida to the royalist-leaning Afonso de Albuquerque in 1509. The Crown had grown intolerant of independent-minded commanders like Almeida (and da Gama) and was attempting to monopolize the India trade for itself. The fifty-six-year-old Albuquerque was loyal, competent, and a known commodity. If the gubernatorial transfer was less than smooth—Almeida refused to be replaced after his great victory at Diu and even imprisoned his successor for a brief spell—Manoel's blessing and a solid relationship with the Florentine banking community won the day for Albuquerque. The Crown would not be disappointed.[64]

Albuquerque understood Manoel's economic interests and had his own ideas about proper Portuguese strategy in the Indian Ocean. Having served in East Africa and India, he determined that the great expanse would have to be invaded selectively. Portugal had limited capital, ships, and manpower to do otherwise. His plan was to seize and fortify four carefully chosen ports—Mozambique Island, Hormuz at the entrance to the Persian Gulf, Goa on the Malabar coast of India, and the fabled entrepôt of Malacca along the straits between Sumatra and the Malay Peninsula. If executed successfully, the plan would add two of the Indian Ocean's most powerful trading centers (Hormuz and Malacca) and provide four secure bases for the Portuguese fleet. The capture of Hormuz and Malacca would be major challenges, but Mozambique Island was already in Portuguese hands and the third-tier port of Goa was likely to be easy prey. Almeida's destruction of the Mamluk fleet off Diu had given Portugal a virtual command of the western Indian Ocean.[65]

The selection of Goa as Portugal's Indian headquarters, rather than friendly Cochin or powerful Calicut, must have surprised the

Muslim trading communities of western India. Hindu-leaning Goa had changed hands regularly. The port was conquered by Muslim Bahmanis in 1356, recaptured by Hindu Vijayanagar in 1379, and retaken by Muslim Bijapuris in 1471. Goa had always been a minor factor in the Indian spice trade. The prime spice routes between Malacca and Hormuz (and Aden) were dominated by Gujarati textile traders and a string of Malabar ports like Calicut, Cochin, and Quilon. Goa's claim to fame was her importation of around one thousand horses per year from Persia and Arabia. In fact, the Muslim sultans had conquered Goa mainly to prevent the horses from reaching their Hindu rivals. Vijayanagar's recent revival under Krishnadeva Raja may have convinced Albuquerque that a Portuguese Goa would have at least one potential friend along a Muslim-dominated coastline.[66]

Albuquerque was less interested in Goa's horse-trading capabilities than its location, harbor, religious leanings, and ease of access during chaotic monsoon seasons. After initial attempts were thwarted, Portuguese troops and local mercenaries captured the forty-eight-square-mile island-port from the Sultan of Bijapur in November 1510. Albuquerque liked Goa's defensible position. The Western Ghats protected the port's eastern flank and provided a water supply for the surrounding agricultural territories. After a Lisbon-styled arsenal and dockyard were in place, the Portuguese would learn to manage the southwesterly monsoons that suspended traffic in and out of Goa between May and October. Even better, the Muslim influence in Goa, as in Malindi, was relatively weak. Jesuit missionaries would manage the unusual feat (in India) of converting thousands of Goans to Christianity by midcentury.[67]

The more ambitious military target was Malacca, arguably the single greatest entrepôt in Asia. Malacca had somehow reached this lofty position in a single century and was viewed by many, including future Portuguese visitors like Tome Pires, as heir to the former glory of Srivijaya—the Kingdom of Srivijaya had controlled the Malacca Straits from Palembang (in southeast Sumatra) between 700 and 1100 CE. Malacca's early history is somewhat murky. One legend has it that the Prince of Palembang had relocated to Singapore after an attack by the Majahapit Kingdom of Java in 1377, murdered a local Siamese ruler there in the 1390s and escaped to found Malacca in around 1400. A more likely scenario is that Malacca, bounded by defensible tropical forests, emerged from one of the many pirate lairs that sprouted along the Straits of Malacca during the late fourteenth century. The location was convenient to the tin mines at Perah, Tenasserim, and other sites on the Malay Peninsula. The existence of Malacca was first noted by Chinese sources in 1403, probably around the time that the pirates who founded the port had solicited naval protection from China (against Siam).[68]

It had been the Mings' shift to isolationism in 1436, however, that converted an erstwhile tin depot and pirate haven into a trading colossus. Malacca's fortune was made by unofficial trading with China, access to Indonesian spices, Sumatran pepper and gold, Malay tin, and a golden location. The Malay-side, deep-water port commanded the straits at its narrowest point and guarded the principal shipping lane between the China Sea and the Indian Ocean. By 1450, Malacca had extended its commercial reach to both sides of the straits, gaining access to Sumatra's prized commodities and establishing trading relationships with virtually every port between East Africa and Indonesia. These relationships delivered a steady stream of customs duties to the Sultan of Malacca in the same fashion that the toll-takers at Aden blocked the Red Sea, Istanbul blocked the Dardanelles, and Copenhagen blocked the Danish Sound. The spectacular wealth of Malacca was eventually introduced to Europeans by Tome Pires, a Portuguese adventurer who documented his jaw-dropping experiences in *Suma Oriental*.[69]

Split into halves by the Malacca River, Malacca was subdivided into districts based on the nationality of her many merchant communities. The bridge that spanned the river functioned as the center of town. Malacca's architecture was not to be confused with Beijing, Cairo, Istanbul, Venice, or even Goa. Most of the homes were wood and thatch affairs, and only the wealthiest citizens had a tile roof. While Malacca's deep, sheltered harbor was accessible twelve months of the year, travel plans were affected by the same May–October monsoon that plagued Goa. Malacca's largest trading partners, the Gujaratis, were forced to adjust. Otherwise, unofficial Chinese trading junks typically arrived with the northeast trade winds and departed in late June to exploit the southeast monsoon. Other than fishing and some shipbuilding, no industries were ever developed in Malacca. Rice, foodstuffs, metals, and timber were all imported. However, a succession of Muslim rulers and businessmen had assembled a vast, multinational trading network that the Portuguese could not have created on their own. Pires, who resided in Malacca between 1511 and 1515, determined that eighty-four different languages were spoken in town. And Malacca had become vulnerable. She had defended herself well since the departure of her official Chinese protectors in the 1430s, but the application of military force had been rarely needed or applied.[70]

In 1509, the forward-thinking Albuquerque had sent a four-ship fleet under Diogo Lopes de Sequeira to actually find Malacca. When Malacca was indeed found, Sultan Mahmud Syah greeted the "barbarian" invaders with hundreds of armed troops. Scores of Portuguese were killed or captured. The survivors were fortunate to escape. The barbarians returned the favor in April 1511 when Albuquerque's heavily armed fleet of eighteen ships and 1,200 men blockaded a sultanate that was not used to being attacked. Malacca fell after a four-month siege. Albuquerque reorganized Malacca's fortifications and naval defenses for the inevitable counterattack and then set off for Goa—barely surviving when his spice-laden ship sank off the coast of Sumatra. However, Albuquerque lived to see the value of his defensive preparations at Malacca. When the sultan's counterattack came in January 1513, the Portuguese navy destroyed a fleet of underarmed Javanese junks completely. The victory ensured that Malacca would remain in Portuguese hands for over a century.[71]

The Sultan of Malacca had at least received support from Java and some of his other neighbors. When the Portuguese analyzed the three thousand pieces of weaponry that had supported Malacca's defense, they discovered that most of the small arms had been man-

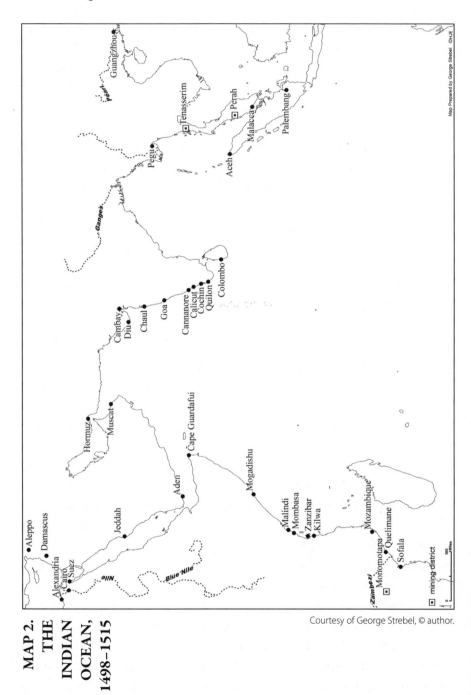

MAP 2.
THE
INDIAN
OCEAN,
1498–1515

ufactured in Pegu (south Burma) and Siam and a bronze cannon had been sent by the always-willing Zamorin of Calicut. The superiority of Portuguese ships and artillery was not lost on the Asians. Anyone fighting with swords, lances, shields, bows, and poisoned arrows could see the potential of firearms and heavy artillery. European ships may have been too expensive (and complicated) for a local sultan, but European cannon and firearms were well within one's reach. In later years, the Achinese of northern Sumatra would import weaponry from Istanbul, and Japanese warlords would acquire thousands of European muskets for their upcoming campaigns.[72]

Portugal's seizure (and retention) of Malacca would never have occurred if China had not removed herself from regional affairs. China was a nation of well over one hundred million people in 1500, twice the size of Europe, and was capable of crushing any threat to her regional hegemony. The Chinese ruled historically by fear, rarely having to resort to bloodshed (or plunder) to achieve their objectives. With technical capabilities ranging from printing with moveable type to artillery, silver smelting and iron making, China had been the world's most technologically advanced nation when the Mings removed the Mongols from power in 1368. It was a temporary surge in annual silver output to thirty-four tons, however, that allowed Emperor Yongle (1403–1424) to build the magnificent Forbidden City in Beijing, assemble the world's most powerful naval fleet, and send Admiral Zheng He (Cheng Ho) and 1,350 heavily armed Chinese junks to the far reaches of the Indian Ocean. The subsequent depletion of China's always-fragile mineral supplies (to a mere two tons annually) prompted the Mings to turn inward and isolationist in 1436. Emperor Zhengtong (1436–1449) and his successors banned (or restricted) foreign trade and limited the Chinese navy to coastal protection. While their naval artillery was still very capable, the Mings had removed China from the world scene.[73]

Emperor Zhengde (1506–1521) declined to assist Malacca during the siege of 1511, possibly because he was as curious about the "barbarian" Portuguese as they were of him. Beijing's official policy of isolationism was even relaxed for a brief spell. A handful of Portuguese traders were allowed to visit the Chinese coastline around Macao and Tamao (Hong Kong) in June 1513, followed by an official call by Jorge Alvares, the new Treasurer of Malacca. Alvares received permission to establish a trading post on Neil-ingding, an island located near the mouth of the Pearl River, and added the fabled port of Guangzhou (Canton) to the list after Rafael Perestrelo managed to sail up the Pearl unimpeded in 1515. These official exchanges were insignificant financially and would be suspended in 1521. However, news that the Portuguese were trading directly with China was shocking to the gnomes of Europe. If Guiliano de Medici was receiving occasional reports from his agents in Malacca, so were the other Italians and the Bavarians.[74]

The capture of Malacca also opened the door to the Spice Islands (the Moluccas)—once the Portuguese could find them. Moluccan spices had been exported to the Mediterranean for over three thousand years, but the precise whereabouts of Tidore, Ternate, Ambon, and the Bandas had remained a mystery. In contrast, the western Indonesian kingdoms of Sumatra, Malaysia, and Java were reasonably well known—Java had been mentioned in Ptolemy's *Geography*, civilized by Hindu-Buddhism during the 400s, and invaded by Muslim traders after around 800 CE. The East Indonesian islands had a very different history. They were colonized in waves from Papua New Guinea prior to 2500 BCE, civilized only barely since then, and isolated from the main Asian trading routes until around 1300 CE. However, someone had directed their clove output to faraway Syria in 1700 BCE, to the Han emperors of China during the 200s BCE, and to a succession of Roman emperors. Chinese rulers especially prized the medicinal qualities of Moluccan cloves and relied on intermediaries from the Malacca Straits, like the Kingdom of Srivijaya after 700 CE, to provide them.[75]

An Indonesian map of 1500 CE would have presented the Moluccas as a chain of tiny islands. However, the economic reality was very different. Large fortunes were generated from the mace and nutmeg produced on the Bandas and the cloves cultivated on Ternate, Tidore, and (later) Ambon. It was a series of middleman price mark-ups and transportation costs, not the products themselves, that made spices so expensive in Europe. Spice cultivation was relatively simple. A clove tree needed twelve years to generate its first dried flower-buds (cloves) but continued to produce for around twenty-five. Nutmeg required ten years to bear fruit but produced for at least sixty. The key point was that a quintal of cloves that could be purchased for one or two ducats in Ternate or Tidore could be sold for at least ten ducats in the spice markets of Malacca. By the time the cloves reached Europe, assuming that the quality hadn't been spoiled by prolonged exposure to salt water, they would fetch prices in excess of twenty.[76]

The Spice Islanders were stuck with Malacca. The feisty Bandanese maintained a trade representative in town and operated a small fleet of spice ships, but they rarely made the voyage to the straits. They preferred to exchange their mace (around two thousand quintals per year) and nutmeg for the rice and textiles delivered by traveling Malays and Javanese. Ternate and neighboring Tidore depended entirely on foreign freighters, reserving their own primitive fleets of *kora-koras* for local military actions. According to Pires, who was prone to exaggeration, Ternate exported twelve thousand quintals of cloves and boasted a harbor that could handle as many as three cargo ships at a single time. Pires estimated that another 17,000 quintals in annual clove output was contributed by Tidore, with no deep-water harbor at all, and the nearby islands of Motir and Makian combined. If each of the Spice Islands had their own specialty, the bulk of the spice profits flowed to the Sultan of Malacca. The islanders guessed that the Portuguese could not be any worse.[77]

After the conquest of Malacca in the summer of 1511, Albuquerque dispatched António de Abreu, three Portuguese ships, and local pilots to find the legendary Spice Islands. If the tale is true, the adventures of one of the three captains, Francisco Serrão, helped to solve Portugal's geography problem. Serrão was forced to abandon his outbound vessel off the island of Ceram, but he managed to obtain a local junk and load it with cloves, nutmeg, and mace from the otherwise unknown Bandas. Serrão's problems continued when this second (spice-laden) junk sank during the return trip to Malacca. The star-crossed Serrão and six other sailors were rescued at sea by local fishermen and carried to the otherwise unknown island of Ambon. News traveled fast in the Moluccas. When Serrão's ordeal was reported to Sultan Abu Lais of Ternate, an island that was known but not yet located, the sultan immediately sent a ship to Ambon to ferry the stranded Portuguese to Ternate.[78]

The Sultan of Ternate had been anxious to do business with the mysterious Portuguese after hearing about their conquest of Malacca. The sultan had also learned that the ruler of Cochin (India) had been enriched by trading pepper and spices with the Portuguese—that was enough for him. Since Malacca was already the primary market for Ternate's spice output and that of the other islands, the sultan seized the opportunity to interrogate Serrão as a representative of the new regime. Both sides were interested in doing business. In 1513, the sultan permitted the Portuguese to establish a fortified warehouse on Ternate and agreed to maintain his traditional clove supply relationship with Malacca.[79]

The agreement with the Sultan of Ternate prompted a breathless Tomas Pires to take stock of what the Portuguese had achieved:

> Men cannot estimate the worth of Malacca, on account of its greatness and profit. Malacca is a city that was made for merchandise, fitter than any other in the world. [T]he trade and commerce between the different nations for a thousand leagues on every hand must come to Malacca [and] while merchandise favours our faith, [the] truth is that Mohammed will be destroyed, and destroyed he cannot help but be. And true it is that this part of the world is richer and more prized than the world of the Indies. ... Whoever is lord of Malacca has his hand on the throat of Venice. As far as from Malacca, and from Malacca to China, and from China to the Moluccas, and from the Moluccas to Java, and from Java to Malacca [and] Sumatra, [all] is in our power.[80]

The triumph at Portuguese Malacca completed a remarkable period in world history. In seventeen short years, between 1498 and

1515, the Portuguese had conquered Mozambique, Hormuz, Goa, and Malacca, commenced trading activities with China (temporarily) and the Spice Islands and launched a commercial empire in the Indian Ocean. If few of these places could be located on a map in 1498, no European ship had sailed into these waters since Roman times. An apparent Golden Age, facilitated by heavily armed ships, huge volumes of Hungarian copper, and the arbitrage of this copper against Asian pepper and spices, was enriching the coffers of Manoel I, Jacob Fugger, and their business partners—and overshadowing Spain's otherwise exciting initiatives in the Caribbean. Meanwhile, the staggering fifty thousand quintals of copper produced annually by the Fugger-Thurzo Saigerhutten, the key to the entire system, was consolidating Antwerp's position as the most powerful entrepôt in Europe.[81]

# CHAPTER TWO

# THE SPANISH HEAD WEST

E ven the well-informed Jacob Fugger would not have guessed in 1498 that it would be Portugal—not Istanbul, Venice, Genoa, Florence, or any other European state—that would provide a shock to the Muslim economic order in the Indian Ocean. But after da Gama had reached the western (Malabar) coast of India, a foundation was established for the subsequent Western European domination of the globe. The Portuguese triumph hinged on the ability of a mineral-poor kingdom to access vast quantities of precious metals—African gold and, above all, Hungarian copper—to effect trading relationships with a Muslim economic system that placed a very low value on European trade goods. The Mediterranean economy was jolted by the increasingly large volumes of pepper, spices, and other Eastern goods that the Portuguese were exchanging for Fugger metals. By year-end 1515, the Portuguese had established Goa as their Indian Ocean headquarters, captured Hormuz and Malacca, visited China, and cut a clove supply contract with one of the fabled Spice Islands.

Portugal's commercial triumphs in West Africa had already demoted the Kingdom of Castile to junior status on the Iberian Peninsula. Fortunes had sunk so low in 1475 that Castile even had to ward off a brazen military invasion by Afonso V. Castile's independence was preserved, but the message was clear—Portugal viewed herself as the senior Iberian power. While Castile would shortly respond with a host of territorial additions of her own— Murcia, Granada, Navarre, Naples, Sicily, Sardinia, and a few North African ports—the Castilian Crown of the late fifteenth century represented a Christian state that was poor, agrarian, isolated, medieval, and intolerant. So was Portugal, but the Portuguese had managed to overcome these traits with foreign enterprise and ample portions of dumb luck.

The marriage of Isabella of Castile to Ferdinand of Aragon in 1469 had led to the unification of the two Spanish kingdoms in 1479 and ended a series of bloody civil wars. Isabella encouraged energetic Spaniards to participate in the West African trade and sent her privateers against the Portuguese during the campaigns of 1475–1477. This was a mistake. The Castilians were no match for the Portuguese fleet, and as many as thirty-five Spanish ships were defeated by Portuguese commanders in 1478. The resulting Treaty of Alcacovas in 1479 was a clear victory for Portugal—Portugal retained all rights to West Africa, the Madeiras, Azores, and Cape Verdes in exchange for renouncing any future claims to Castile (and the Canary Islands). In short, the treaty freed Portugal to continue her expansion to the south (and to the east).[1]

Isabella and Ferdinand shifted their attention to the last remnants of Spain's non-Christian communities. It had been the collapse of the legendary caliphate of Córdoba (808–1031) that had launched the papacy's first Iberian Crusade in 1064 and allowed southward-moving Christian forces to recapture Toledo and the northern half of Portugal by 1085. After overcoming an invasion by North African Almoravids in the early 1100s, the Christians triumphed at the Battle of Las Navas de Tolosa (1212) and reconquered Muslim Córdoba, Muslim Seville, and the Portuguese Algarve by 1249. Christian Iberians lived side-by-side with two non-Christian groups—an ancient Jewish community and the 300,000 or so inhabitants of Muslim Granada—for the next 230 years. Then Isabella and Ferdinand decided to stir things up. The Crown instituted an Inquisition against suspected heretic Conversos (converted Jews) and Moriscos (converted Muslims) in 1480, invaded Granada in 1482, and extorted large sums from Jews, Conversos, Muslims, and Moriscos to finance these and other royal initiatives.[2]

By 1492, the nearly eight-hundred-year-old Kingdom of

Granada had been subdued, Spain's Jewish population had been forced to convert to Christianity (or leave Spain for good), and Isabella and Ferdinand had unified Castile, Aragon, Andalusia, and Granada into a single political entity. While the reconquest of Granada, the last remaining Muslim kingdom in Iberia, was viewed as a spectacular triumph for Christian Europe, unified "Spain" remained medieval, fragmented, and decentralized. The Crown continued to share political powers with powerful (untaxed) nobles and depended on a very modest stream of tax revenues from the *alcabala* (sales tax), the irregular papal *cruzada*, and the irregular *servicio*. State revenues, expenses, and borrowings were very modest by future standards but still needed to be managed. The Crown controlled her military costs by federalizing local militias under the Inquisition, selling annuities and government certificates (*juros*), and resorting to tribute (and extortion). The Kingdom of Granada paid an annual tribute of twenty thousand gold ducats until 1492.[3]

If the royal marriage was a partnership in most respects, Isabella's Castile held sway over the commercially inclined regions of Ferdinand's Aragon. A leadership role during the Reconquest and an effective control of Andalusia gave Castile the political clout to monopolize the subsequent conquest of the Americas from Seville. This meant that Barcelona, Aragon's traditional (Catalonian) powerhouse in Mediterranean commerce, would be excluded almost entirely from the transatlantic trade. Of course, Castile (including Andalusia) was the largest kingdom in Iberia, accounting for roughly 4.5 million of the peninsula's 8 million or so inhabitants in the 1490s. This was over four times the population of Portugal and Catalonia. Spain's largest cities were Seville (50,000), Granada (50,000), Toledo (30,000), Valencia (30,000), and Barcelona (25,000). Backwater Madrid had less than 5,000 inhabitants.[4]

The ability of Isabella and Ferdinand to unify Spain, capped off by their spectacular conquest of Muslim Granada, provided a platform to invest in Christopher Columbus's latest business scheme in 1492. Columbus had pitched a similar proposal to the royal couple in 1486, just prior to Bartholeméo Dias's expedition to the undiscovered Cape of Good Hope and two years after he had been rebuffed by João II. Columbus must have been frustrated with the Portuguese. He had a commercially minded Genoese background, hands-on experience with the Gold Coast of West Africa, at least two years of residence in Lisbon (and Madeira), and shared Portugal's interest in a possible western route to China (Cathay) and Japan (Cipangu). France and England turned down Columbus as well, and a second proposal to the Portuguese Crown was dashed by Dias's triumphant return from the cape in December 1488.[5]

The Portuguese most probably believed that Columbus underestimated the distance between Europe and Asia. While Columbus apparently relied on the outdated calculations of the Florentine school of cosmography, he may not have known that a new generation of Portuguese cosmographers had nearly corrected Ptolemy's longitude errors of yore. They estimated the distance between Portugal and the Far East at 183 degrees or more, well beyond the Florentine projection (135 degrees) and relatively close to the actual distance of 217 degrees. Armed with these projections, João II must have assumed that the Cape of Good Hope offered a much shorter route to the spice and silk markets of Asia than Columbus's westerly proposal. Not only that, João probably recognized that the closure of China's doors to foreign trade had shifted the business objective from the Far East to closer-to-home marketplaces like India and Malacca.[6]

In 1492, Castile was well behind Portugal in matters concerning cartography, cosmography, and maritime skills. But if timing is everything, Ferdinand, Isabella, and Columbus should all be credited with seizing a high-risk/high-reward business opportunity. The task was to beat the rival Portuguese to the spice markets of South Asia. An introduction from Luis de Santangel, an Aragonese Converso with strong ties to Ferdinand, Genoese investors, and the powerful Santa Hermandad, was all it took for Columbus to scrape up the 2,933 ducats needed to secure three ships and a ninety-man crew. He also received royal privileges and appropriate noble titles based on assurances that he would deliver a new (Atlantic) sea route to the spice and silk riches of Asia. The departure from Palos on August 2, 1492, went smoothly. So did the ten-week sail across the uncharted Atlantic Ocean. Columbus landed in the Bahamas on October 12 and made brief visits to the neighboring islands of Hispaniola and Cuba. The Great Navigator assumed that he had discovered a new island chain off the eastern coast of Japan.[7]

Wherever he was, Columbus was especially thrilled by the inadvertent discovery of gold on Hispaniola. One of his wayward lieutenants, Martin Alonzo Pinzon, managed to find traces of alluvial gold during an unauthorized expedition to the mountainous Cibao country. The Cibao was located fifty miles south of the northern coastline. Otherwise, Columbus observed no unworked gold during his initial Caribbean travels and apparently confused the ornamental *guanin* worn by local Taino chiefs with the genuine article. None of the Caribbean peoples practiced mining and metallurgy, but they traded with mainland cultures in Colombia and Venezuela who did. The better news was that the backward Tainos had left pristine deposits of gold to be exploited by enterprising Spaniards.[8]

The independent-minded Pinzon had made his discovery after anchoring the *Pinta* at Montecristi on Hispaniola's northern (Dominican) coast. Columbus had failed to find any signs of gold during his coastal excursions and had had to abandon the *Santa Maria* after arriving on the island's western (Haitian) side on December 6. Pinzon was more successful. When he inquired about the source of some unworked gold nuggets that were deemed to be worthless by local natives, he was directed to the interior Cibao region. The evidence is murky, but Pinzon and his crew had ample time to visit the Cibao before reporting their discovery back to Columbus (off Montecristi) in early January. The find guaranteed that Columbus would be making a return visit to Hispaniola. After constructing a frontier fort at La Navidad, he sailed for home on January 16. Pinzon died shortly (and possibly even suspiciously) after his arrival at Palos.[9]

Columbus was carrying samples from Spain's first significant mineral strike in nearly 1,500 years. Few if any people knew in 1492 that the Iberian Peninsula had once served as the world's leading source of minerals for over a millennium. In fact, the region surrounding Seville had drawn the first Phoenician metal traders to Iberia in around 2000 BCE. The Phoenicians smelted huge quantities of copper, bronze, and silver through 700 BCE, confirmed by the millions of tons of waste slag left behind at Huelva and Tharsis, and attracted a series of mineral-centric invaders from Greece, Carthage, and Rome. It took the Romans nearly two hundred years (from 218 BCE to 19 BCE) to conquer the entire peninsula, but they mined as they conquered—well over one million tons of copper from Rio Tinto, huge volumes of silver from Cartagena, gold from Asturias, mercury from Sisapo (Almadén), and tin from Galicia. Not much was left behind for the succession of Visigoths, Umayyads, Almoravids, Almohads, and native Christians who followed. In fact, it was stated that if the Romans hadn't exploited a particular mine, the mine had not been worth the effort. Medieval Iberians, like the Italians, were dependent on the Berber caravans for their gold until Portuguese caravels began to visit West Africa.[10]

If Isabella and Ferdinand were ignorant of Spain's remarkable mining heritage, they filed claims to the West Indies immediately after Columbus's return on March 15, 1493. Little did they know, however, that the wily Columbus had made an unscheduled stop at Lisbon en route to Spain. Whether because of unfavorable weather, Portuguese naval pressure, or respect, the Great Navigator briefed Portuguese officials about his "Asian" discoveries—apparently leaving out mention of the gold nuggets—even before he had a chance to inform his Castilian sponsors. João II immediately claimed the new islands as a possible eastern extension of the Portuguese Azores. But he declined to send a follow-up fleet for confirmation. The curious incident suggests that the enigmatic Portuguese Crown was either indifferent to a possible Atlantic trade route to "Asia" or was still unclear about the proper demarcation line.[11]

By year-end 1493, Spanish pope Alexander VI had issued an *Inter Caetera* bull to confirm Castile's claim to Columbus's discoveries and to induce the two Iberian crowns to negotiate their global ambitions. Although Portugal had suspended her invasion of the Indian Ocean since the return of Dias in 1488, Castile inexplicably agreed to Portugal's request for a new demarcation line. In 1494, there were no such things as North and South America, no such thing as the Pacific Ocean, and precise longitude lines would be unavailable for nearly two centuries. Still, Isabella and Ferdinand agreed to move the imaginary north-south boundary meridian from the current line, 100 leagues west of the Azores, to a line 370 leagues (1,200 miles) to the west. The Treaty of Tordesillas, signed in June 1494, divided most of the underexplored globe between the two Iberian nations. The revised line of demarcation "ceded" Africa and Asia to Portugal and most (but not all) of the Americas to Spain. By extending the line westward another 270 leagues, the result of skilled Portuguese cosmography and fishermen's' reports of possible land masses, Portugal managed to deprive Spain of Brazil. No one could have known in 1494 that undiscovered Brazil jutted out from the South American continent just enough to fall inside the revised Portuguese line.[12]

Of course, the division of the world into two halves was laughable to other Europeans, Ottomans, Chinese, and scores of other rulers. In reality, the Treaty of Tordesillas applied mainly to Iberians (and possibly Italians) and would be challenged for decades. Another papal bull, *Ea Quae*, would have to be issued in 1506 to confirm that the demarcation line of 1494 was in fact binding. By then, Portugal was colonizing Brazil and receiving lucrative spice cargoes from Asia. No one had yet sighted the Pacific Ocean (from the east), let alone crossed it, and longitude lines of any kind were still a mystery. Another attempt at precision would be made in 1514, after Balboa had sighted the Pacific Ocean from Panama. The *Praecelsae Devotionis* of that year limited the "Spanish half" of the world to territories bordering the Atlantic Ocean (except Brazil) and confirmed that Portugal controlled the "heathen" lands (Africa and Asia) that were reached by sailing east. The hazy demarcation line would need to be reinterpreted once again after the Magellan party's circumnavigation of 1519–1522.[13]

## COLUMBUS INVADES HISPANIOLA

In the meantime, the *Inter Caetera* bull of 1493 allowed Columbus to leave Cadiz in September 1493 for a second expedition to "Japan." This and most of the outbound Spanish voyages that followed were managed by Juan Rodriguez de Fonseca, archbishop of Burgos and the de facto director of Castile's American enterprise for the next two decades. Fonseca opened the royal coffers to the tune of seventeen ships, 1,200 men—including so-called miners and metallurgists, friars, craftsmen, and agricultural specialists—and enough equipment and provisions (except food) to colonize Hispaniola or any other "Asian" island. The colonization plan was modeled on the scheme that had been implemented on Grand Canary Island between 1490 and 1493. There, the inconvenient Guanches were reallocated to Spanish settlers as *encomiendas*. Fonseca's hopes for Hispaniola were so high in 1493 that he even dispatched a royal financial officer, a custom duties collector, and an accountant to set up a financial reporting system for the new venture.[14]

When the stakes were still low, the financially strapped Columbus had cut an attractive business deal with the Castilian Crown. His financial backers provided virtually all of the capital and left Columbus personally responsible for only one-eighth of the cost of the provisions. Columbus's first two voyages to Hispaniola entitled him to one-tenth of the net proceeds of the venture plus additional incentives. These potential returns compared very favorably with the Crown's 20 percent share (quinto) of the enterprise and would make Columbus a very wealthy man when sizable deposits of gold were finally discovered in the highland streams of the Cibao.[15]

As an employee of the Centuriones, one of Genoa's sharpest merchant banking families, Columbus had sailed to Madeira in the early 1470s and had visited a number of Portuguese feitorias in West Africa. His first voyage to the West Indies may have been focused on "Asian" spices, but he and his crew became obsessed with gold. The word *gold* appeared in his diary over sixty-five times between October 12, 1492, and January 17, 1493. The return visit in early 1494 was centered almost entirely on gold. Columbus was instructed to mine for gold, build settlements, establish trade, and convert the native Tainos to Christianity. No matter that the Tainos refused to cooperate. They destroyed the tiny settlement (La Navidad) placed on the island's northern coast and prepared to defend their territory with their lives. The forceful Spanish response that followed left Hispaniola to be conquered in the same manner as the Canaries and Muslim Granada. The Portuguese and Italians had sought to develop commercial factories in host nations, but Castile, a newcomer to global affairs, was fully committed to territorial conquest and expansion.[16]

The Caribbean islands, almost always deemphasized in any discussion of the European conquest of the Americas, served as a laboratory for Spain's future endeavors. The islands had been settled initially by peoples from Central America and the Yucatán Peninsula prior to 3500 BCE. Much later, between 500 and 250 BCE, the Saladoid people of Venezuela invaded Puerto Rico and gradually spread out to Hispaniola, Cuba, and Jamaica. The Saladoids, predecessors to the Tainos, introduced primitive agriculture and ceramics to the island chain. Progress was slow and lagged far behind the dynamic civilizations that had taken root in Mexico and Central America. Political entities, chiefdoms, and ceremonial centers did not emerge on Hispaniola until around 1000 CE and remained relatively simple until the arrival of Europeans in 1492. A single Taino chief might control as many as one hundred villages, some with over two thousand inhabitants and sizable numbers of slaves, among a native population that numbered in the hundreds of thousands.[17]

The Cibao nuggets delivered to Castile in March 1493 had prompted Fonseca to recruit a number of so-called mining professionals for Columbus's return visit in early 1494. Two officers, forty men, and two Taino guides were dispatched from the island's northside station at Isabela to explore the rocky, pine-forested Cibao region in the interior. The Spaniards would in later years discover nearly identical central cordillera topography in Cuba and Puerto Rico. Columbus himself arrived in the Cibao in mid-March and ordered the construction of a frontier fort (Santo Tomas), sited seventy miles south of Isabela, to support the prospecting activities. The project was a failure. The fifty men stationed at Santo Tomas failed to find any gold, and the Tainos knew even less about prospecting and mining than the incompetent Spaniards. Columbus's men were prospecting the upper streams of the Cibao when placer gold, the by-product of weathering and erosion processes, was usually located in the lower-level slates.[18]

Columbus's brother Bartolomé was somewhat more successful. Led by Taino guides, he was directed to the southeastern San Cristóbal region along the Haina River. By August 1496, he had constructed a second fort (Bonao) and a cross-island road to link Isabela to a new port facility (Santo Domingo) on the south side of the island. Santo Domingo, sited ten miles east of the Haina and twenty-three miles south of San Cristóbal, straddled territory that was discovered to be surprisingly fertile. While the gold fields of San Cristóbal (and the Cibao) would eventually be the most prolific in the Caribbean, this happy outcome was quite unclear in 1496. Fonseca's so-called mining professionals declined to prospect systematically. The Spaniards fanned out in helter-skelter fashion, looking for visible nuggets when the actual placer gold was usually submerged in the downstream riverbeds.[19]

The placement of Santo Domingo on Hispaniola's southern coast began a process through which native peoples (Tainos) and imported slaves (initially from the Bahamas) were forced to provide manual labor for Spain's mining and colonization activities. The Columbus brothers and their successors transferred to the New World the same *repartimiento* system of forced labor that had recently colonized the Canary Islands. It was adapted from the same feudal-style, agriculture-based economic system that had been introduced during the Reconquest and managed by an alliance among the Crown, church, military, and large agricultural landowners. Unfortunately for Spain, there was no evidence of spices on Hispaniola, let alone a spice trade, and the driving force of the colonization effort, gold, was proving to be a disappointment. In addition, the local Tainos were prepared to strike back. A coalition of fourteen chiefdoms had to be defeated by force in 1497. When the island's modest gold-mining sites appeared to falter, Spanish "gold rushers" revolted against the Columbuses and sent the brothers back to Spain (in chains) in the fall of 1500. The family's political powers were suspended until 1509.[20]

Francisco de Bobadilla, the newly appointed governor of the Indies, arrived at Santo Domingo in August 1500. He declared immediately that the mineral deposits of the Indies were now property of the Crown, not the Columbus family, and that Spaniards were now free to prospect and mine at will—as long as the spoils were shared with the Crown. By then, Isabella and Ferdinand had revised Castile's chaotic currency system to improve financial controls and to entice foreign investors. In 1497, the value of the gold *excelente* (worth one Venetian ducat) was set at 375 *maravedis*, a silver real was valued at 35, and the copper vellon-based *blanca* (for small change) was set at one-half. These revisions widened the silver-to-gold exchange ratio to 10.7:1 and encouraged gold prospecting activity of any kind. Taken together, the reforms unleashed an immediate gold rush in the San Cristóbal district— over six hundred pounds of Hispaniolan gold was produced in 1501 with much more to come. No matter that Bobadilla and most of his fleet were lost in a hurricane in 1502.[21]

It had taken the Portuguese over three decades to strike gold in West Africa, through trade rather than through prospecting and mining, but Hispaniola was generating meaningful quantities of gold in 1501. Still, Columbus's dreams of a Caribbean El Dorado or a huge Asian spice venture were disappointed. Columbus was a brilliant navigator, a creative businessman, and a very mediocre administrator. Although he died wealthy in 1506, it would be left to his son, Diego Colón, to reap some of the riches that were uncovered in Hispaniola, Cuba, and Panama after the family was restored to power in 1509. Columbus's first two expeditions were financial busts and his later voyages (1498–1500 and 1502–1504) would be remembered more for their exploration value than their economic returns—the latter were highlighted by discoveries of pearls on the northeast coast of Venezuela in 1499 and traces of gold in Panama in 1503. By then, Columbus and his Spanish sponsors were faced with a new competitor in the Americas. Portugal's Pedro Alvarez Cabral, en route to India in April 1500, landed on a landmass that turned out to be Brazil.[22]

Cabral's accidental discovery of Brazil was not the only expedition catalyzed by the exploits of Christopher Columbus. In search of a possible northwesterly route to Asia, the English Crown had hired the Italian John Caboto (Cabot) to determine if the five-hundred-year-old Viking sailing legends were true. They apparently were. Cabot reached Newfoundland in 1497 and may have reached Nova Scotia (and even Maine) during a return voyage in 1498. No matter that Cabot's efforts were commercial failures. Portuguese navigators followed up almost immediately with expeditions to Newfoundland, Labrador, and Nova Scotia in 1501. Manoel I even claimed these barren territories under the Treaty of Tordesillas.[23]

Amerigo Vespucci, regarded by historian Samuel Eliot Morison as "the most controversial character in the history of discovery," was more successful. Having represented a Sevillian affiliate of the Medici family since 1491, the forty-four-year-old Florentine helped to outfit Columbus's third voyage to America in 1498, sailed with Alonso de Ojeda to the Venezuelan coast (for Spain) in 1499, and joined Goncalo Coelho's three-ship fleet to Brazil (for Portugal) in 1501. After receiving advice from Cabral himself during a layover in Senegal, Coelho and Vespucci landed near Fortaleza on the northern coast in mid-August and hugged the Brazilian coastline south to Cabral's landfall site near Porto Seguro. The territory was dangerous—one unfortunate sailor was ambushed, roasted, and eaten by local cannibals. Two stranded survivors from the Cabral expedition added some tales of their own. Coelho and Vespucci subsequently reached (and named) Rio de Janeiro on January 1, 1502, and then sailed another four hundred miles to the southern border of present-day São Paulo State. From there, following separate expeditions to the South Atlantic islands, Sierra Leone, and the Azores, the two explorers joined up in Lisbon on September 5, 1502. They were already preparing for a follow-up visit to Brazil in 1503.[24]

Whether the actual navigation was directed by Coelho or by the self-promoting Vespucci, the ability of *someone* to trace the Brazilian coastline down to a point south of Rio de Janeiro was as shocking as da Gama's recent achievements in the east. It proved that Columbus's island discoveries were nowhere near Japan or the Azores. A huge continent, rivaling in length the recently discovered coastline of West Africa, was located in the middle of the world. The achievement was so great that cartographer Martin Waldseemuller, half in jest, attached the Florentine's name to the new territorial mass that appeared on his landmark map of 1507. Of course, great excitement was generated in the royal chambers of Castile. An entire Spanish-owned continent (excluding Brazil) was waiting to be explored and stripped of her mineral resources. The lucrative spice markets of Asia could wait. The challenge for Castile was simple— find and extract the inevitable supplies of precious metals that were located elsewhere in the New World. If tiny Hispaniola contained a decent amount of gold, the vast, uncharted mainland territories were bound to contain a whole lot more. Vespucci was eventually recruited to Seville in 1508 to serve as pilot major of the Casa de la Contratación (House of Trade).[25]

The lure of gold (and possibly much more) attracted incoming governor Nicolas de Ovando to Hispaniola in 1502 with a huge armada of thirty ships and 2,500 Spanish colonists. Thousands of repartimiento natives and slaves were organized to work the island's two principal gold fields more aggressively. Ovando also cracked down on hostile Tainos and began a pattern of military conquest that would be repeated many times over in future decades. By the fall of 1503, Hispaniola's last independent chiefdoms were crushed and entire villages were assigned as encomiendas to Spanish settlers. The encomienda system of "entrusted" labor, a more formal version of repartimiento, had been a regular feature of Spanish policy since the Reconquest. Reconquered Muslim villages in Spain were allo- cated to the Castilian military orders for "care," conversion, and forced labor. On Hispaniola, Taino chiefs were charged with the unenviable task of supervising their subjects' four-to-six-month stints in the gold fields and agricultural estates.[26]

The receipt of an encomienda was a reward for entrepreneurial initiative. Spanish explorers and would-be conquistadors were all required to self-finance their expeditions to the New World. The Crown had limited capital resources to fund the ships, supplies, and soldiers that were needed for these high-risk ventures, but it had the exclusive power to authorize expeditions, governorships, and encomiendas. The expeditions themselves can only be categorized as regulated entrepreneurial affairs. The deal was that both the conquistador and the Crown had the same rooting interest in the success of the venture. Most of Castile's lower nobility was comprised of cash-poor *hidalgos*, and only the Sevillian Genoese had access to "venture capital" and a willingness to invest. The Genoese provided personal loans, outfitted vessels, and advanced funds to Sevillian merchants to acquire and transport merchandise across the Atlantic. The conquest of the Americas was a low-budget, stepping-stone process through which each entrepreneurial conquest provided a platform for the next.

Columbus's first two voyages to Hispaniola had been tightly controlled affairs. Private individuals were prohibited from carrying trade goods of any kind across the Atlantic, and all persons and cargo had to be registered by a royal treasurer, comptroller, and factor. While a royal decree of April 10, 1495, had permitted any Castilian subject to participate in the transatlantic trade, this brief fling with open entrepreneurship was ended by Columbus's demotion, Bobadilla's gold strikes, and Ovando's rule as a military dictator. However, the death of Queen Isabella in November 1504 prompted Ferdinand, a commercially minded Aragonese, to restructure the entire transatlantic process. By then, Portugal's spectacular successes in the Indian Ocean and additional Spanish discoveries in the Americas had raised the economic stakes considerably.[27]

Investment capital was a limited commodity in Castile, leaving Ferdinand to rely heavily on Genoese merchant bankers during the early decades of the sixteenth century. Sevillian Genoese had invested less than seven hundred ducats in Columbus's maiden voyage, but even this modest amount had not come easily. The credibility of one Genoese banker, Francisco Pinelo, was critical to Luis de Santangel's efforts to arrange the entire financial package. Of course, it helped that Columbus was a Genoese, had worked for the Centuriones between 1477 and 1479, and had assisted Genoese efforts to develop a slave-based sugar industry on Madeira. Pinelo was energetic. He helped to finance Columbus's second voyage, guaranteed a 13,320-ducat loan to the Duke of Medina Sidonia, and arranged the Genoese syndicate that sent Juan de la Cosa and three ships to the Isthmus of Panama in October 1500. Two of de la Cosa's vessels were lost off Santo Domingo, but the third delivered enough Panamanian plunder to meet plan. Pinelo was rewarded with an offer to serve as the Casa de la Contratación's first royal factor.[28]

Since Genoese merchants had been providing these types of services for centuries, most recently in the Canaries, Madeiras, and West Africa, the business model was relatively clear. The royal monopoly was essentially licensed through the Casa de la Contratación to a guild-like association of Sevillian merchants and the mainly Genoese merchant bankers who financed them. Of course, the Crown attempted to limit these entrepreneurial opportunities to Castilian subjects. Unauthorized immigrants were subject to fines, beatings, or imprisonment. These punishments were rarely enforced, however, and forged document services became a vibrant business in Seville. Inquisition-inspired prohibitions against "infidels and heretics"—Jews, Conversos, Muslims, and Moriscos—were also unenforceable, especially when it was discovered that many "heretics and infidels" possessed badly needed capital, business skills, and contacts. Most of the restrictions were eased by special fees and side-deals. In fact, non-Spanish merchants dominated transatlantic business activities. Other than wool and the cloth trades, which expanded greatly in Burgos and Segovia, the Spanish economy was light in exportable products. Non-Spaniards were needed to arrange the importation of European goods through

Medina del Campo or Seville. A steady inflow of American metals and European trade goods would gradually move Seville into the "major leagues" of European commerce—not yet on the scale of Antwerp, Venice, or even Lisbon, but close enough.[29]

Seville's status was bolstered by the establishment of the Casa de la Contratación de las Indias (House of Trade) in February 1503. No matter that imports of Hispaniolan gold, sugar, and hides were very modest by future standards. The casa was a coup for Seville. The city was located fifty-four challenging miles up the Guadalquivir River and her port, San Lucar, was partly blocked by sandbars. But other candidates had problems of their own. Cadiz had a fine harbor and had been the principal port of departure during the early years of the transatlantic trade, but Cadiz was believed (correctly) to be vulnerable to seaborne attacks. Palos was too small, Malaga and Cartagena were on the Mediterranean side, and Bilbao was miles away from favorable trade winds. Seville had the necessary commercial infrastructure, the political clout, and, most importantly, a convenient location. The new casa was placed in the Alcazar Real, in the former offices of the Admiralty Court, and joined shortly by a new School of Navigation. Cadiz was approved as a second port of departure in 1508, but all incoming vessels from the Americas were required to land at Seville. If weather, warfare, or pirating interfered with the delivery, a ship's cargo would still have to be transported to Seville—even by land.[30]

The Casa de la Contratación, like Spanish trading policies in general, was modeled on the Portuguese system. Portugal had had nearly sixty years of experience managing its West African affairs when Manoel I consolidated a hodge-podge of trade offices into the Casa da Guine e India in 1501 (shortened to Casa da India in 1503). The Portuguese casa was redesigned to manage the Crown's emerging trade monopolies in the Indian Ocean and her network of far-flung feitorias. Although the Spanish casa was supported by a much larger population than the Portuguese version, the commercial objective was more challenging. Spain was dealing with primitive tribal cultures, an actual colonization program, and unknown territories. The number of Spaniards (250,000 or so) who immigrated to the Americas between 1506 and 1600 was over ten times the number that left Lisbon for Portuguese Asia. And while the Casa da India refrained from promoting Christianity until the emergence of the Jesuits in around 1540, the Spanish casa was partnered immediately with the Spanish church. This, in turn, provided a degree of structure to an otherwise brutal series of territorial conquests in the Spanish Americas.[31]

Like the Casa da India, the Spanish casa was intended only to regulate and administer trade between Spain and the Americas—in no way was the casa to participate in this trade directly. Commerce was reserved for Seville's merchant community, especially the local Genoese, and the long arms of the Castilian Crown. Having benefited from Portugal's gold, slaving, and sugar trade with the Atlantic Islands and West Africa, the Sevillian Genoese were very familiar with the Portuguese system. Their economic position in Seville had even been enhanced by the timely eviction of rival Converso, Jewish, Morisco, and Muslim businessmen. However, Genoese merchants had operated in Spain since the start of the Reconquest and had traded with Iberia and North Africa for centuries. Even earlier, they had helped to launch the First Crusade by sending a fleet of twelve warships to Antioch in 1096. The potential of the transatlantic trade, coupled with the loss of Genoa's eastern markets to the Ottoman Empire, attracted a new wave of Genoese immigrants to Seville between 1500 and 1530. Twenty-one of Genoa's twenty-eight noble merchant houses were represented in Seville by 1530.[32]

The coincident interests of the Genoese, the Castilian Crown, and the Casa de la Contratación made Seville the economic center of the transatlantic trade. Since outbound voyages left only from Seville (and Cadiz), Seville and the casa were given a virtual monopoly over the entire American enterprise. The monopoly also provided an economic boost to Spanish producers of wine, grain, wool, and olive oil, as well as the shipyards of Biscay, and the iron foundries of Bilbao and La Coruña. The flip side was that Mediterranean ports like Valencia and Barcelona were excluded entirely from an emerging economic bonanza. There would be no change in Seville's lofty status even after Fonseca's death in 1524. The principal architect of the Sevillian monopoly would have applauded the subsequent transfer of the casa's management functions to a newly created Council of the Indies. The council was intended to enhance the economic power of Seville and tighten royal authority.[33]

The system could be termed as flexible (or chaotic) in the early years. Following the termination of Ovando's autocratic regime in 1509, many Spanish Hispaniolans—including the restored Columbus family—demanded the creation of an independent legal authority to check the power of future governors. Fear and greed were not eliminated, but the Spanish Crown responded with a new system of colonial government. By 1511, Hispaniola was ruled by three separate entities—a governor, a treasury (the casa), and an Audiencia de Santo Domingo modeled on the tribunal system of Valladolid. No one could be trusted. To ensure that the Crown's financial interest in the precious metals industry was not circumvented, a chief notary of the mines was appointed in 1508. He and his fellow non-casa officers were eventually encouraged to participate in the various business enterprises themselves—this ensured that they, too, could be trusted.[34]

## CUBA

In the meantime, Hispaniola served as a stepping stone to Cuba in much the same way that the Canaries had served as a stepping stone to Hispaniola. Cuba had been settled in fits and starts by Saladoids and nomadic Ciboneys between 1000 BCE and 800 CE. The Ciboneys were eventually evicted from the eastern half by the Arawaks, a more advanced Taino-based people who had arrived from Hispaniola, and then by Hispaniolan Tainos themselves in the mid-1400s. When Columbus returned to Cuba in May 1494, focused on gold and still thinking it was a Japanese island, he encountered a native population of around 112,000—roughly 92,000 Arawaks,

10,000 Tainos, and 10,000 surviving Ciboneys. A deadly combination of warfare, overwork, and disease would reduce this total to a
mere 2,000 by 1557, but the start of the devastation process was
briefly delayed by Columbus's failure to find any gold in 1494. Cuba
was placed on the back burner of Spanish interest until 1508. For
that matter, so were Hispaniola and Puerto Rico. The underexploited
Cibao fields of Hispaniola were reworked with only limited success
until a genuine mining boom took hold in the 1510–1518 period.
Puerto Rico's modest amounts of placer gold, located in the same
central cordillera streams that yielded gold in Hispaniola and Cuba,
were exhausted shortly after their discovery in 1508–1509. Jamaica
had no gold at all. Unfortunately, the paucity of the region's mineral
wealth failed to spare native Caribbeans from the demographic carnage that was already underway.[35]

Cuba returned to the front burner after Sebastián de Ocampo
circumnavigated the island over an eight-month period in 1508 and
reported the first traces of gold. The discovery was made along the
Rio Arimao near the Bay of Xagua (Cienfuegos). It was not until
1511, two years after Diego Colón had been restored to the governorship of the Indies (1509–1524), that Diego Velasquez de Cuellar
and a force of three hundred Spanish Hispaniolans were authorized
to "explore for gold"—a phrase that meant "conquer Cuba." The
tentative plan widened into a full-scale invasion after a splinter force
of 150 men under Pánfilo de Narváez joined up with Velasquez's
forces at Baracoa on the island's eastern tip. Within four months,
native resistance was overwhelmed, half of the island was under
Spanish control, and plans were readied to colonize Cuba in the
same manner as Hispaniola (and the Canaries). Baracoa, renamed
Santiago de Cuba, served as Cuba's capital while the enterprising
Velasquez planted crops, raised cattle, and developed the island's
gold fields. Partnerships with Sevillian Genoese investors like Juan
Francisco de Grimaldo and Gaspar Centurione were crucial.[36]

Cuba nearly matched Hispaniola in metallic wealth. Placer gold
had already been discovered near the mouth of Rio Arimao, but
much larger deposits were uncovered in 1512–1513 in the mountain

streams of the central cordillera and those located north of the Sierra Maestra. Within four years, the mining towns of Sancti Spiritus, Sierra de Trinidad, and Bayamo were delivering an estimated 62,000 pesos of gold (74,400 ducats) to the primitive smelting plant at Santiago. Annual gold production expanded to an impressive 100,000 pesos during the 1517–1519 period and attracted as many as 3,000 Spanish prospectors to Cuba's highland streams. This was the peak. The volume of gold processed at Santiago dropped steadily during the 1520s and a short-lived strike at Jobabo in the early 1530s would fail to reverse the general trend. Cuba's economic future would rest mainly on sugar and tobacco.[37]

Sugar turned out to be a very big business. Profits from some of the early gold ventures had been reinvested in sugar cane, a cash crop that Columbus had transplanted from the Canaries during his second visit to Hispaniola. The West Indies sugar industry was built from the timely convergence of Genoese capital, the importation of West African slaves, low-cost government loans, and exemptions from import duties. The Genoese had been involved in the sugar industry for centuries, starting in the eastern Mediterranean and Sicily and, more recently, in Madeira, the Canaries, and West Africa. Genoese investors owned a number of water-powered sugar mills (*ingenios*) in Hispaniola, starting in 1509, in conjunction with a local Spanish partner and a Genoese slaving enterprise. While an ingenio required a much larger front-end capital investment (15,000 ducats) than a horse-drawn *trapiche*, a single ingenio was capable of producing over one thousand long tons of sugar and profits of over ten thousand ducats on an annual basis. The sixty ingenios operating in the West Indies by 1528 were usually sited next to a twenty-five- to thirty-acre sugar plantation and a cattle ranch. Herds in excess of twenty thousand head were not uncommon. Of course, a cash crop like sugar diverted land and labor from basic agricultural needs. Most of the islands' foodstuffs had to be imported.[38]

Santo Domingo, laid out in 1508 along the lines of Granada, would shortly receive more income from Hispaniola's sugar and cattle industries than from gold. The Genoese financed most of the twenty-four sugar mills operating on the island in 1523, supported by encomienda labor and imported Caribbean slaves. Cuba, Puerto Rico, and Jamaica would eventually have thriving slave-based sugar industries of their own. However, the same deadly combination of warfare, overwork, and disease was already threatening the viability of the encomienda system. Hispaniola's native population had plummeted to around sixty thousand in 1509, reflecting the same demographic collapse that would be repeated many times over in Spanish America, and most of the island's ten thousand Spanish residents declined to perform manual labor. Labor problems loomed.[39]

The elation caused by the gold, sugar, and cattle industries of the West Indies was not shared by everyone. One vocal critic was Bartolomé de Las Casas. Having joined the priesthood after participating in the military conquests of Hispaniola and Cuba, Las Casas was so outraged by the ongoing demographic collapse that he sought to end the encomienda system before it was too late. No matter that the encomienda was a pillar of the Spanish colonial system. While Las Casas's estimate that as many as three million Caribbean Tainos had died unnecessarily between 1494 and 1508 was probably six times the actual number, the brutal colonization pattern unveiled in Hispaniola had been extended to Puerto Rico, Jamaica, and Cuba with similar effect. The receding pool of encomienda labor prompted some Spaniards to raid the uncharted mainlands of Central America for slave labor and consider the importation of slaves from West Africa—someone had to work the gold mines, sugar fields, and cattle ranches. Ironically, it was Las Casas, the newly appointed Protector of the Indians, who encouraged the Crown to replace native American workers with West African slaves. The whole horrific process was jump-started in 1518 when Laurent de Gouvenot paid eight thousand ducats for an eight-year license to deliver up to four thousand West African slaves to the West Indies. The wily Burgundian, in turn, tripled his money by flipping the contract immediately to a more experienced Genoese syndicate.[40]

Compared with Portugal's achievements in the Indian Ocean, Castile's initiatives in the Caribbean appeared to be a longer-term project. An estimated fifteen tons of West Indies gold were exported to Seville by 1520, but the gold fields of Hispaniola were nearly exhausted, and those in Cuba had peaked. The corresponding one ton or so of annual gold production was less than Portugal's annual take from Africa. In addition, the Portuguese had not had to invest in costly mining works. However, there was some good news. The Spanish Crown's 20 percent share (the quinto) of the West Indian economy had surged from a mere 8,000 ducats in 1503 to 120,000 ducats in 1518. While this was only one-third of Manoel's receipts from the Estado da India, Portugal's Golden Age was drawing to a close, and the Caribbean sugar, slaving, and cattle enterprises were just getting started—on track to contribute 320,000 ducats to Charles V's coffers by 1535.[41]

The better news was that there were two Spanish-owned continents (excluding Brazil) waiting to be explored, and the West Indies—stripped, depopulated, and recast—provided the platform to do so. Spanish America was being folded into the Castilian system in the same way that Granada and the Canaries had been folded into the system since 1492. Most of the defects (and some of the virtues) had been repeated in the stepping-stone conquests of Hispaniola, Cuba, Puerto Rico, and Jamaica and would be repeated again on the American mainland. This meant that millions of native Americans were targeted for extermination. In 1519, the native population of undiscovered Central Mexico was as high as twenty-five million with millions more inhabiting the Central American strip between southern Mexico and Panama. In South America, the pre-Conquest population of Inca Peru stood at around nine million, followed by that of Colombia (three million) and greater Brazil (five million). All of these native populations would be devastated in due course.[42]

In the meantime, labor shortages and the depletion of the Caribbean gold fields had led scores of Spanish adventurers to Tierra Firme—Central America, Colombia, Venezuela, Guiana, and Brazil—in search of gold and slaves. Columbus had sent exploration parties to the region as early as 1499 and followed up with personal visits to Honduras and Costa Rica between 1500 and 1502. Local gold supplies were disappointing, but there appeared to be large numbers of potential slaves. No matter that Indian slaves that sold for around one hundred pesos each (including a 20 percent commission for the slave hunter) were subject to the same deadly diseases that afflicted their nominally free brethren. While labor needs and rumored gold deposits led Juan de la Cosa back to the Gulf of Darien (Panama) for another inconsequential visit in 1504, a follow-up expedition to the Isthmus of Panama in 1509 was anything but. De la Cosa and most of his one-thousand-man expedition party were killed off by tropical disease, starvation, and attacks by hostile tribes. One of the few survivors, Vasco Nuñez de Balboa, helped to stabilize an extremely shaky situation until Pedrarias de Avila arrived with reinforcements. Panama was proving to be a bust, and Central America did not look any more promising.[43]

However, the thinking was that if Cuba was capable of producing one hundred thousand pesos of gold annually, so could uncharted territories that appeared to be much larger in size. This is what drove scores of Spanish adventurers into the wilds of Panama. Would-be conquistadors also had a powerful ally in the person of Charles I. The death of Ferdinand in January 1516 had catapulted Charles of Habsburg, the erstwhile Duke of Burgundy, to the thrones of Castile and Aragon. If that were not enough, the eighteen-year-old monarch was scheming to have himself elected Holy Roman Emperor. No one seemed to care that the current emperor, Charles's grandfather Maximilian, was still very much alive. The likelihood that Charles would eventually inherit Maximilian's modest assets and huge liabilities made him keenly interested in finding another "Cuba" or maybe even a "Joachimstahl." Like the wily Ferdinand, a monarch whose political skills were lauded by Machiavelli, Charles sought to encourage rivalries among his most capable subjects. In Spanish America, the idea was to uncover the conquistadors with the best "business plans" and the strongest leadership skills. These persons would make Charles rich. This also meant that verbal agreements (and even contracts) could be overlooked when it was necessary to do so.

The tacit encouragement of rivalries would generate huge financial dividends and huge political headaches for the Crown. In Cuba, the fading gold boom prompted a number of local Spaniards to devise their own campaigns of plunder and conquest on the American mainland. Diego Velasquez schemed to be made *adelantado* over all territories (islands and mainlands) that could be discovered from Cuba. He even improved his position by marrying a niece of Juan Rodriguez de Fonseca. Since Diego Colón, son of the Great Navigator himself, was already the governor of the Indies, Velasquez's candidacy threatened to create another (probably deliberate) layer of confusion. As it was, Francisco de Garay, the lieutenant governor of Jamaica, was also allied with Fonseca and two of Velasquez's own subordinates (Pánfilo de Narváez and Hernán Cortés) were apparently scheming to circumvent their superior as well.[44]

## CORTÉS INVADES MEXICO

It would be the least powerful of these rivals, Hernán Cortés, who would ultimately win out and lead the first successful treasure hunt in the Western Hemisphere. Cortés was one of the giants of the sixteenth century. Born in the Extremaduran town of Medellín in 1485, Cortés was an educated son of a less-than-aristocratic hidalgo. He had been sent to the university town of Salamanca in 1499, but if he attended school at all, he dropped out in 1501 to pursue fame and fortune either in Italy or in the Americas. After an accident and illness delayed his maiden voyage to Santo Domingo until the spring of 1504, his early career was marked by a relatively uneventful rise through the bureaucratic ranks. Cortés was apparently content to quietly seize opportunities when they were presented.[45]

After arriving in Santo Domingo, Cortés headed straight to the gold fields. His achievements in mining are unclear, but his family connections with Governor Nicolas de Ovando were invaluable. He was appointed notary public of the small town of Azua, located roughly seventy miles west of Santo Domingo, and he received some farm land and a team of Indian workers. The smooth-talking, twenty-six-year-old Cortés must have done something right because he was hired as Velasquez's secretary in 1511 and then served as the royal treasurer of Cuba. Cortés's relationship with Diego Colón, the thirty-year-old heir to Columbus's governorship, is also unclear. Colón had returned to Santo Domingo in triumph in July 1509 with plans to conquer Cuba (before someone else did) and probably became aware of Cortés's service under Velasquez. According to at least one colleague, the esteemed Las Casas, Cortés was already regarded as a "cunning and cautious" man.[46]

Cortés feuded with Velasquez from the start but not enough to jeopardize Cortés's appointment to the post of Alcade of Baracoa and the accumulation of a modest fortune. In addition to investments in cattle, horses, and sheep, Cortés was receiving as much as three thousand pesos worth of gold and residing in a fine house in Baracoa (Santiago de Cuba) by 1514. A grateful Cortés even named Velasquez as godfather to his own daughter. While Cortés had no apparent interest or involvement in the early expeditions to Mexico, he paid close attention to Velasquez's schemes to gain royal confirmation for his unofficial (if well-earned) position of governor of Cuba—even if the confirmation was likely to circumvent the authority of Diego Colón. Cortés would be guided by Velasquez's example in due course.[47]

The lower Gulf of Mexico had been surveyed only barely by 1517. In February of that year, Francisco Hernández de Córdoba and two partners invested six thousand pesos in a three-ship expedition to the Yucatán "islands" in search of slaves and gold. One of the vessels was a loaner from Velasquez. Córdoba's 111-man party included Bernal Diaz del Castillo and other Spaniards who had served under Pedrarias de Avila in Panama. After a few weeks of cruising the Mexican coast, Córdoba sighted an unusually large native town, future Puerto de Carenas, located on a large "island" that happened to be the Yucatán Peninsula. Córdoba's friendly landing in early March was followed by a firsthand look at the town's stone buildings, temples, and gilded ceremonial objects. Two

natives were captured to serve as translators. Unfortunately for Córdoba, the translators failed to prevent the natives from turning hostile when the visitors began to scour for water (and gold). Fifty
Spaniards were killed in the action that followed. Many others,
including Córdoba (fatally) and Diaz, were wounded during a
retreat to the ships.[48]

Velasquez was less concerned about Spanish casualties than with
the fact that another "island" was waiting to be conquered. He
immediately organized (and invested in) a four-ship expedition to
Yucatán comprising four captains, none of them named Hernán
Cortés, 240 men, Córdoba's two native translators, and assorted
trade goods. When it was learned that the natives enjoyed wine,
Velasquez made sure to stock barrels of high-quality beverage from
Andalusia. The expedition was led by Juan de Grijalva, a possible
relative of Velasquez, and assisted by Pedro de Alvarado. Grijalva
was given the following instructions prior to his departure from
Santiago on May 1, 1518: "to acquire by barter as much gold and
silver as possible and if he saw that a settlement was advisable or
dared to settle [one], to do so; otherwise, to return to Cuba."[49]

The Grijalva fleet sighted the island of Cozumel, the Yucatán
Peninsula, and then a native town that compared in size to Seville.
The discovery would have been less shocking had the Spaniards
known that the region had been dotted with 30,000- to 50,000-
person kingdoms during the Classic Maya period (200–900 CE),
highlighted by a metropolis of 500,000 at Tikal, and that the peoples of the Yucatán were the primary survivors of the former Mayan
Empire. The Spaniards withdrew to Cozumel to make preparations
for a full-scale landing on May 11. The landing was a failure. The
natives advised the Spaniards to leave promptly and when the visitors declined to do so, seven of Grijalva's men were killed. Another
sixty were wounded in hand-to-hand combat. The loss of three of
his own teeth from a native arrowhead was enough for Grijalva—
he claimed the Yucatán for Spain and returned to coastal exploration. The fleet rounded the northwestern coast of the peninsula
and made a second landfall along the coast of Tabasco. The Tabasco

natives also advised the Spaniards to move on, but they tantalized their visitors with hints about large quantities of gold in "Culua" (Mexico). Grijalva moved on.[50]

When the northwesterly moving invaders landed at the mouth of the Tabasco River in early June, they were astonished by the snow-capped peaks that could be seen in the distance. This was no island. Grijalva's third landfall was more successful. Thinking that the Spaniards were associated with Quetzalcoatl, the ancient Feathered Serpent who had prophesied the arrival of white, bearded conquerors from the east, the local natives presented their visitors with gifts, gold, and hospitality. The Spaniards declined their offers of human sacrifices and human hearts, but they remained onshore for ten days—long enough for some interior exploration work and long enough to be monitored by some Aztec agents from Tenochtitlan (future Mexico City). No Spaniard could have known that their sponsor, Moctezuma, had ruled the powerful Aztec Empire as the ninth *Uei Tlatoani* (Great Speaker King) since 1503.[51]

Grijalva declined to build a settlement in Tabasco but sent Alvarado back to Cuba with gold samples, wounded survivors, and a request for reinforcements. When Alvarado returned to Santiago with the exciting news, Velasquez made his move. He again circumvented the authority of Diego Colón, who was conveniently away in Spain, by dispatching his chaplain Benito Martín to Spain with the gold samples and a request to be appointed adelantado of the newly discovered "Rich Island" (Yucatán Peninsula). Velasquez knew that Charles V had recently approved an adelantado for Francisco de Garay, the lieutenant governor of Jamaica, that authorized Garay to explore and conquer the upper regions of the Gulf of Mexico.[52]

Velasquez's more pressing concern was to organize a second expedition to find, rescue, or possibly even stop the Grijalva party in the Yucatán. He secured four of Grijalva's own ships for the enterprise and reluctantly appointed Hernán Cortés to serve as captain general. Cortés was instructed to take possession of all of Grijalva's discoveries (and any future ones) on behalf of Velasquez. If the enterprise hinged on the ability of Martín to win an adelantado for Velasquez in Spain, the written orders to Cortés dated October 23 were left intentionally vague to provide the utmost flexibility—to Cortés. It is probable that Cortés, an on-and-off secretary to Velasquez with a sizable fortune and demonstrated diplomatic skills, had worked out a deal with Velasquez's administrative staff. The deal would have allowed Cortés to draft much of the vague language that was contained in Velasquez's written instructions.[53]

The expedition plan was delayed by the sudden arrival of Grijalva at Matanzas, Cuba, on October 8. Velasquez was outraged. He had dispatched Grijalva to explore and possibly settle the Yucatán, not to focus only on exploration. Native gifts from the Tabasco River area were valued at no more than two hundred pesos and the six hundred gold-surfaced hatchets obtained in trade were actually made of copper. But there was some very good news. Grijalva had also discovered deposits of low-grade alluvial gold along some of the local rivers. These turned out to be worth over fifteen thousand pesos. Together, the Grijalva expedition netted roughly twenty-four thousand pesos of gold—nearly one-fourth of Cuba's annual gold production in 1518—and confirmed that the Yucatán was attached to a larger (Mexican) mainland. Yet Grijalva had failed to establish a Spanish settlement, and Velasquez feared correctly that the exciting discoveries were in danger of being exploited by someone else—rivals like Colón, Garay, Narváez, and possibly even Cortés.[54]

Colón was safely away in Spain, but Garay, a man who had sailed with Columbus in 1493, had been authorized to send Alonso Alvarez de Pineda and three ships into the upper Gulf of Mexico. Pineda left Cuba in late 1518 to explore the vast expanse of coastline between western Florida and present-day Tampico. The voyage was eventful. Pineda was the first European to sail up the Mississippi River (for possibly twenty miles) and he was the first to trace the entire northern coastline of the Gulf of Mexico. Unfortunately for Pineda, he and most of his crew were massacred by hostile natives at the mouth of Rio Panuco (Tampico). One of the surviving ships, commanded by Diego de Camargo, eventually met up with the Cortés party in Veracruz. By then, it was Cortés—not Garay,

Colón, Velasquez, or Narváez—who was in a position to invade Mexico.[55]

Having rewarded Grijalva's achievements with a demotion, Velasquez reconsidered his appointment of Cortés to lead the next expedition to Mexico. Pineda was en route to Florida, Colón and Narváez were indisposed, and Velasquez apparently had no intention of commanding the expedition himself. But on November 18, 1518, five days after Martín had won a coveted adelantado for Velasquez in Spain, the "cunning and cautious" Cortés decided to take matters into his own hands. It is unclear whether Cortés realized that he was being decommissioned, and it is unclear whether he or Velasquez had financed most of the seven-ship fleet that left Santiago that day—Cortés claimed later that he paid for nearly two-thirds of the cost. But it is clear that Velasquez watched helplessly as the fleet sailed out of Santiago harbor. Within days, he would order the arrest of Cortés for treason.[56]

Cortés's invasion fleet was expanded to eleven ships, 518 men, and a sufficient amount of supplies as it sailed along the Cuban coastline. During a ten-day stop at Trinidad, Grijalva's own home was used as a staging area to recruit additional ships, captains, soldiers, and provisions. The expeditionary force included Pedro de Alvarado, 32 crossbowmen, 16 cavalrymen and their horses, 200 Indian servants, royal agents, 10 cannon, and a future chronicler (Bernal Diaz del Castillo) of the entire enterprise. The fleet sailed past the western tip of Cuba on February 18, 1519, and headed for the Yucatán. The flotilla moved slowly. Cortés was careful to repeat Grijalva's stops (with a few skirmishes) during his two-month cruise between Cozumel and San Juan de Ulua (Veracruz). The latter, sighted by Grijalva in 1518, had been inhabited since around 3000 BCE.[57]

Two of the stops were more significant than they might have been. During a layover on Cozumel to repair a damaged ship, Cortés was surprised by the arrival of a canoe filled with three barely clothed Spaniards on March 13. The men had been shipwrecked off Yucatán during a voyage between Panama and Santo Domingo in 1511 and were more than happy to meet some fellow countrymen. One of the three, Father Geronimo de Aguilar, had learned Mayan in the interim and served as a translator when the Spaniards arrived at the Tabasco River on March 22. However, Cortés received an even greater prize during his three-week stay at the port of Potonchan—possibly present-day Frontera. On April 17, the local chief presented him with a slave girl, Malinali, who had been sold into slavery by her own mother. The key point was that the noble-born Malinali spoke both Mayan and Nahuatl, the language spoken in Aztec Mexico. The girl was christened shortly thereafter as Marina but gained fame as "La Malinche"—an interpreter, mistress, and loyal confidant to Hernán Cortés.[58]

Cortés founded Villa Rica de Vera Cruz shortly after landing his armada at nearby San Juan de Ulua on April 21. He immediately had himself elected alcade and captain general of Veracruz to assert his shaky legal authority in Mexico, co-opt the small Velasquez faction, and establish his personal loyalty to Charles I. The "election" was intended to exploit medieval codes, the *Siete Partidas*, that had been adopted by Alfonso X in the mid-thirteenth century. Cortés was betting that an *alcadeship* of a newly established Castilian "municipality" (Veracruz) would supersede the authority of Velasquez (and Diego Colón) in Mexico and allow him to operate as a direct subordinate to Charles I himself.[59]

Cortés's election scheme was bolstered by the gifts of gold, silver, and precious stones that he received from local natives. Few of the "electors" wanted to jeopardize their stake in a possible treasure trove, especially since the natives were continuing to associate the Spaniards with Quetzalcoatl. No matter that one of Velasquez's lieutenants, Francisco de Saucedo, sailed into Veracruz harbor unexpectedly on July 1 and informed Cortés that Velasquez had in fact received his royal adelantado on November 13, 1518. Cortés was undaunted. He prepared to dispatch the initial treasures to Spain immediately. He recognized that he had a very tenuous legal claim to his discoveries and that his only hope was to impress the powers-that-be with actual results. The treasures included a large piece of gold foliage, a large circular piece of silver, silver armlets and shields, collars made of gold and stones, twenty-six gold bells, gold pendants and headpieces, 232 red stones, 163 green stones, various cottons, and featherwork. The gold was carefully inventoried on July 6 and sent back with two "municipal representatives," six native Indians, and the first of five personal letters to Charles I. The letter hoped to persuade the monarch that Cortés was his best bet in Mexico.[60]

Cortés then dismantled his entire invasion fleet in dramatic fashion—there would be no turning back (and no evidence of Velasquez's contributions)—and headed twenty miles north to Zempoala. From there, Cortés sent 150 men back to Veracruz to construct a fortress before heading inland on August 16. He was accompanied by three hundred or so foot soldiers and fifteen horsemen. His exact route to the Aztec capital of Tenochtitlan (Mexico City) is unclear, but crossing the Sierra Madre Oriental must have been a challenge. Local chiefs, having presented the Spaniards with their own gifts of precious metals and stones, urged Cortés to bypass existing Aztec trails and thrash his way through the countryside. The intention was to minimize interactions with native peoples and to avoid Moctezuma's vaunted chain of highway garrisons.[61]

Supplemented en route by four thousand anti-Aztec Indians and one thousand human porters (*tlamemes*)—a native profession that would carry Spanish supplies for the next century—Cortés reached the independent city-state of Tlaxcala in early September 1519 and compared it favorably with Granada, Venice, and Genoa. From there, his westerly moving army sacked Cholula and approached the island city of Tenochtitlan from the south. The Spaniards were astonished by the quality of the stone-paved roads, the bridges, and the fortified causeway that straddled Lake Texcoco. They were welcomed by a passive Moctezuma on November 8, nearly one year after their departure from Cuba, and remained in Tenochtitlan for around six months searching for gold, plundering at will, and holding the Aztec ruler as a virtual hostage.[62]

Moctezuma ruled an Aztec Empire that had been assembled in the early 1400s from the fragmented remains of the Toltec Empire. Unlike their more civilized predecessors—the Teotihuacanos (150

BCE–700 CE), Zapotecs (150 BCE–700 CE), Classic Maya (200–900 CE), and Toltecs (800–1300 CE)—the Aztecs were a nomadic, barbarian people (Chichimecas) from somewhere north of the traditional Tampico-Guadalajara frontier line. Significantly more is known about the Aztecs than anyone else in Mesoamerica simply because they were in power when Cortés arrived in 1519 and could be observed and documented. The Aztecs borrowed heavily from the earlier civilizations but managed their affairs through a ruthless, federated system of trade, tribute, and strategic alliances. One such alliance, the Triple Alliance among Tenochtitlan, Texcoco, and Tlacopan, had served as a platform for Moctezuma's expansion (and centralization) between 1503 and 1519. Cortés guessed that the Aztec Empire between the Gulf (Veracruz) and the Pacific (Guerrero) was nearly as large as Spain herself.[63]

The spectacular island city of Tenochtitlan (translated as "prickly pear tree") had been established in 1344 as a southern extension of Tlatelolco on Lake Texcoco. How Tenochtitlan compared with the earlier capitals of Teotihuacan and (Toltec) Tula is unclear, but it was clearly more sophisticated than the soon-to-be-discovered Inca capitals of Cajamarca and Cuzco in Peru. The unwitting Spaniards had arrived at the precise moment when the imperial capital of Tenochtitlan was at its peak, with a population as high as 200,000, hundreds of tribute-paying city-states, and a fabulous marketplace at the Plaza of Tlatelolco. Local chiefs had carved up the great marketplace into sections and were receiving a 20 percent share (an Aztec quinto) of all business transactions.[64]

The dumbstruck Spaniards compared Tenochtitlan favorably with Istanbul, Rome, and Naples. In his *True and Full Account of the Discovery and Conquest of Mexico and New Spain*, published nearly fifty years after the event, Bernal Diaz del Castillo described the marketplace at Tlatelolco as follows:

> The moment we arrived in this immense market, we were perfectly astonished at the vast numbers of people, the profusion of merchandise which was there exposed for sale, and at the good police

and order that reigned throughout. The grandees who accompanied us drew attention to the smallest circumstance, and gave us full explanation of everything we saw. Every species of merchandise had a separate spot for its sale. We first of all visited those divisions of the market appropriated for the sale of gold and silver wares, of jewels, of cloths interwoven with feathers, and of other manufactured goods; besides slaves of both sexes. This slave market was upon as great a scale as the Portuguese market for negro slaves at Guinea. . . . Next to these came the dealers in coarser wares—cotton, twisted thread and cacao. In short, every species of goods which New Spain produced were here to be found; and everything put me in mind of my native town Medino del Campo during fair time. . . . I had almost forgotten to mention the salt, and those who made flint knives; [plus] instruments of brass, copper and tin [and] gold dust as it is dug out of the mines, which was exposed to sale in tubes made of the bones of large geese. . . . The value of these tubes of gold was estimated according to their length and thickness, and were taken for exchange, for instance, for so many mantles [of] cacao nuts, slaves or other merchandise.[65]

Of course the "tubes of gold" caught everyone's attention and fueled rumors that Moctezuma had stashed a trove of ceremonial treasures somewhere in town. Based on the primitive ornaments encountered in Hispaniola, Cuba, and Tabasco, no Spaniard could have known in 1519 that the Aztecs were guardians of a spectacular metallurgical heritage. This heritage had been started by the hemisphere's greatest craftsmen, the ancient Peruvians, and spread its way north over the centuries to Ecuador, Colombia, Panama, and Costa Rica. When Peruvian-styled metallurgical techniques finally reached the backwater shores of western Mexico after around 600 CE, they were adopted by the Tarascans of Michoacán and the Mixtecs of Oaxaca. The Tarascans fabricated metals from the surface-level deposits of the Sierra Madres—alluvial gold, oxidized copper, and silver—and smelted these metal-bearing ores in Peruvian-styled clay furnaces. Teams of human blowers were enlisted to blast air currents through five-foot-long cane-tubes to reach the extremely high temperatures required for smelting copper (1,083°C), silver (960°C), and copper-silver alloys (779°C). The Mixtecs produced gold metal sheets, lost-wax-cast gold, and gold alloys that are regarded as the finest in Mesoamerica. By the end of the Toltec period (800–1300 CE), even the remote Huastec people of the Rio Panuco valley were mining, smelting, and trading copper and copper-tin objects.[66]

Precious metals and ceremonial objects had been sacred to the Toltecs and the later Aztecs—they rarely (if ever) left town. By 1300 CE, when the Toltec Empire was on its last legs, Mixtec goldsmiths were producing their greatest works and the Tarascans were smelting, casting, and gilding bracelets, masks, and other ceremonial objects from copper-gold and copper-silver alloys. Even utilitarian copper-tin-bronze products like needles, hooks, axes, awls, and tweezers, were reserved for the religious elites. While many of these treasures eventually found their way (as tribute) from Michoacán and Oaxaca to Aztec Tenochtitlan, the Spaniards would discover shortly that objects appearing to be pure gold or silver were actually gilded metallic surfaces—achieved by a process of hammering, annealing, and burnishing that been invented centuries earlier by Peruvian craftsmen.[67]

Six months of plundering Central Mexico had convinced Cortés and his fellow Spaniards that the wealth of the Aztec Empire was many times more significant than that of Hispaniola, Cuba, and Panama. Moctezuma and his predecessors had had limited experience with vaults—no native Mexican would have dared to even think about stealing a single nose-ring—but he might have considered safeguarding Mexico's treasures had he fully understood the Spaniards' obsession with gold. He had been forced to dispatch messengers to the far corners of the Aztec Empire after being taken hostage by Cortés and being informed that Charles I "had need of gold for certain works." The results were astonishing. Together with the plunder seized to date, the Aztec provinces delivered a jaw-dropping 700,000 gold pesos worth of ceremonial objects, precious metals, precious stones, and featherwork. This was over seven times the value of Cuba's annual gold production, more than double Manoel's annual take from the Estado da India, and more than the annual revenue stream from the single largest silver mine in Central Europe (Joachimstahl). It also meant that Charles I was in line to receive a huge quinto.[68]

There was no time to celebrate. Before Cortés had a chance to pocket his plunder or send a second treasure ship back to Spain, he was forced to deal with a powerful new threat. Velasquez had been enraged by Cortés's rebellion, and his anger only intensified when he learned that Cortés's initial treasure ship to Spain had even stopped for supplies at Marien (Cuba). The adelantado responded in early March 1520 by sending Pánfilo de Narváez and a thirteen-ship fleet to halt Cortés's unauthorized power play in Mexico. By then, Cortés's own father, Martin Cortés de Montoy, had guided two of Cortés's envoys and assorted Aztec treasures to Charles V's chambers at Tordesillas. The recently elected Holy Roman Emperor was thrilled with the treasures, the implication that the new territory held exciting economic potential, and even Cortés's performance. The emperor rejected Fonseca's recommendation that Cortés be arrested for treason and ordered on May 10 that the remainder of Cortés's confiscated cargo be released by the Casa de la Contratación in Seville. Little did Charles know that Cortés had just uncovered an Aztec treasure trove that would dwarf the items received in his initial dispatch.[69]

Charles V probably had no idea that Velasquez had sent Narváez and nine hundred men to crush his insolent rival. News of Narváez's arrival in late April forced Cortés to suspend his treasure-hunting activities in order to deal with a Velasquez-financed army at Veracruz. The recently constructed fortress was not even needed. Cortés had always been more popular than Velasquez, so it was easy to persuade Narváez's men that their future was brighter with the man who had just plundered Tenochtitlan than with a man who just sat back and watched. Narváez was taken prisoner after losing an eye to a Spanish pike on May 27. Cortés followed up this triumph by making peace with Tlaxcala, an independent chiefdom that had always had an uneasy relationship with the Aztecs, and by marching his expanded army back to the capital in style. An otherwise happy return on June 25 was greeted with hostility. The plunder of Tenochtitlan had been one thing, but Pedro de Alvarado had shocked the local citizenry by murdering (and looting) a group of feasting Aztec nobles in Cortés's absence on May 16. Alvarado claimed that the massacre had been necessary to ward off a rumored conspiracy among Moctezuma, Aztec nobles, and Narváez. The evidence is murky. The massacre poisoned relations between the two cultures for good and led to the sudden (unexplained) death of Moctezuma on June 30—possibly at the hands of his own subjects.[70]

Cortés managed to escape the counterattack, but hundreds of Spaniards and thousands of allied Tlaxcalans were slaughtered by Aztec warriors on June 30. Even worse for the would-be conquerors, large amounts of Aztec plunder had to be abandoned during the retreat to the coast. The treasures abandoned during the so-called *Noche Triste*—estimated at 457,000 gold pesos worth of "gold and silver and jewels of many kinds, shields, plumages, necklaces and many other things of gold and stones of much value"— were reported painfully to Charles V on September 4, 1520. Another 100,000 gold pesos of tributary trade goods were also lost in the flight. While these losses were staggering, it must be assumed that some (or even most) of the treasures were eventually recovered by Spanish hunters in later months (or years).[71]

Cortés was not through. Reinforced on the coast by allied Spanish ships (and more Tlaxcalan warriors), he decided to mount a third invasion of the Aztec capital in the spring of 1521. This time, the otherwise diplomatic Cortés was out for blood. Hostile villages were sacked en route—between five thousand and ten thousand Cholulans were massacred within two hours—and alliances were struck with some of Tenochtitlan's enemies. By late May, the staging area established on the east side of Lake Texcoco was home to an invasion army of around nine hundred Spaniards, seventy-five thousand-plus native allies, three large iron guns, fifteen smaller bronze cannons, and a fleet of thirteen brigantines. The boats had been constructed at Tlaxcala and hauled west. This formidable armada proved to be too much for the Aztec Empire. Coordinated assaults by three captains—Alvarado, Cristóbal de Olid, and Gonzalo de Sandoval—led to the surrender of Tenochtitlan on August 13, 1521, after a grueling eighty-five-day siege. The siege left over 200,000 dead Aztec troops in its wake and a smoldering pile of rubble. When news of the triumph reached Spain, Fonseca immediately issued another order for the arrest of Cortés and dispatched Cristóbal de Tapia in early December 1521 to assume the governorship of Mexico. The overmatched Tapia fared no better than Saucedo and Narváez. He was flattered, bribed, and sent back to Santo Domingo.[72]

The treasures recovered initially from Tenochtitlan were valued at only 130,000 gold pesos, or less than one-quarter of the original hoard. If the total was extremely disappointing to the Spaniards and their Indian allies, it was still more than Cuba's annual gold production. One-fifth (a quinto) of the spoils was reserved for the Crown, one-fifth went to Cortés, and the remaining three-fifths were splintered off to Cortés's senior officers, Indian chiefs, and hundreds of foot soldiers. Since the meager fifty pesos awarded to each foot soldier was equivalent to the value of a sword, the Spaniards had to be paid off with encomiendas and repartimientos in order to prevent an insurrection. Few cared that the royal quinto never even reached Spain. Two of the outbound treasure ships, laden with precious metals, jewels, pearls, turquoise, and jade, were hijacked off the Azores by the French pirate Jean Fleury in May 1523. The heist prompted a change in the royal accounting system. Eight separate inventories were taken between 1524 and 1526 to account for the Aztec treasures that did reach Spain—including those recovered from the *Noche Triste* and a cache salvaged from Lake Texcoco after the lake was drained. These priceless ceremonial objects were stripped, melted down, and converted into bullion.[73]

Unlike the Portuguese, who had subdued the coastlines of Africa, India, and the Straits of Malacca with a minimum of hand-to-hand combat, Spain's conquest of the primitive Americas had not been so easy. The conquest of the West Indies had required the application of land-based military force, and the invasion of Mexico was marked by a series of brutal battles with warriors armed with obsidian-bladed clubs, axes, spears, slingshots, bows and arrows, tightly quilted cotton armor, and shields fabricated from bark and cane. The Aztec armies that greeted Cortés were formidable. They would have been even more formidable had they had access to metal weaponry, horses, wheeled vehicles, guerrilla warfare techniques, or professional military training. They were also overwhelmed by the modest-sized Spanish cavalry and the quality of the Spanish "soldiers," gnarly, cutthroat types who were armed only with swords and crossbows. Otherwise, muskets (and ammunition) were in limited supply, available firearms provided more of a psychological blow than an actual one, and cannons had to be dragged across mountains and rivers.[74]

Cortés recognized that the conquest of the Aztec tribute system would be a challenge. After the fall of Tenochtitlan in August 1521, he constructed a fortress on the city's ruins, dispersed scouts into the Aztec hinterlands, and began to build roads. Plans were readied to cut a *camino real* (royal road) parallel to an existing Aztec highway in order to link Veracruz with the soon-to-be-established capital at Mexico City. The new road would have to be sturdy enough to carry large Spanish wagons (*carros*) and an anticipated heavy volume of inbound and outbound cargo. The first of many caminos reales, like the public works projects, were built mainly with repartimiento labor—that is until smallpox, measles, and plague began to take their toll. Otherwise, the sophisticated Aztec Empire was a very different story than the West Indies. None of the Caribbean peoples had practiced mining or metallurgy until the arrival of the Spaniards. In Mexico, the challenge was to locate additional treasure sites and the underlying deposits of precious metals from which the brilliant ceremonial objects had been made. Guided by Moctezuma's tributary maps, Cortés prepared to lead expeditions into Michoacán (west) and Panuco (north) in search of mineral wealth. He also directed lieutenants to the southwesterly Isthmus of Tehuantepec to identify a possible Pacific-side seaport.

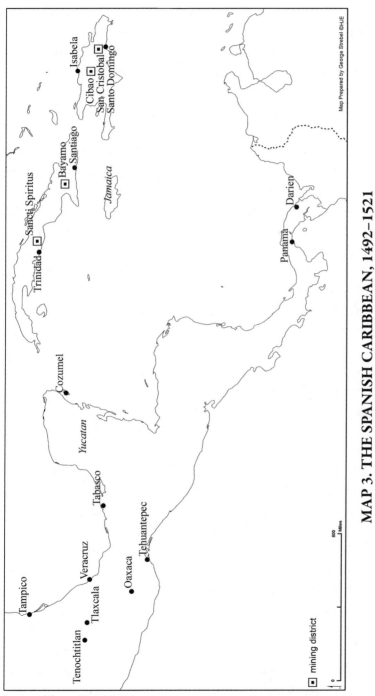

MAP 3. THE SPANISH CARIBBEAN, 1492–1521

Courtesy of George Strebel, © author.

Cortés established his conquests as "New Spain" after receiving word that a binding royal decree had been issued by Charles V on October 15, 1522. Pushed through by Mercurio Gattinara, the emperor's principal adviser and an arch-foe of the Fonseca faction, the royal decree confirmed Cortés as governor and captain general of New Spain and reconfirmed the controversial encomienda system in Mexico. If Cortés's extended family in Spain had provided valuable political services to his cause, Cortés had not been thrilled about the unexpected arrival of his wife of eight years, Catalina Suarez, along with various family members and friends in July 1522. They all settled into the family household in Coyoacan. No matter that everyone knew that Cortés had a number of mistresses, including his confidant and interpreter Marina, and that Marina was just about ready to deliver a son (Martin). It was sometime in November 1523, within weeks of the arrival of the royal decree, that Catalina died from either a heart attack, suicide, or murder. More than a few people, including the fellow interpreter Geronimo de Aguilar, suspected Cortés of foul play.[75]

The royal decree did not relieve Cortés of some serious threats. He had to recruit 420 Spaniards and 40,000 allied Indians to head off a challenge by Juan de Garay in the Rio Panuco region, make plans to crush a rebellion by Cristóbal de Olid in Honduras, and then prepare for the arrival of four royal inspectors in October 1524. The royal inspectors were in for a few surprises. They learned that the undermanned Garay had died suspiciously while visiting the Cortés household in December 1523. They learned also that the wily Cortés had funded his "pacification" campaigns with sixty thousand gold pesos that actually belonged to Charles V (as the quinto) and planned to cover up the scheme by sending the "borrowed" funds back to the emperor in the form of a small silver cannon and some other gifts. If that were not enough, they learned that Cortés had even threatened to arrest Velasquez in Cuba for conspiring against him with Olid. The last piece of news placed Hernán Cortés in deep political trouble with the emperor. So deep that Francisco de los Cobos, Fonseca's successor, was authorized to conduct a formal investigation of Cortés's activities under the auspices of the newly formed Council of the Indies.[76]

In the meantime, Cortés had some work to do. Having incurred thirty thousand pesos in personal debt to commission the silver cannon and other outbound gifts, he was forced to deal with his disloyal lieutenant in Honduras. When it was discovered that Olid had been in league with Velasquez and was claiming the rights to uncharted Honduras after Velasquez's death in 1524, Cortés marched into the jungles himself to crush the nascent rebellion during the summer of 1525. Olid was captured and beheaded by Cortés's lieutenants. Honduras turned out to be light in gold, but Cortés organized an ambitious road-building project to link the region with Mexico City. Conditions were harsh. Two Franciscan missionaries, recently arrived from Flanders, died en route, and Indian attacks were pervasive. However, Cortés cleared hundreds of miles of jungle brush and built scores of bridges. A one-thousand-beam structure across a western Campeche lagoon was constructed by allied Indian chiefs in less than a week, an engineering feat described by Cortés as masterful and remarkable. Many of these roads and bridges evolved into segments of the Pan American Highway. The town of San Cristóbal de las Casas was founded shortly in Chiapas to honor the famous priest.[77]

Having had enough of the jungles of Central America, a haggard Cortés returned empty-handed to Veracruz in May 1526 to reassert his authority in New Spain—most people had thought he had perished in Honduras. The next steps were to handle the four royal inspectors in the same fashion that Saucedo, Narváez, and Tapia had been handled, accelerate mineral prospecting activities in Mexico, upgrade agricultural methods, and continue the reconstruction of Mexico City. Cortés was a builder. He had begun an ambitious urban renewal project on the ruins of Tenochtitlan in March 1523, based on the Santo Domingo model, thanks to a huge supply of repartimiento labor. The massive effort deployed as many as 400,000 unfortunate Indians, hundreds of mule teams, and hundreds of wagons—native Americans had functioned to date without wheels and had only domesticated dogs (and llamas). The plan was to build one hundred thousand homes, drain Lake Texcoco, and extend ancient roadways into newly reclaimed agricultural lands. If the implementation of this plan was bittersweet to the site's former inhabitants, Mexico City would serve as the capital of Spanish North America for the next three hundred years. No matter that the city's energetic founder, Hernán Cortés, would have little time (or ability) to enjoy it.[78]

# CHAPTER THREE

# CONVERGENCE WITH THE HOUSE OF HABSBURG

The discovery, conquest, and reorganization of the Aztec Empire capped off a remarkable period of overseas activity between 1492 and 1521. The fledgling commercial empires assembled by Spain in the Americas and by Portugal in the Indian Ocean had catapulted two minor kingdoms into positions of prominence. After a somewhat rocky start, Spain was importing sizable quantities of gold, sugar, and hides from the West Indies and preparing to exploit the mineral riches of Aztec Mexico. The Portuguese had jolted the Venetians, the Mamluks, and scores of other Muslim trading enterprises along the Indian Ocean with their ability to exchange increasingly large amounts of pepper, spices, and other eastern goods for increasingly large amounts of Central European copper and silver. They established Goa as their Indian Ocean headquarters, captured Malacca as a gateway to China, cut a clove supply contract with one of the fabled Spice Islands, and even gained title to Brazil. Better yet for the new Iberian powers, there were two huge American continents waiting to be explored and plundered.

Coupled with a powerful revival in the Central European mining industry, economic growth had returned to Western Europe after 150 years of stagflation. This growth, in turn, created fresh demands for precious metals to finance new business ventures, invest in emerging markets, and exploit arbitrage opportunities. The value of precious metals was rising despite an expansion in the money supply. The purchasing power of precious metals, especially gold, was so attractive that investors began to fund high-risk exploration ventures that, twenty years earlier, would have been impossible.

Capital starved for centuries, and ignited by newly discovered opportunities in the Indian Ocean and the Americas (with much more to come), European trade began to explode on all fronts.

While all of these commercial activities were benefiting Lisbon and Seville, it was Antwerp in the Habsburg Netherlands that emerged as the foremost trading center in Europe. It was Antwerp that came to manage the metals-for-spices trade with the Indian Ocean, the emerging transatlantic trade with the Americas, and the perpetual financing needs of the House of Habsburg. These economic forces were combining to create a new type of economic power—capitalism. It wasn't pretty, but the results were becoming very persuasive. The great irony was that the Iberians, despite their spectacular overseas triumphs, were still locked into the medieval tradition of royal trading monopolies, aristocratic connections, and regulated merchant guilds. Half-hearted attempts were made to replicate the commercial networks developed by the Muslims, Italians, and even Bavarians, but the profit potential overseas was simply too lucrative to be left entirely to private enterprise. While Maximilian of Habsburg had allowed copper to become a Fugger monopoly, the Iberians were attempting to apply an exclusive royal monopoly on gold, silver, mercury, pepper, and spices. This medieval mind-set opened the door to a free-wheeling marketplace like Antwerp.[1]

Despite Portugal's remarkable achievements to date, Manoel I had discovered a painful fact—Portugal's business and financial capabilities were far inferior to those of the Italians and Bavarians. The great Italian city-states had introduced (and adapted) Muslim forms of corporate organization to Europe between the thirteenth and fifteenth centuries and created a global network of private merchants and investors in the Mediterranean and much of the Eastern world. The trailblazing Italians had made sure to retain strong financial relationships with the papal treasury as well. These networks generated business intelligence, controlled risk, and provided valuable lessons for ascendant financial powerhouses like the Fuggers and Welsers. Supported by their massive copper- and silver-mining interests in Central Europe, the Bavarians had begun to organize private trading networks of their own. There was a need to address the growing financial requirements of the Crowns of Portugal, Castile, and the House of Habsburg.

Merchant bankers of all stripes were eager to tap into the Iberians' royal monopolies, but the Central European mining industry had created an entirely new type of economic power in the case of Fuggers. The Fuggers' ruthless allocation of capital to the most profitable markets was shifting more of the economic decision making from inept monarchs to professional investors. Not only that, the intersection of Bavarian capital, Central European copper and silver, African (and Caribbean) gold, and Portuguese Asian pepper and spices was occurring in one principal location, Antwerp, and Antwerp was set to explode when the evolving Spanish American market was added to the mix. Supported by Lisbon, Seville, the House of Habsburg, and the House of Fugger, Antwerp consolidated the commercial and financial trading capabilities that had been handled previously by Venice, Genoa, Florence, Lyon, and Bruges. It emerged as the world's first quasi-capitalist marketplace.

Antwerp's astonishing rise could not have been predicted. Located in Brabant near the mouth of the Scheldt, a navigable river that could handle large Portuguese carracks, Antwerp and her cloth merchants had traded with English wool traders since the eleventh century. But there was little else until the Duke of Brabant established (and protected) a series of official trade fairs in the late 1300s and incorporated Antwerp's twice-yearly fairs into the wider Brabant system. It was Bruges, dominated by energetic Italian and Hansa merchants and regarded as the "Venice of the North," that served as the textile and financial capital of the Netherlands. The marketplace at Bruges was driven by the daily trading activities of merchants—rather than monarchs, town councils, or princes. The term "bourse" was even derived from Bruges' main square.[2]

The otherwise unremarkable Antwerp Fairs of the late 1300s offered one capability of note—access to Italian-style financial instruments like "letters obligatory" and negotiable "bills of exchange." While Europe's economic depression since 1350 had devastated trade of all types, the shortage of precious metals prompted the development of new financial instruments to supplement conventional barter sales. Long-distance trade of the Italian (Muslim) variety required access to credit, unless buyers and sellers were able (or willing) to transport large quantities of hard-to-get precious metals and other trade goods across dangerous territories and waters. As textile trading volume expanded at the great medieval fairs, like those in Brabant, Italian-style paper transactions began to appear more frequently. Smaller-scale merchants utilized the "letter obligatory," a nonnegotiable letter stating that payment was postponed but backed by a "payable to bearer" clause (or "bond"). After the Antwerp Fairs were recognized officially in 1415, Antwerp's "paper capability" separated it from its peers during the European "bullion famine" and allowed it to supersede Bruges as the leading financial center in northern Europe.[3]

Antwerp's growth after 1450 was driven by free trade, especially with Portugal and England, the acceptance of bills of exchange, and the importation of precious metals from Central Europe and West Africa. After a merchandise-oriented Merchant Bourse was established in 1460, based on the Bruges model, the convergence of bills, metals, and textiles made the Antwerp Bourse something unique. The biannual trade fairs gave way to daily transactions that were free from the tightly controlled, feudal-styled restrictions that operated even in Bruges. Antwerp was open to anyone with capital and energy. Maximilian's adoption of a pro-English, anti-Flemish posture helped Antwerp as well—in June 1488, he ordered (and guaranteed) that all Habsburgian commerce in the Netherlands be transferred from Bruges to Antwerp. The decision was fatal to Bruges and induced the Merchant Adventurers of London to more than double their annual cloth shipments to Antwerp to eighty thousand pieces by 1500. This represented nearly two-thirds of England's overseas trading volume. If that were not enough, Antwerp's status was bolstered by the election of Maximilian to the Holy Roman Emperorship in 1508 and the proclamation of fifteen-year-old Charles of Habsburg as Duke of Brabant in January 1515. Within four years, this otherwise insignificant heir to the Burgundian Netherlands would become the sole ruler of Habsburg Europe, Imperial Spain, and the Holy Roman Empire.[4]

Antwerp was inextricably tied to the House of Habsburg, the House of Fugger, and Portuguese India. Utilitarian copper worked well in Africa and served as Portugal's most important trade item in Malabar in the early years. But silver joined copper in the outbound fleets after the copper markets became saturated and after the Portuguese discovered that silver was valued more highly in the Indian Ocean than it was in Europe. The Central European mining boom had created an intriguing arbitrage opportunity between Europe's 10:1–11:1 silver-to-gold exchange ratio and the 9:1 ratio received in the Indian Ocean. Central European and Balkan silver that had been shipped east (via Venice and Cairo) for centuries was now being redirected to Antwerp. The new trading pattern raised the demand (and price) for imported silver. Of course, a large chunk of the Central European silver industry was in the hands of the Fuggers, who were keenly aware of the arbitrage opportunities between the Eastern and Western markets (and even between Antwerp and Lisbon). The Fuggers ensured that Antwerp received as much European silver as Venice (nearly thirty thousand marks) during the early 1497–1504 period with much more to come. By 1508, Antwerp was receiving a whopping sixty thousand marks of Central European silver (nearly fifteen tons) to finance the Portuguese spice trade.[5]

As Antwerp eclipsed Venice as the leading metals-for–spices emporium in Europe, the Portuguese and Fuggers were trading prized commodities on an unprecedented scale. The Portuguese simply out-bought the small-scale Muslim traders who had purchased Malabar pepper over the centuries. The annual volume of Malabar pepper unloaded at Antwerp more than tripled (to eight thousand quintals) between 1504 and 1511 and facilitated Manoel's establishment of a full-scale feitoria (factory) at Antwerp in affiliation with the Casa da India. While pepper accounted for less than one-third of Portugal's total spice imports, volumes and profits were two different things. A subsequent doubling of spice imports between 1510 and 1520 reduced spice prices significantly, pressuring what had been spectacular profit margins, devastating a number of commodities traders in Antwerp, and nearly destroying the Venetian spice industry. However, Antwerp was insulated by a diversified economy and the outbound copper and silver shipments that had made the spice trade possible.[6]

## FUGGERS AND HABSBURGS

The temporary spice glut between 1510 and 1520 was largely the creation of the House of Fugger. The spice-laggard Fuggers had seized their opportunity after Manoel had placed a royal feitoria in Antwerp in 1508—this allowed them to match their copper monopoly to the growing needs of the Portuguese Crown and to consider entering the spice business themselves. The Fuggers were soon operating a wildly successful factory in Antwerp. Between 1508 and 1514, their huge inventory of Hungarian copper prompted the Affaitati spice syndicate to purchase 636 tons of Malabar pepper on behalf of Manoel's factor in Antwerp. It was these purchases that led to the general spice glut in Europe and the near-destruction of Venice's spice business. The glut only worsened when Filippo Gualterotti, the Affaitatis' agent in Antwerp, attempted to arrange a formal five-year copper supply contract between the Fuggers and the Portuguese Crown in 1516. The contract called for a staggering 694 tons of Fugger copper to be delivered annually to the Crown's factor in Antwerp.[7]

Jacob Fugger had his own ideas about supply contracts. He had analyzed the Italians' spice business practices very carefully and determined that he no longer needed them. Betting that spice demand would eventually catch up with the glut in supply, that copper was the Crown's crucial export item, and that the Fuggers' fifty thousand quintals of annual copper production gave them a powerful monopoly, Fugger lobbied the Portuguese factor, Rui Fernandez de Almada, for a new supply contract in July 1517. Portugal agreed to purchase six thousand quintals of Fugger copper and allowed the Fuggers to purchase ten thousand quintals of Malabar pepper for the first time. A series of follow-up meetings in Augsburg and Munich led to a doubling of the quotas for the 1519–1523 period—Portugal agreed to purchase a whopping twelve thousand quintals (six hundred tons) of Fugger copper at fourteen cruzados each in exchange for the Fuggers' right to purchase fourteen thousand quintals of Malabar pepper at twenty-four cruzados per quintal.[8]

Almost everyone made money. The Portuguese Crown was happy to pay fourteen cruzados per quintal for Fugger copper because this raw copper could be sold at a very profitable twenty-four cruzados in Malabar. If shipping costs were around 4.5 cruzados per quintal, the Crown stood to realize 5.5 cruzados per quintal in profits—a 39 percent return. Of course, profits were much higher when the copper was bartered for Malabar pepper. Pepper could be purchased in India for as little as three cruzados per quintal and resold to European spice distributors for over twenty-four. In this case, the Crown stood to profit from both sides of the transaction. The Fuggers would profit from both sides as well. They received a copper price that was triple their production cost (4.5 cruzados) and could mark up the retail price of their imported pepper purchases in Antwerp. The Sultan of Gujarat, whose textile-centric markets valued European copper even more highly (eighteen cruzados) than those in Malabar, had estimated India's annual copper demand at as much as forty thousand quintals. Unfortunately for him, textiles did not (as of yet) provide the same mark-up opportunities in Europe as pepper and spices.[9]

Assuming that the Portuguese Crown was earning tidy profits on both sides of the copper-for-spice transactions, Manoel probably received annual profits in excess of 200,000 cruzados during the 1510–1520 period. These impressive returns helped to sustain his Golden Age. Of course, smaller-scale European and Indian merchants were outraged by the ability of the Fuggers and Affaitatis to cut exclusive deals with the Portuguese Crown. The Fugger (copper) and Portuguese (spice) monopolies constrained every entrepreneur's ability to purchase enough volume of either commodity to make an otherwise very risky ocean voyage worth the effort.[10]

The Venetian economy was nearly destroyed by these developments. Prior to 1500, ten Venetian galleys had returned annually with as much as 1,750 tons of cargo from the Mamluk spice depots at Alexandria and Aleppo. Thanks to Portugal, however, Venetian spice imports plummeted to around five hundred tons between 1502 and 1505—less than half of the volumes imported by Portugal during this period—and declined some more between 1508 and 1514. If that were not enough, Venice suffered through the ramp-up of the Fuggers' massive copper-for-spice deals with Portugal and the loss of her precious metals trading business to Antwerp. It was left to the Ottoman conquest of Mamluk Egypt and Syria (1517) to bail out the reeling (but opportunistic) Venetians in due course. The Ottomans needed Venice's European distribution network.[11]

While the Fuggers schemed to dominate the global spice market, their metals-intensive business activities were becoming increasingly entwined with those of the House of Habsburg. The simple formula had been repeated many times over since the 1490s—when Maximilian defaulted on his loans, as he usually did, the underlying security was transferred to the Fuggers. By 1508, newly elected Holy Roman Emperor Maximilian had already pledged his entire silver and copper mining revenues to the Fuggers. The transfers extended the family's copper monopoly and facilitated the huge long-term supply contracts with the Portuguese Crown. When another round of military campaigns in Italy forced Maximilian to secure another round of loans with Habsburgian real estate, Jacob Fugger received enough property to join the landed aristocracy. The newly ennobled count invested a total of ninety-two thousand florins to purchase the Duchy of Kirchberg in Swabia, additional Swabian properties, and even a castle between 1507 and 1514. By then, Jacob had gained complete control of the House of Fugger. He rechristened the firm as Jacob Fugger & Nephews in 1512, following the deaths of his brothers George (1506) and Ulrich (1510), and took in George's sons Raimund and Anton for a six-year "trial period." The firm's business capital had nearly quintupled to 269,000 florins since 1494, two-thirds of which comprised trade goods, cash, and valuables, and one-third fixed properties, homes, and mining interests.[12]

Metals and spices were not enough for the House of Fugger. They were also receiving sizable contributions from the sale of papal indulgences and the collection of church fees. This was an ancient business, but the Fuggers had raised the most controversial of these practices to a higher scale. Fugger agents like Johann Tetzel earned large fees from the "retail" sale of indulgences, leaving the Fuggers to focus on the interest income generated from the collection funds, the transfer fees, and the foreign exchange arbitrage. Between 1494 and 1510, the Fuggers transferred 152,931 florins on behalf of Cardinal Melchior von Meckau, financed the church Jubilee of 1500, managed at least one hundred thousand ducats of Pope Julius II's personal funds, and even sold an eighteen-thousand-ducat diamond to the pontiff. The Fuggers controlled papal coinage between 1510 and 1534 and lent twenty-nine thousand florins to Albrecht of Brandenburg, the archbishop of Mainz, so he could enter into an indulgence partnership with Johann Tetzel and the Holy See in 1515. These types of transactions placed the Fuggers very much in the mind of Martin Luther when he nailed his ninety-five theses to the doors of Wittenberg (Saxony) Cathedral on October 31, 1517—the devoutly Catholic Fuggers returned the favor by orchestrating Luther's transfer to Fuggerhaus (Augsburg) for questioning in 1518.[13]

The Fuggers also began to provide direct financial services to the House of Habsburg. If the family was a latecomer to the Italian-dominated public-finance business, it had been a latecomer to the mining and spice businesses as well. No matter that conventional public finance was a very different business than extending credit to monarchs and securing these loans with tangible mining properties. The Fuggers entered the fray nevertheless. In 1508, Jacob Fugger managed the remarkable feat of raising 170,000 ducats from Maximilian's European allies and then transferring these funds, as bills of exchange, to a single Augsburg account in less than six weeks. The speed of this large-scale transaction was shocking. Of course, the wily

Fuggers profited handsomely by arbitraging the bills of exchange from one bourse to another. When Maximilian requested 300,000 ducats to finance an election bid for the papacy in 1511, knowing full well that he had already pledged his mining properties and much of his real estate, he proposed to secure the election loan with his "four best caskets" of royal jewels and other valuables. The Fuggers stood to receive a commission, interest, and exchange fees totaling 100,000 ducats—33 percent of the value of the entire loan. Fortunately for Rome, Maximilian decided to abandon his papal quest.[14]

Maximilian was up to his eyeballs in debt. Inconclusive campaigns in Italy forced him back to the Fuggers in 1514 for a loan of nearly 340,000 florins secured by all of Schwaz's silver output through 1521 plus an additional eight years of reduced-price Schwaz copper (through 1523). It was these pledges that facilitated the Fuggers' long-term copper supply contracts with the Portuguese Crown. Given the fact that the Fuggers' business capital stood at only 300,000 florins and annual Schwaz silver production had fallen temporarily to around seven thousand marks, Jacob Fugger was placing a very large bet on copper. The bet paid off. The Fuggers realized huge profits on the two-sided Portuguese deal and benefited from a continuation in Maximilian's financial woes. In 1515, the emperor was forced to extort one hundred thousand florins from his own grandson, Charles of Habsburg—handled, of course, by the Fuggers—in return for recognizing that his grandson "had come of age." Holy Roman Emperor Maximilian of Habsburg was essentially flat broke when he died on January 2, 1519.[15]

Maximilian's death cleared the way for Charles of Habsburg. The career of one of the true giants of the sixteenth century was launched by a remarkable series of accidents. Charles was born in Ghent on February 24, 1500, to Joanna of Castile and Philip of Burgundy, educated as a French-speaking Burgundian noble, and destined to rule in the Duchy of Burgundy, the Burgundian-dominated Netherlands, or in one of the Habsburg domains in Germany and Austria. These assorted kingdoms and duchies had comprised the Habsburg Empire since 1477. But Charles also happened to be the grandson of Isabella (Castile), Ferdinand (Aragon), and Holy Roman Emperor Maximilian of Habsburg (Austria), and their deaths, coupled with the unexpected passing of his father and the incapacity of his mother, would matter greatly. Charles was made ruler of the Netherlands in January 1515, then king of Castile (after Ferdinand's death in January 1516), and then ruler of the entire Habsburg Empire (after Maximilian's death in January 1519). The assumption of six separate crowns—Castile, Aragon (including Naples and Sicily), Burgundy (including the Netherlands), Austria, Hungary, and Bohemia—gave the erstwhile Duke of Burgundy title to the largest territorial entity in Europe since Charlemagne.[16]

If that were not enough, Charles felt obligated to get himself elected Holy Roman Emperor in 1519. The bid presented a formidable political challenge and would test mightily the financial clout of the House of Habsburg. As everyone knew, the archbishops who elected candidates to positions of high office had to be bribed for their vote. This had been an accepted practice since 1300. Interest was banned, but the financiers who brokered these bribery deals were compensated handsomely for their "trouble, danger and expense." Since two powerful monarchs, Charles I of Spain and François I of France, were both vying for the largely ceremonial position of Holy Roman Emperor, the contest was expected to break the all-time record for bribes—even prior to the death of the incumbent, Maximilian of Habsburg.[17]

The election of 1519, fascinating in itself, had important ramifications. As early as August 1517, as Charles I was sailing south to his new home in Valladolid, it was estimated that over 500,000 florins would be required to win a majority of the seven German electors responsible for electing the next Holy Roman Emperor. If Maximilian was still very much alive, it never hurt to plan ahead. The seventeen-year-old king of Spain was already consulting with a variety of merchant bankers to raise the necessary funds. The principal advisers were the Florentine Filippo Gualterotti, a syndicate of Genoese bankers, and the Augsburg branch of the House of Welser. The all-powerful Fuggers were curiously absent from these initial discussions—Jacob Fugger may have demanded the lead role or no role at all. It was even rumored that François I of France had approached Jacob Fugger for a 300,0000-*ecu* loan to advance his own candidacy. The Fuggers declined either out of loyalty to Charles or because the French king had a demonstrated knack of not repaying his debts in a timely fashion. François still had access to the friendly capital market of Lyon, but most of the "smart money" appeared to side with Charles of Habsburg.[18]

When Maximilian finally passed away in January 1519, Charles's financial advisers had raised or guaranteed nearly 300,000 florins in bills of exchange payable to either Charles I's representative in Germany or directly to the electors in Augsburg and Frankfurt. Of course, the bribes were to be paid only if Charles was elected Holy Roman Emperor. Unfortunately for Charles and his advisers, the absence of the Fuggers turned out to be a potential deal-breaker—the German archbishops demanded to be bribed in cash (rather than in bills of exchange) and stipulated that the unimpeachable Jacob Fugger be employed as Charles's representative in Germany. Charles reluctantly agreed to the cash requirement but preferred to keep the Fuggers on the sidelines.[19]

Like a modern-day corporate bidding war over a prized acquisition target, the twenty-five-year-old François I was not yet ready to surrender the Holy Roman Emperorship to a nineteen-year-old Habsburg. He and his mainly French advisers had access to the venerable capital markets of Lyon. For their part, the various financial advisers were operating like modern-day mergers and acquisitions bankers who earn a fee even if their sage advice results in an idiotic transaction. The Florentines were backing Charles because they were already stuck with 262,500 ducats in unpaid claims against the less-than-responsive François I. The Genoese were also backing Charles. They had been prohibited from trading in Lyon since 1463 and recognized that Lyon, the primary financial marketplace for the French monarchs, was far less dynamic than the commodities-and-merchandise markets of Antwerp. Lyon would be nearly destroyed by a rival Genoese fair at Besançon in 1535, the French state bankruptcy of 1557, and the emergence of the public finance markets of Paris.[20]

With the gnomes of Lyon pitted against Charles's minions in Florence, Genoa, and Augsburg, the German electors—including the same archbishop of Mainz who had recently purchased the right to sell papal indulgences in Saxony—were delighted. The value of the bribes rose to 500,000 florins in March, then to 720,000, and then to a whopping 850,000 during April. It was then that the Fuggers finally (and inexplicably) entered the financial fray. The Fuggers responded to a demand by the electors that only they could be trusted to transfer the massive bribe monies to their secret accounts, a practice that the family had employed skillfully for decades. Jacob Fugger managed to parlay a minor role in the process to one that was front and center. Of course, the Genoese were outraged by the power play and refused to subordinate themselves to the House of Fugger—the Welsers had to guarantee the Genoese contributions to Charles's election fund. But if the prize was a dubious credit risk to the Fuggers at 500,000 florins, how could it be less so at 850,000? The answer lay in security, a Fugger specialty. The Fuggers agreed to lend 543,000 florins to the election campaign only when Charles agreed to sign the bills and notes personally and to specify that the funds be payable in the Netherlands (not Castile) in gold *gulden*. The remaining funds (and guarantees) were provided by the Welsers and the Italians. Charles was elected Holy Roman Emperor on June 28, 1519.[21]

The 850,000 florins needed to bribe the seven German electors, equivalent to 602,026 ducats, placed Charles I of Spain in severe debt even before he assumed the Holy Roman Emperorship. In fact, he had been in semi-severe debt since paying 100,000 ducats to Maximilian in 1515 and having been stuck with nearly 300,000 ducats in Maximilian's obligations in early 1519. Jacob Fugger needed Charles's personal signature for the election loan because Maximilian had already pledged virtually all of the Habsburg copper and silver mining revenues in Tyrolia to the Fuggers (through 1523), gold shipments from the West Indies had peaked, and

Hernán Cortés had only just landed at Veracruz. Charles won large annual subsidies (servicios) from the hopeful governments of Castile and Aragon shortly after his arrival in Castile, but these generous grants were insufficient to service his accumulating debts to the House of Fugger. These same debts helped to trigger a populist revolt by Castilian Comuneros in 1520 that was put down with additional funds (and debts) near Valladolid in April 1521. However, Charles's surprisingly brutal military response to the Comuneros endeared him to Castilian nobles and church officials.[22]

The Fuggers were less concerned about their election loans to Charles V than they might have been—they had just cut a lucrative copper supply contract with the Portuguese Crown, were scheming to break the increasingly fragile Portuguese spice monopoly in the Far East, and recognized that Habsburg Europe was in the midst of a powerful mining boom (that would last until around 1540). While Jacob Fugger was less optimistic about the uncharted Spanish Americas, he was one of the principal investors in Magellan's ambitious expedition of September 1519 and was waiting patiently for the outcome. Given the above, it didn't hurt to be owed a large amount of money by an emperor who controlled the most valuable mining properties in the world, an unexplored Western Hemisphere, and the means to exploit an emerging global economy.

Jacob Fugger also recognized that the eroding Portuguese spice monopoly was likely to erode the powerful Portuguese presence in Antwerp. The Portuguese were being attacked commercially by the ascendant Ottomans, Istanbul's newfound partners in Venice, and scores of Muslim spice traders who had been displaced by the loss of Malacca in 1511. Many of them, including the ex-sultan of Malacca, Mahmud Syah, had relocated to friendly Muslim ports on the Malay Peninsula or to Aceh on the northern tip of Sumatra. The Muslims were not going to give up the wealth of the Spice Islands, the Moluccas, without a struggle. The name *Moluccas*, in fact, had been coined by Muslim traders as *Jazirat al-Muluk* (Land of Many Kings) to honor the rulers of Ambon, Ternate, Tidore, and the Bandas. The sultans and rajas who ruled these island kingdoms had varied in their allegiance to the Portuguese—Ternate and Ambon were friendly; Tidore and the Bandas were not. Ternate would eventually become even more hostile than Tidore.[23]

## MAGELLAN

When Fernão de Magalhaes (Magellan) shifted his loyalties from Portugal to Spain in 1518, Portugal had another potential rival in the mythic-turned-real Spice Islands. Born in northern Portugal in 1480, Magellan had sailed to India with Francisco de Almeida in 1505 and was a member of the four-ship Portuguese fleet that was dispatched to locate Malacca in September 1509. Magellan was wounded in a skirmish with the sultan of Malacca's troops, a battle that cost one Portuguese ship and sixty lives, but he returned to conquer Malacca with Albuquerque's nineteen-ship armada in the summer of 1511. After volunteering to find the famous Moluccas in 1513, Magellan reached only as far as Ambon and the Bandas. He returned to Portugal in 1517, joined a military campaign against Morocco, and then angrily renounced his Portuguese citizenship—he did not like being charged with pocketing an excessive amount of local Moroccan plunder.[24]

Magellan's Asian experience had been gained via the cape route. But his bold proposal to Charles I of Spain (and probably Jacob Fugger) in early 1518 was to sail in a westerly direction to the Moluccas. Balboa had sighted the Pacific Ocean in 1513 but no one had attempted to cross a vast body of water that was beyond comprehension. For that matter, Iberian sailors had reached no farther south than Rio de la Plata (Uruguay) in February 1516, and no one had come close to locating the southern tip of South America. Magellan's ambitious proposal was an easy sale. With the Treaty of Tordesillas still in limbo, due to the absence of any information about the Pacific Ocean or longitudinal demarcation lines, Magellan and the Spanish Crown apparently assumed (hoped) that the Moluccas were located within the Spanish zone. Having previously visited Malacca and the Moluccas, Magellan was prepared to openly defy both the Portuguese Crown and the papal bulls.[25]

The contract drafted by Juan Rodriguez de Fonseca, and backed partly with Fugger capital, called for Charles to contribute five heavily armed, but modest-sized ships (of 75–125 tons) in return for all of the Asian spices that Magellan could find and deliver. Magellan would receive 5 percent of the profits, the governorship of lands discovered and conquered, plus a coveted knighthood with the Order of Santiago. On September 20, 1519, within weeks of Hernán Cortés's initial visit to Tenochtitlan, Magellan left Seville with five ships, 260 men, and trade goods en route to the Pacific Ocean via an unidentified southwest passage through southern Argentina. One of the crew members, a young Venetian noble named Antonio Pigafetta, chronicled the fleet's hair-raising adventures and lived to tell the tale.[26]

Of course, locating a southwest passage through Argentina was not so easy. After sighting (but not entering) the Rio de la Plata, Magellan pushed south to a winter ("summer") anchorage at San Julian Bay in Patagonia. At this frosty latitude, Magellan was forced to crush a mutiny by over forty of his men, including Juan Sebastián del Cano, and suffer through the loss of one of his ships. Matters only worsened when the expedition resumed in August 1520. A two-month sail was required to identify a possible passage through the treacherous straits (the future Straits of Magellan) in late October, followed by seven harrowing weeks to negotiate a 370-mile channel strewn with barren islands, sandbars, and rocks. A diet of shipboard rats was all that sustained Magellan's nearly starved crew members. When the fleet finally reached the Pacific, it was reduced to three ships—the large supply vessel had been abandoned en route.[27]

From there, Magellan enjoyed excellent winds and weather as he sailed west to an inadvertent landfall on the Ladrone Islands in early March 1521. Replenished with local supplies of rice, fruit, and water, the one hundred or so survivors were informed that they were only a few days away from the unchartered Philippines. Leyte was sighted in mid-March, followed by Mindanao, an island visited by wandering Portuguese sailors as early as 1512, and then Cebu. Magellan named the islands for San Lazaro, in honor of his death-defying, Lazarus-like feat of survival, and received a warm welcome from Cebu's Chief Humabon. Humabon's village of three hundred bamboo huts and two thousand inhabitants was one of the largest in the archipelago. While the chief and his friendly subjects agreed to convert to Christianity in late April 1521, the Europeans discovered that the neighboring islanders on Mactam, including Chief Lapu Lapu, were happy with their pagan rituals. Six days later, the overconfident Magellan decided to attack Mactam's primitive 1,500-man army with just 60 Spaniards. In this case, numbers, spears, and arrows triumphed over musket-and-crossbow-bearing Europeans. Magellan and forty of his men were killed during the forced retreat to the ships.[28]

Humabon only backed winners. He conspired to have as many as twenty-two of the surviving Spaniards murdered after a feast—an inadvertent native payback for Alvarado's massacre of feasting Aztec nobles in May 1520—and nearly prevented del Cano and the remaining crew from leaving Cebu. Del Cano somehow directed the *Vittoria* and the *Trindad* to the island of Tidore by November 1521. There, he and his fellow Spaniards were greeted warmly by the sultan of Tidore, an age-old rival of the late sultan of Ternate, and were informed that they could assist his efforts to break Ternate's alliance with Portugal. Ternate was in chaos following the recent death of the ruling sultan. The Spaniards were permitted to trade for cloves and establish a local warehouse on Tidore. The wily sultan hoped to play the Spaniards against the Portuguese.[29]

The Portuguese responded to this unexpected Spanish incursion by capturing the *Trinidad* in local waters. Del Cano escaped with the eighty-ton *Vittoria*, however, and kept alive a possible circumnavigation of the globe. His return route followed the traditional Portuguese sea lanes through the Banda Sea, across the Indian Ocean, and around the Cape of Good Hope. The Spaniards were not home free. The clove-laden *Vittoria* lost twenty or so men (from disease) en route and was impounded by Portuguese officials at

Cape Verde. The Treaty of Tordesillas had prohibited the cape route to Spain and other foreign flags. But eighteen men, including del Cano and Pigafetta, managed to flee their Portuguese captors at Cape Verde and return to Seville as conquering heroes on September 8, 1522—this was nearly three years after the Magellan expedition had departed.[30]

The first circumnavigation of the globe was a coup for Charles V, enhanced by the Tidorean cloves aboard the *Vittoria* and the recent death of his archrival Manoel I in 1521. If the emperor's financial position was somewhat shaky, the achievement pushed Jacob Fugger and the always-thinking Hernán Cortés to finance a series of possible follow-up expeditions to the Far East. The Great Conquistador informed the emperor in mid-October that he was constructing four ships at one of his Pacific bases in preparation of a planned expedition to the Moluccas in 1525. Immediately after the conquest of Tenochtitlan, Cortés had sent an expedition party through Michoacán to identify an interior route to the Pacific and to scout for rumored deposits of Tarascan gold—the route was marked by crosses. However, Cortés's Pacific launch was scuttled when the Council of the Indies commissioned an investigation of Cortés's activities in New Spain and Charles V authorized Juan Garcia Jofre de Loaisa and Sebastian Cabot to retrace Magellan's route in 1525 and 1526, respectively. An opportunity to bypass Portugal's increasingly vulnerable spice monopoly prodded Jacob Fugger to invest 10,000 ducats in one (or both) of the expeditions. Loaisa's seven-ship fleet and 450 men (including del Cano) left La Coruña, Spain, in late July 1525 to revisit the Magellan Straits and to rescue possible survivors in the Far East.[31]

The rescue mission was a failure—three ships were lost in the treacherous straits, two more were lost or wrecked en route, the *Santiago* was lucky to struggle back to Mexico, and Loaisa and his crew were stranded on the island of Gilolo in November 1526. Del Cano and 329 other unfortunate sailors failed to survive at all. But there would be some intriguing news. After a stop at Mindanao in September, the Spaniards learned that native Filipinos were conducting trade with Chinese junks, exchanging gold and pearls for silks and other goods. This was valuable intelligence because the untapped Chinese market was still viewed as the world's most lucrative business opportunity. In the meantime, the unexpected arrival of the *Santiago* on June 1 persuaded Charles V that he needed the services of Hernán Cortés after all. Cortés was authorized to launch a three-ship rescue fleet from Zihuatanejo (Guerrero) under the command of his cousin, Alvaro de Saavedra, on October 31, 1527. However, Saavedra fared as badly as his predecessors. Two of his ships were lost off the Marshall Islands in December, the flagship accomplished little after reaching Tidore in late March 1528, and Saavedra apparently died en route to Mexico. The Spanish survivors, including those from the earlier Loaisa and Magellan expeditions, were left to be captured by Portuguese officials and held in limbo until 1535.[32]

Spain's failures in the Pacific were inconsequential. News of the Spanish threat to the Moluccas had prompted the Portuguese into action. António de Brito and three hundred men were dispatched to Ternate to complete a long-delayed fortress project in February 1523 and to sign a commercial treaty with the new sultan of Ternate. The Portuguese were given the right to purchase the island's entire clove crop at a fixed price of slightly less than one-half a cruzado per quintal, paid in trade goods, and were authorized to halt any unwelcome foreign ships. A combined force of Portuguese and Ternatens invaded Tidore in late 1524 and burned down the palace of the recently deceased sultan. The action was intended to send a message to both Tidore and Spain.[33]

By then, the Muslim merchants and sultans who had controlled the regional spice trade for centuries had become increasingly outraged by Portuguese brutality and corruption. Traditional business partnerships were restored among the Javanese, Malaysians, and Sumatrans to circumvent the Portuguese and to take advantage of the surge in spice demand brought about by the Portuguese-Fugger partnership. Native islanders replaced wild clove trees with large-scale orchards and planted cloves in underdeveloped islands like

Ceram and Ambon. The local sultans typically retained a share of the harvested spice crop as tribute and sold the rest to a favored buyer—whether it was the reviled Portuguese, friendly Muslims, or the Chinese merchants of Java. The Chinese traders at Grise benefited from China's refusal to purchase spices directly from Indonesia during the entire sixteenth century.[34]

Rising spice production, increased hostility to the Portuguese, and a growing Ottoman naval presence revived the traditional Red Sea spice trade with Ottoman Egypt and Syria (and Venice) during the late 1520s. This was bad news for Lisbon and Antwerp. Even if Spain had been more successful in the Pacific, the global spice business was too fragmented to be monopolized by any one power. Venice's commercial treaty with Istanbul was cutting into Antwerp's profits and giving second thoughts to Jacob Fugger about his future role in the spice business. Rival spice routes to Ragusa (Dubrovnik) and Marseilles were bypassing both the Portuguese and the Venetians. Lyon received nearly 85 percent of her imported spices from Marseilles by 1534. While these rival spice centers never matched the scale of Antwerp (or Venice), they prompted the spice merchants of Antwerp to consider new lines of business.[35]

Portugal was still a powerful player in the spice trade, but dreams of a global monopoly were over. The basic problem was that the costs to outfit and defend the royal spice fleets were rising, while spice profit margins were falling. The massive Portuguese-Fugger contracts had disrupted the entire marketplace. Even if a private Portuguese trader could generate spice profits of around twenty thousand cruzados from the exchange of one hundred thousand cruzados worth of merchandise between Goa and the Moluccas, the cost to carry merchandise around the always-dangerous Cape of Good Hope could wipe out most of the profits. The Crown finally permitted private Portuguese enterprises to purchase Moluccan spices directly in 1535, but only if one-third of the cargo was transferred to royal authorities at official (discounted) prices of around one-half a cruzado per quintal. Few entrepreneurs took advantage of this less-than-generous offer.[36]

Antwerp and her two primary sponsors, the House of Fugger and the House of Habsburg, adjusted to the steady erosion of Portugal's spice monopoly. They could do so because mining investments catalyzed by the Duke of Saxony, Venetians, and Bavarians had launched a Habsburgian mining boom of epic proportions in Central Europe. Antwerp was its primary outlet. While Thurzo's separation process and drainage techniques had raised copper output significantly, the silver mines of Tyrolia, Bohemia, and Saxony were approaching record levels. The nearly fourteen tons of silver produced by the Falkenstein mines at Schwaz (Tyrolia) in 1523 may have matched their record output in 1485. This achievement in turn was eclipsed by an unexpected bonanza on the eastern slope of the Erzgebirge range in Bohemia. Since the great Bohemian silver mines at Kuttenberg had peaked at nearly twenty-two tons in the early fourteenth century, few were prepared for the massive silver strike made at Joachimstahl in 1516. Thanks mainly to Welser capital, Joachimstahl reached an impressive 56,437 marks (nearly fourteen tons) of annual production and made Bohemia accountable for nearly one-fifth of Europe's total silver output. A grateful Ferdinand of Habsburg, who would shortly receive the crowns of both Bohemia and Hungary in 1527, had a new source of funds with which to defend against the Ottoman Empire. It would get even better. By the early 1530s, Joachimstahl would be producing as much as 87,500 marks per year (twenty-one tons) and serve as the single largest silver mine in Europe.[37]

In Saxony, silver production was recovering from an insignificant one ton per year during the 1511–1520 period to a substantial 43,000 marks (over ten tons) in 1520. There would be much more to come. A new round of investments at Annaberg, Marienberg, and Freiburg would raise annual output to a record 88,240 marks (over twenty-one tons) in 1540. Even the ancient Rammelsberg mine in the Upper Harz mountains was producing two tons of silver—less than the others but roughly equivalent to Rammelsberg's peak output back in 1300. Freiburg didn't top out until she reached eight tons in 1572. By then, interested parties had become familiar with the Saxon mining secrets revealed with the publication of *De Re Metallica* by Georgius Bauer (Agricola) in 1550.[38]

Another source of silver was derived from the transfer of Thurzo's Saiger separation process to the Fugger copper works in Thuringia and Carinthia. Annual copper output at Mansfeld (Thuringia) reached 2,100 tons in 1526, more than double the volume produced in 1506 and roughly equivalent to that produced at Neusohl, plus a bonus supply of separated silver. Mansfeld and Neusohl may have generated forty thousand marks (nearly ten tons) of separated silver combined. The mining and processing of Central European copper, silver, and iron were also bolstered by fresh investments in large-scale smelting and forging facilities, rolling mills, blast furnaces, and the introduction of bellows, tilting hammers, winches, hoists, and water-powered pumps. Productivity improvements were nonsectarian. They raised profit levels at the massive iron works in Upper Palatinate and Schmalkalden (Thuringia), in the process enriching the coffers of the Schmalkaldean League of Protestant princes, and promoted a gradual shift in the European copper industry to Sweden.[39]

Charles V and Jacob Fugger both knew that the reversion of the Tyrolian mining properties back to the House of Habsburg in 1523 would restore the emperor's annual revenue base to nearly one million ducats. Roughly one-fourth of this total was generated by Charles's unpledged mining interests in Saxony, the single most lucrative imperial domain in Europe, and one-fourth from the duchies of Austria, Tyrolia, and Bohemia. The other half (500,000 ducats) was received from the assorted Habsburg kingdoms of Castile, Aragon, Burgundy, Milan, Naples, and Sicily. The more encouraging economic news was that Joachimstahl was looking like a potential bonanza, Spaniards were becoming familiar with the Pacific Ocean, and the ability of Hernán Cortés to find (and lose) over five hundred thousand pesos of Aztec plunder in 1520–1522 suggested that there was more work to do in the Western Hemisphere.[40]

Backed by the emerging mineral wealth of Habsburg Europe and the economic promise of the Spanish Americas, Charles's one million ducats in postelection debt didn't look so bad. The Cortes of Valladolid had granted their French-speaking monarch a generous servicio of 181,333 ducats per year for the 1519–1521 period under the condition that Charles remain in Spain, learn Spanish, and marry a suitable wife. The Spanish Cortes, established in 1188 to provide an occasional forum for representatives (*procuadores*) of Castile's leading city-states, figured that this was plenty of money for a monarch to conduct himself as king of Spain—even if Charles had already incurred seventy thousand ducats in debt to cover the cost of his lavish tours and processions. Spain was financially healthy, and returns from Spanish America were likely to follow an upward path.[41]

Unfortunately for Spain, the Spanish Cortes failed to appreciate the fact that Charles V was a Habsburg, Spain had been converted into a Habsburgian domain, and the emperor was already planning his next military campaigns on the backs of Castile's tax-paying citizenry. The generous Castilian servicio would be insufficient to dent the emperor's one million ducats in outstanding debt and the millions more that would follow. Of course, managing the finances of an emerging global empire was not easy. The far-flung Habsburg Empire of the early sixteenth century operated less in the traditional mode of the trading-centric Venetians (and Portuguese) than in the portfolio management style of present-day multinational conglomerates. As "chief executive officer" of Habsburg, Inc., Charles V relied heavily on "chief operating officers" (military commanders) to wage war and on "chief financial officers" to manage imperial strategy, finances, and administration. In the first half of the sixteenth century, Charles was served by four prominent advisers—Archbishop Fonseca, Mercurino Arborio de Gattinara, Nicholas Perrenot (Lord of Granvelle), and Francisco de los Cobos—but it was the esteemable Cobos (d. 1547) who dominated the age.[42]

Born sometime between 1475 and 1480 in the tiny village of Ubeda on the upper Guadalquivir River, Cobos had gained prominence from his skillful management of the Crown's mining and smelting operations in the Gulf of Mexico. He managed to parlay a November 1519 appointment as *fundidor y marcada mayor* for

Diego Velasquez's unrealized rights in Mexico into the directorship of Spanish smelting and assaying activities between Florida and Panuco and between Panama and Venezuela. Although these territories excluded virtually all of New Spain, Cobos's 1 percent participation in their gold and silver output made him a wealthy man. By 1524, he was presiding over the newly created Council of the Indies, investigating Hernán Cortés's past and present activities in New Spain, and exploiting the political problems of Mercurino Gattinara. Despite the receipt of 7,000 ducats in annual income from the Duke of Milan and a one–time 14,628-ducat gift from Charles V himself in late 1524, the esteemed Gattinara was complaining about being underpowered, underappreciated, and underpaid. The passing of Gattinara in June 1530 would split the imperial chancellorship between Cobos (finance) and Granvelle (foreign affairs).[43]

Cobos managed the Habsburg Empire "on the fly"—without the benefit of a centrally managed accounting system. He simply transferred funds, cash flow–generating assets, and debts from one Habsburg realm to another to suit the emperor's needs. Fund transfers were an adventure. Carrying cash in late medieval Europe was always risky, and unpaid troops were likely to mutiny. In a typical wartime transaction, Habsburg troops stationed in the Netherlands were paid only after Cobos had shipped thousands of ducats to Genoa to create a bill of exchange that could then trigger the remission of gold and silver coinage in Antwerp. The associated transfer charges, exchange fees, and interest costs were excessive but less than the cost of changing money in the Netherlands and then having to hire an armed convoy to carry the funds directly to the military front.[44]

Cobos moved the undertaxed Kingdom of Castile to the center stage of the Habsburg financial system. His initial action, however, was to transfer Maximilian's unpaid debts and most of Charles's election loans to mineral-rich Habsburg Austria in 1521. The transfer relieved the Castilian treasury of an unpopular burden and freed up additional Habsburg borrowing capacity in Castile. Cobos could get away with this because the Tyrolian mines were scheduled to revert back to Habsburg Austria in 1523 and the entire region was in the early stages of a spectacular mining boom. The move also reflected the peculiarities of the Castilian tax system. The powerful Castilian nobility, claiming nearly 11 percent of Castile's five million inhabitants, had been untaxed for centuries (and would remain so). The landed aristocrats (*grandees*) conspired with the lower-level horsemen (*caballeros*) to avoid tax payments at all costs in return for a legal obligation to defend the Crown. The lowest-level hidalgos, poor in both cash and land, provided most of the noble manpower for the colonization of Spanish America. It was the remaining 89 percent of the population, the commoners, who did all of the work, served in the military, and paid most of the general taxes.[45]

Cobos schemed to raise government revenues wherever he could. He created the three-person Council of Finance (Consejo de la Haçienda) in 1522, based on the successful Burgundian model in Brussels, and then the Council of the Indies in 1524. Next he converted the alcabala (sales tax) into a form of tribute payment, pressured the Cortes at Valladolid to award a second servicio of 136,889 ducats per year for 1523 and 1524 after Charles had prepledged his projected revenue stream for 1524 to meet his expenses for 1523, and farmed out the collection of *maestrazgo* (mastership) funds from Castile's ancient military orders. The venerable orders of Santiago, Alcántara, and Calatrava were out the difference. Charles always needed money for something. The costs to support his trappings of power continued to shock, especially after the relatively frugal spending habits of Isabella and Ferdinand, and worsened with time. The expenses associated with royal persons, palaces, visits, trains, and the arts represented an estimated 10 percent of state income.[46]

Cobos's greatest contribution to imperial finance was his restructuring of government debt. When traditional Castilian sources of revenue—the alcabala, servicio, cruzada, and quinto—had failed to match royal expenditures, the Castilian treasury had bridged these cash flow deficits by issuing long-term certificates (juros) to private investors. Juros ("I swear") were secured by specific streams of government revenues collected at the local level. Unlike a medieval-style annuity, a juro gave an investor a right to receive an annual interest payment from the Crown's ordinary revenues (*rentas ordinarias*). Interest was still prohibited by law so the implied interest rate had to be embedded into the juro principal itself. The key point for Cobos in 1522 was that only one-third of Castile's one million ducats in ordinary revenues had been pledged to service the interest on the Crown's outstanding juro obligations. Since Castile faced no direct military threat and controlled a potentially huge source of revenues from Spanish America, the other two-thirds of unpledged revenues were available to Charles V as security.[47]

Castile became the heart of the imperial treasury. Cobos's skillful transfer of much of Charles's outstanding debts to Habsburg Austria in 1521 paved the way for a surge in future juro borrowings. No matter that these juros supported Habsburg military campaigns waged in Italy, France, and Germany (not Castile) and overlapped a new federal tax system (*Romermonate*) that had been established to finance the emperor's military campaigns on a regular basis. Yet these and other machinations would still be insufficient to meet the emperor's many obligations. As had been the case with his hapless grandfather Maximilian, the more money Charles had, the more he spent. Potential security from Castile, Spanish America, and the Central European mining boom would only make matters worse. The emperor's increasingly ambitious military campaigns against France, the Ottoman Empire, and German Protestantism would soak up ever larger amounts of capital. Unless something could be done with Portugal, a tantalizing possibility after Charles married Isabella of Portugal in 1526 and fathered a son (Felipe) on May 21, 1527, Charles would have to make do with Castile's unpledged revenue stream and the unfulfilled promise of the Americas.[48]

Even the Fuggers had to grapple with Cobos's machinations. Their lead role in the imperial election of 1519 had given them a very powerful friend but a very large unpaid claim. The more surprising discovery was that even Jacob Fugger was capable of making a mistake. He discovered painfully that the value of Charles's personal security was negated by his overestimation of the emperor's personal net worth. Cobos's transfer of most of the Fuggers' massive unpaid claim to Habsburg Austria was inconsequential. As unpaid interest accumulated, the claim stood at 415,000 florins in Austria and 198,121 ducats in Castile in early 1523. In an otherwise insulting letter to Charles in April 1523, Jacob Fugger reminded the emperor of his family's great service to the Habsburg cause. A lesser person would have been executed for treason.[49]

The emperor caved in by awarding the Fuggers a no-cost contract to collect the maestrazgo (mastership) funds for the 1523–1527 period. The predictable stream of revenues from Castile's three military orders—consisting of pasture lands, thousands of sheep, and the dues paid to graze them—had been just recently transferred to royal authority. Jacob Fugger was happy to receive it. The five-year maestrazgo contract reduced his outstanding claims in Castile and became a virtual Fugger franchise until 1614. The exceptions were the 1527–1532 period, when the Fuggers were outbid by a Genoese syndicate willing to pay 145,000 ducats for the rights, and the crisis years of 1557–1562. The value of the contract rose after Charles agreed to include the Order of Calatrava's mercury mines at Almadén and an unexpected silver strike was made at nearby Guadalcanal. The Fuggers were paying an annual advance of over 250,000 ducats on the Almadén lease in 1600.[50]

Cobos's transfers reduced the Fuggers' outstanding claims in Castile but not in Austria. Of course, no one grieved for the House of Fugger. The family was envied, feared, and detested. Martin Luther, himself the son of a Mansfeld-based copper metallurgist, was outraged by the Fuggers' involvement in the sale of papal indulgences and their role in crushing the German Peasant Rebellion of 1524. In Hungary, where the Fugger-Thurzo partnership had gained control of the entire copper industry, a cash-strapped King Ludwig defended his kingdom against the Ottoman threat by confiscating 140,000 ducats of Fugger property in 1525 under the false pretense that the Fuggers were delivering low-quality metal to the Royal Mint. Public opinion was clearly on the side of the Hungarian Crown. As the Fugger name became a generic term for evil activities, Charles V was forced to issue the Edict of Toledo in May 1525 to place the family under imperial protection. King Ludwig had his own problems. He was killed at the epic Battle of Mohacs in 1526, an event that transferred the Crown of upper Hungary to Ferdinand of Habsburg and left lower Hungary in the hands of the Ottoman Empire for the next 150 years. The vested interests of the Habsburg brothers, Charles and Ferdinand, ensured that the Fuggers would regain their Hungarian copper operations in due course.[51]

By then, the Fuggers had diversified into merchant banking, commodities trading, and the provision of advisery services to the House of Habsburg. They were the most powerful financial entity in Europe when Jacob Fugger passed away (childless) on December 30, 1525. He left a family business that was approaching 1.5 million ducats in value—a staggering amount that was over seven times the family's net worth back in 1511—and a blessing from Charles V. With one eye on the Fuggers' political problems in Hungary and another on his own accumulating debts, the Edict of Toledo had confirmed the legality of the Fuggers' mining monopolies throughout Europe. The heirs to this massive concentration of capital, Jacob's nephews Raymond (1489–1535) and Anton Fugger (1493–1560), had been partners in Jacob Fugger & Nephews since 1510. It was the talented Anton, however, who gradually assumed full control of the enterprise, renamed the firm Anton Fugger & Nephews in 1538, and followed his uncle's lead into Habsburgian real estate. Ennobled in 1530, Anton eventually added seventy-nine thousand florins worth of properties in Bavaria, Swabia, and Wurtemberg.[52]

Anton Fugger's most fateful decision was to move the family more aggressively into the Antwerp money market, a line of business that had been largely avoided by his hard assets–oriented uncle. Anton intended to put his own stamp on the family business. After gaining valuable experience at the Fondaco dei Tedeschi in Venice and at the family's Nuremberg office in 1510, he relocated to Breslau between 1512 and 1514 and served stints at the Fuggers' copper-centric stations at Cracow, Vienna, Budapest, and Schwaz. Anton was sent to Rome for a six-year term in 1517 and received an appointment as a papal judge (and tax collector) from Medici pope Leo X in August 1519—following the emperor's successful election bid. Like Charles V, Anton was friendly with the great humanist theologian Erasmus of Rotterdam and expanded the family's contributions to various churches and the arts. Albrecht Dürer painted Anton's portrait in Nuremberg in 1525.[53]

The Fuggers' financial statements of December 31, 1527, valued the family enterprise at 2.1 million florins (nearly 1.5 million ducats). This mind-boggling sum was highlighted by the estimated value of the Fuggers' mining assets and metal inventories (650,000 florins), private assets held by the Fugger partners as individuals (430,000), and a vast array of real property, farms, offices, and cash. However, careful observers would have noted the inclusion of nearly 1.2 million florins in "book debts" to Ferdinand of Habsburg (651,000) and the Castilian treasury (507,000) that were less than certain. There was also an "unrecoverable" of over 200,000 from the political chaos in Hungary. In short, the solidity of the family fortune appeared to depend heavily on getting repaid by the House of Habsburg.[54]

## THE EMPEROR'S MULTI-FRONT WARS

Anton Fugger would have to be patient with Charles V because the emperor was planning a hugely expensive round of military campaigns against the Ottoman Empire, France, and German Protestantism. He left Habsburg Austria, mineral-rich and largely protected by mountainous terrain, to fend for itself against the vaunted Ottoman army. This despite the fact that Sultan Suleyman the Magnificent had taken lower Hungary in 1526 and intended to invade a mining region that was even more valuable than the Ottoman Balkans. Charles V's more pressing concern was Habsburg Italy, a collection of kingdoms that had been seized by the Kingdom of Aragon during the early 1440s. The Ottoman navy was threatening the Kingdom of Naples (including Sicily) from the east; France was challenging Charles's territorial claim to the Duchy of Milan

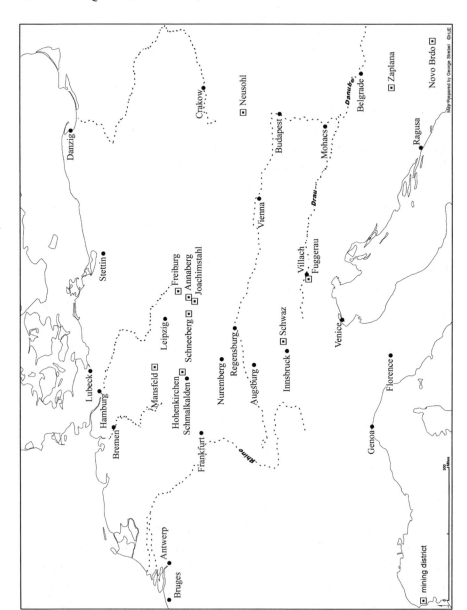

**MAP 4.
CENTRAL
EUROPE,
1490–1540**

Courtesy of George Strebel, © author.

from the west; and the papacy was attepting to manage the entire messy situation to Rome's own benefit.

Habsburg Italy had also inherited Aragon's enemies. A new generation of Barbary pirates, fueled in part by exiled Iberian Muslims and Moriscos, was focused on the upturn in Mediterranean shipping activity and Spain's tenuous possessions in North Africa. Aragon, Genoa, Naples, and Sicily were occasionally attacked as well. The Barbary threat to Italy was elevated after the Barbarossa brothers, Baba Oruc and Hayreddin, evicted the Spanish from Algiers in 1516 and joined the Ottoman cause. The Ottoman conquest of Mamluk Syria and Eqypt in 1517 began a process through which the fragmented Muslim navies, and even the Venetians, were all coordinated by Istanbul. Venice had been nearly destroyed by the Portuguese spice monopoly and was willing to change sides. While these developments were frightening to Christian Europe, Christian naval responses had been failures. A Genoese-Neapolitan fleet was storm-wrecked off the coast of Sardinia in May 1519 and a twenty-six-ship fleet was destroyed off Algiers in August 1523.[55]

Suleyman the Magnificent (1520–1566) maintained the pressure. Armed with Balkan mining revenues, huge tribute payments from the Mamluk Empire, and a commercial treaty with Venice, he expanded the Ottoman Empire by adding present-day Iraq, Cyrenaica (eastern Libya), and the island of Rhodes by January 1522. The defenders of Rhodes, the Knights Hospitaller of the Order of St. John, withdrew to the island of Malta. Suleyman subsequently assembled a powerful Ottoman navy from the ragtag Barbary corsairs, privateers, and pirates who operated along the North African coast. The sultan's offensive even raised fears about a possible Fifth Column of Moriscos in Spain herself after thousands of unhappy Moriscos revolted against royal authority in 1526. The revolt occurred in the Sierra de Espada after the "convert or leave" decree of 1502 was extended to Valencia. Most of the Mediterranean world, especially Naples, Sicily, and Malta, were targeted by Ottoman war galleys.[56]

A possible alliance of convenience among Istanbul, Venice, France, Barbary corsairs, and Iberian Moriscos was frightening to

Charles V. The emperor had been happy to add the kingdoms of Naples and Sicily to his imperial empire, making him the first Holy Roman Emperor to rule these lands since the first half of the thirteenth century, but the cost to manage and defend Habsburg Italy usually exceeded the realm's revenue contribution. Habsburg Italy was chronically unprofitable. While Charles was in no position to challenge the formidable Ottoman Empire, he was not about to watch Habsburg Italy fall into the unfriendly hands of France, Venice, or an occasional hostile pope. With Italy in mind, the emperor had even signed the Treaty of Windsor with English king Henry VIII in June 1522 to gain another ally against France.[57]

Despite Cobos's machinations, the Habsburg treasury was ill prepared to handle a major military campaign in Italy (or anywhere else) in 1525. When a high-quality borrower like the Netherlands state treasury was forced to pay interest of as high as 22 percent in Antwerp, lesser-quality borrowers had to pay much more—when they could borrow at all. Charles V's relatively solid credit rating was becoming increasingly dependent on dumb luck. In February 1525, he was heavily in debt and the Habsburg army in northern Italy hadn't been paid in three months when a likely retreat (or mutiny) turned unexpectedly to the emperor's favor. General Antonio de Leyva somehow persuaded his unpaid Habsburg troops to attack the French position at Pavia in an all-or-nothing quest for spoils. Not only were the French routed at the Battle of Pavia, but Leyva managed to capture François I himself and force the shell-shocked monarch to leave his two sons as hostages. The ransom payment turned out to be a staggering 1.2 million ducats. While the windfall was not received until 1531, it provided badly needed financial relief and a new source of security for Charles's next campaigns. The unpaid Fuggers even agreed to lend one hundred thousand crowns to the emperor's Italian war chest in 1527.[58]

Unfortunately for all concerned, great victories were almost always temporary, and treaties were almost always made to be broken. While Charles's smashing victory at Pavia forced François to renounce all French claims to Italy (and Flanders) in January

1526, the renunciations were unlikely to hold. François I (1515–1547) was content to leave his two sons in Castile as hostages for the time being. Charles recognized French fears of a possible Habsburg domination of Western Europe and knew that François, financed by a huge taxpaying population, would always return to battle. However, the emperor was outraged by Rome's willingness to side with the French position. The papacy feared the Habsburg Empire as well. The emperor responded in May 1527 by doing the unthinkable—he sent his underpaid troops to Rome and allowed them to plunder the city at will. The audacious action warned the French-leaning papacy to stay neutral in European affairs.[59]

The Italian campaigns left huge debts to be paid by all sides. By September 1527, the Habsburg army was due 373,000 ducats in back wages, 514,000 was owed by Pope Clement VII (Giuliano de Medici) and the Republic of Florence, and the already-devastated French treasury was on the hook for a massive ransom payment. Charles followed up his brutal sacking of Rome by shifting his attention to France herself. Already blessed with the finest army in Europe, he recruited the brilliant Genoese admiral Andrea Doria in August 1528 to defend against the increasingly powerful Ottoman navy. Doria was available because Genoa's French-leaning government had just been replaced by a group of twenty-eight Habsburg-friendly nobles. Since these same Genoese families dominated Spanish sugar and slaving enterprises in the West Indies, Charles was willing to defend their Mediterranean interests against Istanbul, Venice, and even France.[60]

By the fall of 1528, the emperor had supplemented Doria's twelve-ship Genoese fleet with sixteen Spanish vessels, eight thousand sailors, ten thousand German mercenaries, and two thousand cavalrymen. He was preparing for a full-scale invasion of France. No matter that Charles was heavily in debt, the Fuggers and Welsers were tapped out, and a loan of 200,000 *scudi* from the Genoese Ansaldo Grimaldi was insufficient to support Doria's fleet. Margaret of Austria, Habsburg Regent of the Netherlands, had dispatched the German troops to the front without adequate financing. It was then that the resourceful Cobos persuaded Charles to cash in on his recent marriage to Isabella of Portugal—Charles's brother-in-law, King João III of Portugal, was willing to pay 350,000 ducats for a renunciation of Castile's tenuous (but defensible) claim to the Spice Islands at the Treaty of Zaragoza (1529). The Portuguese funds were carried by mule-train to Barcelona and then on to the Italian front.[61]

Charles's invasion plan would have paid off spectacularly if not for the actions of Suleyman the Magnificent. The opportunistic sultan took advantage of the brewing Habsburg-Valois conflict by invading the Austrian frontier in the summer of 1529. As a huge Ottoman army neared the gates of Vienna, the threat posed to the world's most valuable mining properties shocked Charles V and Medici pope Clement VII into a peace treaty in July 1529. Charles agreed to restore the Medicis to power in the Republic of Florence. The papacy agreed to restore cruzada payments to Castile, lean on François I to sign the Treaty of Cambrai on August 3, 1529, and (finally) anoint Charles as Holy Roman Emperor. Catalyzed by Suleyman's advancing army, the Treaty of Cambrai called for the withdrawal of all French forces from Naples and for François to make good on the 1.2 million ducats in ransom funds required to release his two sons from captivity.[62]

Suleyman the Magnificent was the most feared man in Europe. Born in the eastern Black Sea port of Trebizond in 1495—a cosmopolitan, Greek Orthodox–dominated town linked to the nearby silver mines at Gumushane—Suleyman assumed the Ottoman throne on October 1, 1520, with a huge war chest. He was receiving massive tributary funds from the Mamluk Empire and as much as fifty tons of silver annually from the refurbished mines of the Ottoman Balkans (Rumelia). The crown jewel of Novo Brdo (Kosovo), so revered by Mehmed II when he reconquered the town in 1455 that he enslaved everyone except the miners, was in decline, but the ancient works at Sidrekapsi, Srebrenica, and Serez had been expanded. Ottoman mining engineers were also focused on the underexploited deposits at Trepca and Zaplana. An outdated Genoese report of 1475 had provided only sketchy details on the great Balkan mines, so most Europeans were unfamiliar with the fact that Sidrekapsi (east of Thessalonika) was producing as much as ten tons of silver per year. A French visitor in 1554 estimated that Sidrekapsi's six hundred furnaces and six thousand miners were generating around three hundred thousand ducats in annual revenue. Some Ottomans smelted their silver-bearing ores with a somewhat unusual flux material. Rather than using traditional lead, vinegar, or even dried wine as flux, at least one operator was adding 1.3 grams of dry human waste to the molten mix to "ensure that [the] silver will be of pure standard."[63]

The capture of Belgrade in 1521 had freed Suleyman to march up the Danube River uncontested against the Kingdom of Hungary and threaten the Fugger-Thurzo copper works in Slovakia, Joachimstahl in Bohemia, and even the prolific Habsburg mines in Tyrolia. The subsequent Ottoman triumph at the Battle of Mohacs on August 28, 1526, gave Suleyman control over a large chunk of lower Hungary, including an annual tribute payment of around fifty thousand ducats, and a platform to move against the mineral wealth of Habsburg Austria. The Ottomans even had their appeal. Based on their track record in the Balkans, they had managed their subject peoples in an effective and even tolerant manner. The five million or so residents of the Ottoman Balkans (Rumelia) south of the Danube River, four-fifths of whom were Christian, were ruled in virtually the same fashion as the citizens of Istanbul and Egypt. The tributary *timar* (fife) system was less onerous to the common citizenry than the abusive tax policies of Habsburg Europe and France. And there were no forced conversions to Islam. Mehmed II had converted Constantinople's famous Hagia Sophia into a mosque but allowed the Holy Apostles Church to continue as a Christian church. Compared with the religious policies practiced against Protestants, Jews, Conversos, Muslims, and Moriscos in Habsburg Europe, the Ottoman Empire didn't look so bad. Charles, Ferdinand, and the papacy all had to be concerned about the fate of upper Hungary, Austria, Italy, and their neighbors.[64]

Suleyman, of course, was focused on the Habsburg mining

industry. After decades of stalemate, the sultan had seized lower Hungary, killed Hungarian king Ludwig, and induced the crowns of Hungary and Bohemia to be transferred to Ferdinand of Habsburg. This meant that the next campaign would be directed at the Habsburg Empire itself. The attraction was obvious. Central Europe was on track to produce nearly one hundred tons of silver on an annual basis—twice the output of the Ottoman Balkans—and could help to finance the sultan's next campaigns against Western Europe and Persia. In addition, an assault on the mineral wealth of Habsburg Europe would follow the same strategy that had been applied by Suleyman's illustrious ancestors against the Christian kingdoms of Macedonia, Thrace, and Serbia in the fifteenth century and complement the spectacular seizure of Mamluk Egypt and Syria in 1517. Ottoman Egypt alone was contributing as much as 500,000 ducats in annual tribute to the imperial coffers.[65]

The sultan was not rich enough. He was personally receiving as much as half of the state's ten million ducats in annual revenues in 1528. The sultan's personal hoard included 3.2 million ducats in annual tribute from Muslim vassal states and another 750,000 ducats in annual poll taxes from non-Muslim subjects. While the gold coins sitting in the Istanbul treasury were multinational—Ottoman *sultanis*, Venetian ducats, Egyptian ashrafis, and even Hungarian issues—most of these funds were allocated to the sultan's personal army, the Janissary corps, and the construction of fortresses, ships, and public works. Janissary garrisons were located in major cities to protect the caravan trade and the carriage of tribute funds. The elite Janissaries consumed more than one-third of state income in the form of timar (land grant) distributions. Like the Spanish encomiendas, the Ottoman timar system helped to conserve cash in Istanbul and decentralize imperial power.[66]

Vienna had withstood the first Ottoman invasion in 1526 thanks to timely reinforcements and Suleyman's caution—the sultan feared that Charles's recent triumph at Pavia had freed up thousands of Habsburg troops to possibly meet him. However, Charles had refused to provide any direct assistance to his brother Ferdinand, had saddled him with a massive load of debt, and had opted instead to lead his next invasion against France. That left Ferdinand, ruler of Habsburg Austria since 1522 and Hungary and Bohemia since 1527, to manage the Ottoman threat on his own. He did not have to wait long. A powerful Ottoman force of around eighty-seven thousand took Buda (Hungary) in early September 1529 and moved north to besiege Vienna. Regent Mary of Hungary scrambled mightily to raise funds to defend the city. A very reluctant Ambrosius Hochstetter agreed to advance 200,000 florins at 37 percent— assuming that there was a Habsburg Austria left to pay the bill. But Habsburg Austria was spared by Charles's timely peace treaties with François I and Pope Clement VII and Suleyman's continued caution. The Ottomans abandoned their three-week siege of Vienna on October 15, 1529.[67]

When Charles V was finally anointed Holy Roman Emperor in Bologna on February 14, 1530, things were looking up. The emperor's shaky financial position was set to be bolstered (temporarily) by the receipt of the French ransom funds. Naples was back in the Habsburg fold after twenty-one thousand potentially mutinous troops were finally paid. A successful siege of Florence had restored Cosimo de Medici to power. And Charles's brother Ferdinand (1502–1564) was on schedule to receive the largely ceremonial title of King of the Romans in January 1531. If that were not enough, the Treaty of Cambrai with France, on top of Charles's timely recruitment of Andrea Doria, had prompted Suleyman to think twice about invading Habsburg territories by land or by sea.[68]

The geopolitical results were positive, but Cobos had been challenged to cover the 1.3 million-ducat cost of the Italian campaigns between 1528 and 1530. The Castilian treasury had already been drained of 1.7 million ducats in transfers between 1522–1528, leaving the 1.2 million-ducat windfall from France to be spent nearly as soon as it was received in 1531. In the meantime, Cobos had financed the Italian campaigns from a combination of sources— Portugal's "purchase" of the Spice Islands; reluctant advances from the still-unpaid Fuggers, Welsers, and Genoese; Brussels's approval of a five-year "coronation" subsidy to Charles for the 1529–1534 period; a resumption of the papal cruzada; and a four-year *donativo* from the Parlamento of Naples. Since Charles still retained garrisons in Naples, Sicily, and Lombardy, the new donativo was not likely to go very far.[69]

Charles's financial position was bolstered significantly by the financial windfall from France, the resumption of the papal cruzada, and a long-awaited increase in the ordinary servicio in Castile—to nearly five hundred thousand ducats over a two-year period. So when Suleyman decided to make a third assault on Vienna in 1532, the emperor was ready. The emperor was an army man. He inherited an excellent Spanish army that had been battle-tested in the long campaign against Muslim Granada, refined in North Africa, and improved during an endless series of Italian campaigns. Famous commanders like Gonzalo de Córdoba had experimented with new arrangements of pikemen and arquebusiers. One successful arrangement placed the 3,000-man regimental unit, the *tercio*, into a square formation of 1,500 pikemen, 1,000 swordsmen, and 500 arquebusiers. While the French had countered with formations of their own, the emperor was confident that he could handle Suleyman's underarmed military machine. No matter that Suleyman had him beat when it came to numbers—the most men and the most money.[70]

The temporary peace with France induced Charles to go for broke in 1532—he would personally lead a massive Habsburg flotilla down the Danube River and crush the Ottoman army once and for all. Since the latest Ottoman invasion threatened Christians of all denominations, the Protestant princes of Germany were even willing to suspend their own grievances to support the pan-Germanic cause. The French and Venetians stayed on the sidelines. By early September 1532, Charles, Ferdinand, and the Protestant princes had assembled one of the largest Christian armies in history—an estimated 112,000 infantrymen, 26,000 cavalrymen, 3,000 trench diggers, and 10,000 boatmen—to float down the Danube (from Regensburg) and hopefully reverse a century of Ottoman achievements in southeastern Europe.[71]

Fortunately for both sides, Suleyman opted for caution once more. He ordered his 136,000-man army back to lower Hungary and assumed correctly that Charles would decline to pursue. With Safavid Persia lurking on his eastern front, the sultan was not about to risk his resources on a less-than-sure thing—he even failed to take the Slovakian fortress at Guns after a costly three-week siege. Charles had a similar concern with respect to France. So what would have been one of the epic battles in world history was averted by Charles's timely intervention and Suleyman's prudence. Of course, the aborted campaign carried a huge financial cost. The Habsburg treasury was saddled with another 530,000 ducats in campaign costs and 100,000 more to cover the expense of the emperor's always-lavish retinue. As Suleyman returned to his naval campaigns, Charles entered Vienna on September 23, 1532, as the savior of Christendom.[72]

# CHAPTER FOUR

# THE GREAT AMERICAN TREASURE HUNTS

Charles V paraded through Vienna in September 1532 as the savior of Christendom. His successful campaigns against the Ottomans (in Austria) and the French (in Italy) had preserved the Habsburg Empire for the time being, but he had just spent two million ducats simply to maintain the political status quo. The emperor's heavy debt burden would have been even heavier if not for the receipt of a huge one-time ransom payment from France, the creative financial skills of Francisco de los Cobos, and the spectacular mining boom in Central Europe. While the unsettled military situation in Europe would require even larger sums in the immediate future, the imperial cupboard was nearly bare. The Fuggers and Welsers held huge unpaid claims against the House of Habsburg, a series of Pacific expeditions had been inconsequential, and the promise of Spanish America remained unrealized. The Sevillian Genoese were working wonders with the West Indian sugar industry, but the reopening of transatlantic trade to all subjects of the Holy Roman Empire in 1524 had been uneventful. The Fuggers were sitting on the sidelines, and the Welsers were struggling to make money on the American mainland.

While the Welsers were investing some of their Bohemian silver profits in mining ventures in Mexico and Venezuela, the Fuggers' direct investments in Spanish America were next to nil. Jacob Fugger had preferred to concentrate on metals-trading opportunities in Europe and schemes to circumvent the Portuguese in Asia. Prior to his death in December 1525, he had agreed to finance the westerly expeditions of Loiasa and Cabot only because they were aimed at the Portuguese spice business and the murky demarcation line in the Pacific Ocean. Jacob and his nephew Anton Fugger had no apparent interest in the uncharted stretch of Portuguese territory that ran from northern Brazil down to the so-called Rio de la Plata (River of Silver) off present-day Uruguay. If the Welsers wanted to dabble in high-risk places like Venezuela, that was their business.

Brazil had been largely ignored by the Portuguese Crown as well. Following Cabral's accidental discovery in 1500, Manoel had simply farmed out the rights to Terra de Vera Cruz to a private company organized by Fernão de Noronha and other Portuguese Conversos. In 1501, the company paid four thousand ducats annually for an exclusive eleven-year trading and exploration contract modeled on Fernão Gomes's earlier license in West Africa. Noronha's contract was focused on brazilwood, a reddish dye that could be marketed to the European textile industry, and possibly sugar. Financial incentives helped to establish Pernambuco in 1502, followed by Porto Seguro (1503) and São Vicente (1508). While all three of these establishments were just tiny frontier posts, the forty thousand cruzados worth of brazilwood shipped in 1506–1507 had persuaded the Crown to change the territory's name to Brazil. Royal contracts issued between 1512 and 1516 were focused almost entirely on brazilwood.[1]

Vespucci had returned to Brazil in November 1503, the only one of Coelho's six-ship fleet to do so that year, but he explored the coastline only as far south as Cabo Frio (just north of Rio de Janeiro). Noronha and other brazilwood traders operated no further south than São Vicente (downriver from São Paulo). This left the discovery of the Rio de la Plata to two Portuguese brazilwood freighters, Esteban Froes and João de Lisboa, in 1511. While the news was ignored by a Fugger agent in Cadiz in 1512, Spanish interest in the disputed territory heightened after Balboa sighted the Pacific Ocean in 1513. The discovery prompted Ferdinand to dispatch a three-ship fleet to southeastern Brazil in October 1515 under Juan Diaz de Solis, the casa's new pilot major, in the hopes of finding a cross-continental passageway to the Pacific. The fleet reached (and possibly named) the Rio de la Plata by February 1516 and made landfall somewhere in Uruguay. The Portuguese-born Solis should have stayed at home. He and five crew members were abducted by hostile natives, roasted, and eaten. Eighteen other sailors were shipwrecked off Santa Catarina Island. The survivors returned to Seville in September with less-than-glowing reports about the region.[2]

Things heated up in the Rio de la Plata after Magellan stopped near the entrance in 1519. In the spring of 1524, one of the ship-wrecked members of the Solis expedition, the Portuguese Aleixo Garcia, led three or four fellow Iberians and a large number of Guaranis on an unbelievable inland trek between southeastern Brazil and the wilds of uncharted Bolivia. The Guaranis of Santa Catarina had excited the stranded Iberians with tales of a silver mountain range, a Sierra de la Plata, and the existence of a fabulously wealthy kingdom that turned out to be the Inca Empire. The tales seemed to jibe with Cortés's recent discovery of a spectacular inland kingdom in Mexico. Western affiliates of the Guarani tribe had raided the Incas' Bolivian frontier for decades, but no Guarani had ever attempted an invasion on the scale of the proposed Garcia expedition.[3]

Garcia and the Guaranis traveled past the spectacular Iguazu Falls at the junction of the Paraná and Uruguay rivers and across the southern plains of Paraguay. At the future site of Asunción, home to a friendly Guarani village in 1524, the invasion force was expanded to as many two thousand before proceeding up the Paraguay River to a point near the southeastern corner of Bolivia. From there, the westerly moving invaders marched another five hundred miles across the Bolivian Chaco and plundered local villages along the Inca frontier. They were within fifty miles of Chuquisaca (the future site of Sucre) when the Incas responded. Thousands of Inca warriors pushed Garcia and his Indian allies back into Paraguay in late 1525. At a site possibly 150 miles north of Asunción, Garcia dispatched twelve messengers back to Santa Catarina with two or so *arrobas* of silver (twelve pounds) and requests for reinforcements.[4]

Garcia was never heard from again. He and many of his allies were massacred on the west bank of the Paraguay River by either his erstwhile friends, the Incas, or by another tribe. The Guarani threat prompted the Incas to construct a chain of garrisons along the Bolivian frontier and to alert their ruler, King Huayna Capac, about the presence of white-skinned Iberians. These actions, on top of Garcia's silver shipment and letters to Santa Catarina, added credibility to the wild rumors of silver mountains and Inca treasures that had circulated for over a decade. News of Garcia's discoveries could not have reached Spain until late 1526, but a would-be conquistador like Francisco Pizarro was sure to have heard about it. If Garcia reached as far north as Chuquisaca, near the headwaters of the Pilcomayo, he had come within one hundred miles of the greatest silver bonanza in American history.[5]

The next major expedition to the Rio de la Plata was an inadvertent visit by Sebastian Cabot, the new pilot major at the Casa de la Contratación. No fewer than sixty-five investors had raised a total of 56,185 ducats for the project, including contributions from seventeen Sevillian Genoese families, the Fuggers, and even the Crown. By March 1525, the Venetian-born Cabot was armed with enough investment capital to outfit a fleet of at least three ships for a two-year voyage to Asia. The objective was not the Rio de la Plata but another follow-up to Magellan's westerly voyage to the Moluccas. With increasingly large volumes of Central European silver on hand, the investors were interested in tangible Asian spices and silks—not rumored rivers and mountains of silver.[6]

A dispute delayed the fleet's departure until April 3, 1526—the Genoese had demanded that two of their representatives be allowed to join the expedition (as *veedors*) to monitor the independent-minded Cabot. The pilot major was not happy about it. Finally, Cabot headed his fleet and a two-hundred-man crew in a southwesterly direction to a landing site at Pernambuco on the Brazilian coast. Pernambuco had been founded in 1502. It was there that Cabot learned that one of the few survivors from the Solis massacre of 1516, Francisco del Puerto, had reported the existence of fantastic treasures in the upper Rio de la Plata region and that Aleixo

Garcia had disappeared in a very recent attempt to find them. Since Cabot was probably anxious about crossing the treacherous Magellan Straits, he jumped at the opportunity. He immediately canceled the Molucca project, headed south toward Rio de la Plata, and imprisoned the two Genoese veedors on nearby Patos Island when they dared to complain.[7]

Cabot was on his own. The mood darkened after his primary supply ship ran aground off Santa Catarina in October and most of his crew began to die off from tropical disease. Yet when Cabot encountered two other survivors from the Solis party and two more from the Garcia de Loaisa expedition of 1525, he was presented with firsthand reports about the rumored "mountain of silver." He learned that Garcia had reached as far north as Bolivia, had sent back tantalizing quantities of silver, and had not been heard from since. Cabot was undaunted. He recruited one of the Solis survivors to serve as his guide and sailed up the Rio Paraná to a point thirty miles north of present-day Rosario, Argentina. After constructing a frontier fort, he continued north to the confluence of the Paraguay River only to face a series of Indian attacks near Asunción. Two lieutenants were sent back to the ships with some gold, silver, captured Indians, and instructions to solicit aid from Charles V. Stranded in the deep interior of Paraguay and operating in clear violation of his investors' instructions, Cabot explored the region for three long years. He came up empty but at least escaped with his life. When his modest gifts arrived in Toledo in November 1528, Cabot was branded as an outlaw—his investors had sought Asian spices and silks, not an adventure in Paraguay, and refused to reinforce an unauthorized expedition.[8]

The destruction of Cabot's frontier fort by local natives was the final straw. The pilot major returned to Seville in late 1529 with nineteen surviving crewmembers and little else. Cabot was arrested immediately by casa officials in July 1530 and tried by the Council of the Indies. Charles V must have admired Cabot's spunk and his tales of fantastic treasures. He pardoned Cabot, restored him to his position of pilot major, and advised him that his intriguing Paraguay project might be relaunched at a later date. The Genoese were outraged. They

refused to participate in another Spanish expedition until 1545—
Francisco de Orellana's disastrous eight-month voyage up (not down)
the Amazon River—and returned to their sugar, slaving, and trade
finance businesses. As for Cabot, he was eventually recruited to lead
an expedition for Edward VI of England in 1547.[9]

If the Rio de la Plata had yet to live up to its tantalizing name,
there was plenty of uncharted territory to be investigated in Spanish
America. Investigators were encouraged to find new sources of revenue to finance the emperor's next campaigns against France,
Istanbul, and German Protestantism. Spanish American mining
activities to date had been highlighted by a twenty-year run of alluvial gold in the Caribbean and one-shot plunders of native temples,
palaces, and gravesites in Mexico and Central America. After native
ceremonial objects had been looted, the Spaniards had simply melted
down the gilded (copper-gold) and silvered (copper-silver) alloys into
bullion and exported the metal back to Seville. Conventional mining
ventures were limited to the occasional excavation of surface-level
silver ores in Mexico. Compared with Central Europe and even Portuguese Africa, Spanish American mining activities had been disappointing and the native labor pool was being devastated by disease.

An optimistic Hernán Cortés had compared the Aztec Empire
favorably with Muslim kingdoms and even German states in the
hopes of catalyzing investment and colonization in New Spain. In
Michoacán, the cradle of Mexican metallurgy, Cortés and his fellow
Spaniards became skilled at finding, seizing, and melting down
many centuries worth of Tarascan artistry during the 1520s. Gilded
and silvered shields, bracelets, cups, and other ceremonial objects
from the Tarascan capital on Lake Patzcuaro were reduced to large
slabs (*plata baja*) for shipment back to Spain. But Cortés and his
fellow Spaniards struggled to pinpoint the precise location of the
mineral deposits from which these ceremonial treasures had been
made. Even with the assistance of Aztec guides and Moctezuma's
own tributary maps, the Spaniards failed to discover the primary
Tarascan deposits at Tamazula and wasted their energies on minor
silver works in Jalisco and Compostela.[10]

The failure to discover any meaningful silver deposits in metallurgically advanced Michoacán suggested to Spanish officials either that Mexico was poor in precious metals or that professional prospectors and mining engineers needed to be imported from Central Europe. In 1529, the Welsers and Ehingers were called upon to recruit over fifty German mining engineers to train some of New Spain's would-be mining magnates. Flush with their spectacular profits from Joachimstahl, the Welsers were already supplying four thousand black slaves to the sugar plantations of Hispaniola and Cuba and preparing to colonize Venezuela. The debate over Mexican geology was subsequently tabled in 1531 when sizable quantities of surface-level silver were finally discovered in Zumpango (Guerrero) along the Sierra Madre del Sur escarpment. This was hundreds of miles south of the Tarascan region of Michoacán. Like virtually all silver strikes in the Americas, the discovery at Zumpango was led by native prospectors and exploited by royal officials. Zumpango was developed by a partnership (*compania*) headed by Juan de Burgos, Juan Alonso de la Sosa, and the irrepressible Hernán Cortés.[11]

While entrepreneurs like Cortés were prospecting for Mexican silver and Cabot was wandering around Paraguay, other energetic Spaniards were focused on the unexplored territories of Tierra Firme—the Spanish name for the mainland territories of Central America and Colombia. The focus was on gold. Iberians were especially interested in the yellow metal because European gold supplies had become scarce again. Between 1500 and 1520, the West Indies had shipped an average of around one ton of gold to Spain on an annual basis before entering into a period of decline. The depletion of the major Caribbean fields, the diversion of West African gold to Portuguese Asia, and the recent surge in Central European silver production all combined to raise Europe's bimetallic exchange ratio to a lofty 12:1. This encouraged profit-hungry Spaniards to invade Panama, Nicaragua, Honduras, and Mexico in search of gold. While they would be largely disappointed—these four locales produced no more than forty thousand pesos of gold annually during the first half of the sixteenth century—their efforts would lead to major discoveries in the uncharted south.[12]

No matter that most of the Spanish explorers and conquistadors spent as much time fighting among each other as they did with local Indian tribes. Based on the plunder strikes at Tenochtitlan, the stakes had become simply too high to allow a potential rival to gain an upper hand—especially with the added risk of a premature death at the hands of some barbarian Indian tribe or a political foe, an old-fashioned shipwreck, or starvation. It was easy for common soldiers and the conquistadors who commanded them (and the Crown itself) to shift sides at a moment's notice. All of this was on display during the feverish metal rushes that were set to take place in Mexico, Bolivia, and Colombia. In the race to achieve wealth and glory, loyalty was a short-term commodity in the service of unbridled profit opportunity.

Spain's tenuous foothold in Panama had solidified in 1512 after Vasco Núñez de Balboa assumed command of the surviving members of the disastrous de la Cosa expedition of 1509. Balboa founded the city of Darien and other settlements, promoted the isthmus as Castilla del Oro to market the region's very modest gold supplies, and even discovered the Pacific Ocean in September 1513. Francisco Pizarro was among the sixty-seven Spaniards who participated in the initial sighting. Unfortunately for Balboa, his entrepreneurial achievements had been made without royal authority. Governor Pedrarias de Avila, a feisty veteran of the Granada campaign, ordered Pizarro to arrest (and behead) Balboa for treason in 1517. Balboa's great achievement, discovering the Pacific Ocean, was demystified very shortly by the Magellan expedition of 1519–1522.[13]

Pedrarias shifted Panama's capital to Pacific-side Panama (City) in 1519 and dispatched a prospecting expedition to Nicaragua in 1523–1524. The intention was to beat the increasingly long arms of Hernán Cortés to Central America. The Nicaragua campaign included future notables like Francisco Hernández de Córdoba, Hernán de Soto, Sebastián de Benalcázar, Diego de Almagro, and possibly Francisco Pizarro. Benalcázar, already in possession of an encomienda in Panama, had a falling out with de Soto that would cost him a lot of money in later years. In the meantime, very limited quantities of gold were uncovered in Central America. The disappointment prompted Pizarro, the loyal executioner of Balboa, to gain permission to explore the uncharted Pacific coastline to the south of Panama.[14]

## THE PIZARROS INVADE INCA PERU

The man who would conquer the otherwise unknown Inca Empire, Francisco Pizzaro, had landed in Hispaniola as a twenty-four-year-old in 1502. Although he was an unrecognized illegitimate son of Gonzalo Pizarro, a hidalgo who had participated in the conquest of Granada, an introduction from his uncle Juan Pizarro in Hispaniola guided Francisco through the ranks. He served as a lieutenant to Balboa in 1513, was listed (after Balboa) as the second Spaniard to have sighted the Pacific Ocean, and was then ordered to execute Balboa in 1517. The loyal Pizarro participated in Pedrarias's founding of Panama (City) in 1519 and was rewarded with one of the largest encomiendas in Panama. It was in 1522, after striking an informal mining partnership with Diego de Almagro, Fray Hernán de Luque, and Diego de Moria, that Pizarro began to earn a substantial amount of money. He was making at least 3,500 pesos per year, if his dispatch of 705 gold pesos to Seville in 1522 represented the royal quinto, and had received an encomienda of 150 Indians to exploit the island of Taboga opposite Panama (City). Shortly thereafter, the forty-four-year-old Francisco learned that his father Gonzalo had been killed during the Navarre campaign and had left a modest estate in Trujillo to a single legitimate son—the twenty-year-old Hernán. Two other illegitimate sons, Juan and Gonzalo, had at least been recognized in their father's will of 1522.[15]

The uncharted territory south of the Isthmus of Panama had excited a new generation of plunder-seeking conquistadors after Cortés's astonishing success in Mexico. Most of the Isthmus-based

Spaniards were off salary and forced to fend for themselves. This converted their status from soldiers to *compañeros*. In 1522, Pascual de Andagoya and Licenciado Gaspar de Espinosa received an apparently nonexclusive royal *capitulación* (agreement) to explore the Pacific coastline south of Panama. Their efforts were inconsequential, but they prompted Francisco Pizarro to form a 25,000 gold peso partnership (*compaña*) with Almagro, Espinosa, and Luque to retrace Andagoya's route in 1524 and again in 1526–1527. These efforts were also inconsequential. Constrained by disease, hunger, and mosquitoes, the Spaniards reached only to the northern border of Ecuador before they withdrew to the relative safety of Gorgona Island in 1527. From there, off the southwestern coast of Colombia, Pizarro and fourteen Spaniards readied plans to make a third expedition to the south in 1528.[16]

This third expedition would be more successful. After stopping at a few coastal villages, Pizarro landed at Tumbes on the southern (Peruvian) side of the Gulf of Guayaquil and shortly captured two Indian youths to serve as translators. The nearby Rio Viru apparently became the source of the word *Peru*. The Spaniards received gifts of silver and gold jewelry (and some llamas) from the natives of Tumbes and learned two important bits of information—there was an opportunity to make friends with the local Chimus, an otherwise legendary empire that had been overthrown by the Incas, and that the ruling Incas possessed an enormous amount of treasure. This was enough for Pizarro. Leaving Almagro behind to recruit more men and ships from Nicaragua, Pizarro decided to sail back to Spain in late 1528 to obtain a signed capitulación from Charles V himself. If approved, the agreement would give Pizarro's Compaña del Levante the exclusive right to conquer the region and provide Francisco Pizarro with the adelantado and governorship of Peru. Pizarro also hoped to recruit some trusted adventurers from Extramadura, including his three half-brothers in Trujillo.[17]

Of course, the likelihood that an illegitimate and unrecognized son of a lower-class noble would obtain a prestigious capitulación from Charles V was slim. Pizarro had a strong track record in Central

America, but his primary connection to the Crown was limited to a chance association with the brother-in-law of one Lope Conchillos during one of his many campaigns. The key point was not that Cochillos had worked closely with the late Juan Rodriguez de Fonseca in the early days of the casa, which he had, but that he had supervised a young clerk by the name of Francisco de los Cobos—Charles's principal adviser and arguably the second-most powerful man in Spain. If that were not enough, Cobos was extremely familiar with American metals thanks to his service as *fundidor y marcada mayor* in the early 1520s. In short, Pizarro had a very good contact.[18]

Pizarro managed to convert a tenuous connection to Lope Cochillos into a formal capitulación—a medieval-style reward to those who served the Crown at their own expense. The capitulación was received from Empress Isabella (not Charles) on August 17, 1529, on the condition that Pizarro carry along three officials from the royal treasury to supervise his activities in Peru as well as three or more Dominican friars. In return, the agreement gave Pizarro's compaña the exclusive right to discover, conquer, and colonize Peru. Pizarro's concurrent request for vassals and encomiendas was deferred, but the Crown agreed to contribute some weapons and artillery. The status-conscious Pizarro was also issued his own coat of arms (the first of three) and admission to the prestigious Order of Santiago (a virtual Fugger subsidiary during the sixteenth century).[19]

Pizarro still had to be concerned with Hernán Cortés. No matter that Cortés's dream of a Pacific spice empire was in limbo and he was in danger of being stripped of his virtual kingdom in Mexico. The energetic Cortés had participated in a series of inconsequential silver mining ventures prior to the strike at Zumpango in 1531 and launched a few naval expeditions from his Pacific bases at Tehuantepec, Zacatula, and Zihuatanejo. His biggest challenge, however, was a political one. While he had apparently mollified the four royal officials who essentially ruled New Spain during his regular absences, the arrival of Luis Ponce de León in early July 1526 was cause for concern. Ponce de León, whose more famous brother Juan had died in 1521 from wounds suffered in Florida, was instructed to investigate matters (in *residencia*) on behalf of the Council of the Indies. However, Cortés dodged another "bullet" when Ponce de León and thirty of his colleagues succumbed to a plague (or food poisoning) on July 20. A relieved Cortés was left to deal with the unexpected arrival of one of Loaisa's wayward ships, the *Santiago*, and a subsequent royal authorization to dispatch Saavedra and a three-ship rescue fleet from Zihuatanejo.[20]

The Great Conquistador's wealth, talent, and shipbuilding capabilities made him a credible threat along the untapped Pacific coastline. After launching Saavedra's rescue fleet in October 1527, the forty-three-year-old Cortés left for Spain in March 1528 in an attempt to regain confirmation of his Mexican realm. The now-legendary figure arrived in Palos (with Pedro de Alvarado) with great fanfare and was received warmly by Charles V. But the emperor had already banned the publication of Cortés's personal narratives—four personal letters to the emperor (1519–1524) were outlawed in March 1527, and a fifth, dated September 3, 1526, was "misplaced"—and was bound to reject his claims to Mexico on April 1, 1529. Cortés's triumphs had been achieved by flaunting the bureaucratic rules. Yet his commercial efforts and a 200,000-peso donation made in 1525–1526 were very much appreciated. Cortés was awarded the title of Marques del Valle de Oaxaca, a coat of arms, an honorary captain generalship of New Spain, vast agricultural territories, twenty-three thousand vassals, and even a new commission. On October 27, 1529, following Charles V's renunciation of all Spanish claims to the Moluccas, Cortés received a royal capitulación for future discoveries that he might make in the Pacific Ocean. No matter that Pizarro had just received a Pacific capitulación of his own on August 17.[21]

The Pacific capitulación and other royal honors provided Cortés with at least some hope that he would make money in the Pacific spice business. But when he returned to Veracruz in July 1530, he was faced immediately with a string of lawsuits. The five-man *audiencia* prohibited the new marques, the so-called captain general of New Spain, from stepping within ten leagues of Mexico City. The director of the audiencia—Don Sebastián Ramirez, the bishop of Santo Domingo—was even preparing to occupy Cortés's former residence in town. Although Cortés was allowed to assume a new feudal headquarters at Cuernavaca, prospect for silver, and make preparations for his next Pacific enterprise, he would have no role in the post-Cortés regime in New Spain.[22]

Stripped of his primary political power in New Spain, Cortés refocused his energies on various business enterprises—silver mining, sugar, silk, wine, cattle, horses, cotton—and the construction of a new Pacific fleet at Tehuantepec. Tehuantepec, which had employed at least forty Spaniards as shipwrights, woodcutters, blacksmiths, and sailors since 1521, was a logical staging area for Cortés's Pacific capitulación. The general plan was to regain some of the disputed northern Mexican domains, explore uncharted lands located even farther to the north, and scheme of ways to participate in the exploration of Ecuador and Peru. Of course, Cortés knew full well that he would have to compete with the formidable Francisco Pizarro in the south.[23]

Pizarro had returned to Panama in January 1530 with a royal capitulación in hand and his three half-brothers in tow. He immediately renewed his compaña agreement with Almagro, readied plans to invade Peru, and agonized over the strength of his capitulación since the formidable Cortés was still lurking around. He would eventually have to send his brother Hernán back to Spain to seek a more binding royal *cedula*. In the meantime, the invasion force that left Panama in January 1531 consisted of the four Pizarro brothers, three ships, 180 Spaniards, and no more than thirty horses. After landing at Tumbes in April, the party headed south to Coaque and immediately seized an estimated 22,500 pesos worth of gold and silver from local tribes. The plunder was sent back to Panama and Nicaragua to entice additional recruits. The ploy worked. Within a few months, one ship under Pedro Gregorio arrived with supplies, three treasury officials, twenty men, and thirteen horses. Sebastián de Benalcázar, who had organized his own expedition after Pedrarias's death in March 1531, followed with two ships, thirty men, and twelve horses.

Yet another recruit, Hernán de Soto, arrived in December 1531 with two more ships, one hundred men, and twenty-five horses. De Soto was an educated hidalgo who had led dangerous scouting parties in Panama and Nicaragua. One of the few odd men out was Pedro de Alvarado, the current governor of Guatemala and Cortés's former right arm in Mexico. Pizarro was so wary about Cortés that Alvarado was not summoned to the scene (Quito) until August 1534.[24]

Pizarro's enlarged expedition party was reorganized on Puna Island, on the Gulf of Guayaquil, in the face of occasional attacks by local natives. From Tumbes, the Spaniards marched one hundred miles inland to Piura by February 1532 and observed their first Inca highways, warehouses, and llama herds. The Spaniards missed the great Sican metallurgical center of Batán Grande (near Lambayeque) by around 150 miles, but they collected modest nuggets of pure gold en route and melted them down into ingots as they marched into the interior. The ad hoc invasion scheme was enhanced by the knowledge that the Inca Empire had been engulfed in a power struggle since the death of King Huayna Capac in 1527. The timely rift between the two successors to the Inca throne, Huascar (1527–1532) from Cuzco, and Atahualpa (1532–1533) from Cajamarca, could be exploited by the opportunistic Spaniards. Cajamarca was located three hundred miles south of Piura.[25]

Pizarro would discover shortly that the Incas, culturally unsophisticated compared to the Aztecs, were the hemisphere's greatest metallurgists and civil engineers. Without writing capabilities or scientific knowledge, the Incas had constructed a highly engineered fourteen-thousand-mile road network between Quito (Ecuador) and Santiago (Chile) to link their two primary capitals—Cajamarca in the north and high-altitude Cuzco in the southeast—and provide a three-thousand-mile route along the treacherous Andes. Inca roads may have been pedestrian oriented and unpaved, but they featured retaining walls, drainage systems, cantilevered bridges, causeways, and death-defying rope suspension bridges across mountainous gorges. The famous span at San Luis Rey would influence a host of future bridge builders. These and other structures were enhanced by a messenger system between Cuzco and Tomebamba (Cuenca, Ecuador) that was more efficient than anything the Aztecs (or any other empire) had to offer. With transfer stations located every 1.5 miles, Inca runners could deliver a message from one end of the 1,250-mile route to the other in just five days.[26]

The powerful Inca armies that emerged from a Cuzco-based chiefdom in 1200 CE were equipped with wooden shields, spears, axes, clubs, cotton armor, and mace slings. It was the work of kings Pachacuti (1438–1471) and Tupa Inca (1471–1493) that had conquered as many as nine million native Peruvians and Bolivians and created the largest empire in South America. Local puppet rulers were appointed to collect tribute, organize rotating gangs of tributary labor (the *mita*), and manage the construction of highways, irrigation systems, aqueducts, and adobe houses. The Incas (1200–1533) domesticated llamas, produced a variety of wools (cotton, llama, and alpaca), and crafted some of the world's most sophisticated weavings. They were not as religious and ceremonially oriented as the Aztecs— human captives were deployed in public works projects, not sacrificed to the gods—but, like the Aztecs, the Incas benefited greatly from the achievements of earlier civilizations. These included the coastal Moches between Ecuador and Lima (100–800 CE), the Nazcas to the south of Lima (100–800), the Waris in the Ayacucho Valley of central Peru (600–1000), the Tiahuanacos of highland Bolivia (500–1100), and the Sicans (900–1100) and Chimus (1150–1470) in Ecuador and northern Peru. The Chimu adobe capital at Chan Chan boasted one hundred thousand inhabitants at its peak.[27]

Then there was metallurgy. If the Spaniards had been astonished by the craftsmanship of the Aztec treasures at Tenochtitlan, they would be absolutely shocked by the quality (and quantity) of the Inca treasures collected at Cajamarca and Cuzco. Peruvian metallurgical skills were the richest in the hemisphere. Copper had been collected, fabricated, and exchanged in the Americas in stops and starts for possibly 2,500 years before highland Peruvians began to place copper disks in the mouths of their dead in around 1500 BCE. By 1000 BCE, formal metallurgical centers were separating surface-level deposits of copper, silver, and gold, deploying goldsmithing tools in and around Cuzco, and trading copper and copper-tin-bronze metals with tribes on the Pacific coast. These technical capabilities may have been two thousand years behind those practiced in ancient Mesopotamia—the Sumerians were smelting gold, silver, copper, and copper-tin (bronze) alloys by 3000 BCE—but they predated the establishment of the first Phoenician mining colonies on the Iberian Peninsula. By 400 BCE, gold-oriented metallurgical centers like La Tolita were operating near the present-day border with Ecuador.[28]

The metallurgy practiced by highland Peruvians was taken to much greater skill levels by the coastal Moches after 100 CE. Native silver was rarely mined or fabricated in pre-Pizarro Peru, but the Moches were skilled at finding, excavating, and smelting surface-level deposits of silver-lead and copper-silver ores. Moche mining trenches, rarely deeper than one hundred feet, were viewed as sacred places (*huacas*). Their craftsmen learned to produce silver-copper alloys via reduction smelting in charcoal-fueled furnaces. The smelted ingots were then stone-hammered and annealed into thin sheets of metal that could be welded into ceremonial masks, bracelets, or headgear for the royal and religious elites. The hammering and annealing process enhanced the metals' toughness and left a shiny golden or silvery exterior, even with small concentrations of gold and silver. While the Moches were ignorant of processes like cupellation, liquid metallization, or the "lost-wax" casting technique developed independently in Colombia, they were superior metallurgists in everything else.[29]

The later Sicans refined many of the metallurgical practices that had been pioneered by the Moches. By around 1000 CE, while Muslim Iberia was enjoying a "Golden Age" under the Umayyad Caliphate of Córdoba, Sican craftsmen at Batán Grande were smelting copper-arsenic (and some copper-tin) alloys and casting Peru's first bronze ornaments, tools, chisels, and needles. The production of arsenic bronzes involved the partial roasting of copper, followed by a smelting process to separate unwanted sulfur from the arsenic. The Sican smelting process was less efficient than the

Sumerian version, but it was reasonably effective. Teams of human blowers, equipped with five-foot-long cane tubes with ceramic tips, raised the heat in the charcoal-fueled smelting bowls to a temperature that generated small droplets of metal ("prills"). The prills could then be separated from the slag and remelted into an ingot. This "prill process" made Batán Grande the premier copper alloy smelting center in Peru between 900 and 1500 CE. The Sicans were also skilled with gold. Eighty objects uncovered from the nearby tombs of Huaca Loro demonstrate their capabilities with gold sheets (for the elites) and depletion-gilded *tumbaga* (for lesser nobles).[30]

The later Chimus and Incas made their own contributions to Peruvian metallurgy. The Chimus applied primitive acids to remove (deplete) the base metals from gold-copper and gold-silver-copper alloys before the remaining gold (or silver) layer was hammered and burnished into tumbaga sheets. Chimu gilding techniques were extremely persuasive. The Spaniards discovered that the gold content of the otherwise spectacular treasures of Cajamarca and Cuzco was only one-third of that plundered from Tenochtitlan. The gilded mummy masks and wall coverings fabricated by Peruvian metalsmiths contained as little as 12 percent gold. However, it could have been even worse for the Spaniards—had the Chimus and Incas been familiar with cupellation, they could have reduced the gold content of their fabricated treasures to as little as 2 percent. This would have reflected the actual gold-to-silver content of their regional mineral deposits. The major contribution of the Incas was to switch Peru's entire bronze-making system from copper-arsenic to copper-tin. The switch exploited the vast quantities of copper-tin located in the southeastern altiplano and removed the deadly arsenic fumes that were killing off their workers. Scores of Chimu metallurgists were transferred to the Inca capital of Cuzco to supervise the transition.[31]

Cuzco presided over one million recently conquered Bolivians who inhabited the 13,000-foot-high altiplano east of Lake Titicaca. The Incas had reorganized a federation of Quechua-speaking Aymara tribes into the tributary state of Kollasuyo, one of only four Inca provinces, and colonized the region aggressively after around

1500 CE. The altiplano, enclosed by the 16,000- to 21,000-foot-high Cordillera Occidental to the west and the 14,000-foot-high Cordillera Oriental to the east, was carved by a series of river valleys like the Yunga (near La Paz) and the Cochabamba and Chuquisaca (near Potosí). In addition to copper-tin, the Incas were attracted to Bolivia's agriculture, llamas, and sheep (alpaca and vicuña). Little attempt was made to conquer the rugged frontier country between Santa Cruz and the southeastern border with Paraguay and Argentina. The so-called Llanos de Chaco was controlled by the same backwater hunting and gathering tribes who had terrorized Aleixo Garcia and some of his successors. The wild region had remained independent from Inca control for good reason.[32]

The Incas had only scratched the surface of Bolivia's mineral deposits. Virtually all of the region's mineral wealth was located east of the Cordillera Oriental along a strip that included the silver deposits of Porco, Cerro de Potosí, and Oruro. There were also modest supplies of alluvial gold, copper, tin, lead, zinc, bismuth, and antimony. The native Aymaras who worked some of these sites under Inca supervision left behind little evidence of mines, smelting furnaces, or slags. The availability of weathered (oxidized) surface-level ores meant that there was no need to excavate (or smelt) deeper sulfide ores. If and when smelting was required, the Incas deployed the traditional reduction method.[33]

This remarkable three-thousand-year-old metallurgical heritage, the foundation of the hemisphere's single greatest economic prize to date, lay in the direct path of an advancing Spanish army in February 1532. The empress Isabella had unwittingly awarded a capitulación in 1529 to a man, Francisco Pizarro, who would create more wealth (and more headaches) for the Crown than even Hernán Cortés. Her husband, emperor Charles V, was a busy man in early 1532. He was readying a Danube flotilla to safeguard the mines of Habsburg Austria from an assault by the advancing Ottoman army. An estimated 860,000 ducats of the 1.2 million windfall from France had been transferred from Medina del Campo (via Genoa) to the flotilla at Regensburg and points east during the summer of

1532. The emperor made his point. His adversary, Suleyman the Magnificent, decided to reverse course and retreat back to lower Hungary. While the Danube campaign added another pile of debt to the Habsburg treasury, it would have been much worse for Charles V if the prudent Suleyman had decided to become engaged.[34]

When Charles entered Vienna triumphantly on September 23, 1532, the French ransom funds had been exhausted, Habsburg Austria was heavily in debt, and the Castilian treasury was tapped out. Treasure imports from the Americas were insignificant, and there was no Inca Peru to worry about. But Charles had always found a way to raise capital when he needed it and he had inadvertently backed an extremely talented man in the person of Francisco Pizarro. Few of the Spanish invaders, not even veterans like Belalcázar and de Soto, possessed the hands-on experience that Pizarro had accumulated in the Americas. Pizarro's campaigns in the Isthmus of Panama had been fought on foot, not on horseback, and his Extramaduran style was simple and straightforward. He was also an avid gambler and was happy to plunge himself into projects that required manual labor, like the two dams that he constructed later in Lima. Pizarro was also smart enough to involve his trusted (and literate) half-brother Hernán in the decision making and to leave the most dangerous missions to younger men like de Soto and Almagro. The flipside was that Pizarro promoted his inexperienced but trusted brothers over superior men and was capable of breaking promises to loyal supporters (Almagro) and unfortunate captives (Atahualpa).[35]

Pizarro had almost certainly learned en route to Piura that the two heirs to the Inca throne—Huascar and Atahualpa—had been mired in a war of succession since 1527 and that Aleixo Garcia had come within fifty miles of Chuquisaca (Bolivia) in 1525. Coupled with the Chimus' traditional hatred of the Incas, Pizarro's timing could not have been any better. After a base of operations was organized at Piura in February 1532, he and 167 men marched farther inland in search of the rumored cache of Inca treasure. When the Spaniards eventually reached Atahualpa's adobe capital at Cajamarca on November 15, 1532, they walked into an abandoned city.

Atahualpa and thousands of Inca warriors had withdrawn to the outskirts of town. Borrowing a page from Cortés's manipulation of Moctezuma at Tenochtitlan or, for that matter, Antonio de Leyva's capture of François I at Pavia, the new plan was to find and capture the unsuspecting Atahualpa and hold him for ransom.[36]

As the Spaniards occupied Cajamarca, Atahualpa was somehow persuaded to meet the invaders in the town's main square. Both sides were scheming against the other. The trap laid by Pizarro included the placement of sixty cavalrymen under Hernán Pizarro, de Soto, and Belalcázar; twenty-five infantrymen under Francisco himself; seventy more held in reserve; and a few artillerymen and musketeers positioned atop Cajamarca's abandoned walls and rooftops. Atahualpa had his own plan. Thousands of Inca troops, armed with wooden shields, spears, axes, clubs, and mace slings, were placed outside of town to cover the arrival of Atahualpa and his two thousand or so military escorts. The two sides met in the late afternoon of November 16, 1532. The Dominican Fray Vicente de Valverde and the two Indian translators were sent forward to converse with the last significant king of the Inca Empire.[37]

When it became clear that the parley was going badly, the Spanish cavalry charged into the crowded square. The confusion provided the opportunity for Francisco to grab the unsuspecting Atahualpa and drag him to "safety." Most of the Spanish guns proved to be useless, but Pizarro's action was a spectacular success. The dumbstruck Inca forces regrouped outside of town and were subsequently routed by the superior Spanish cavalry. A Spanish horse at full gallop was terrifying to a civilization accustomed to llamas. When it was all over, thousands of Incas lay dead, not a single Spaniard was killed, and Atahualpa agreed to fill an entire room with gold and silver objects in exchange for his freedom. Pizarro occupied Cajamarca in wait of the deliveries and the arrival of Spanish reinforcements from the north.[38]

The wait was extended into the spring of 1533. Three Spaniards were dispatched to the southern Inca capital at Cuzco to accelerate the process and shocked everyone when they returned a few months later with a llama caravan of gold. Meanwhile, an anxious Hernán Pizarro and twenty men spent four months looking for a rumored Inca temple at Pachacamar, a village located south of the future site of Lima, but came up empty. However, the Spaniards somehow managed to capture a large treasure caravan that was passing through Jauja to the northeast. When the three Spanish treasury officials arrived from Piura to inspect the captured treasure, they were astonished by its scale and quality. Preparations were made to send the royal quinto back to Spain.[39]

The treasures that streamed into Cajamarca were assayed and melted down by mid-May and distributed to the 167 invaders (and to Charles V) in July. Francisco's generous thirteen-share allotment was challenged by Almagro when he and two hundred Spaniards arrived from Panama. Almagro, a partner in Pizarro's compaña from the very start, was outraged by Pizarro's decision to cut him out of the plunder entirely. He had been absent from Cajamarca only because Pizarro had ordered him to stay with the ships. Almagro's financial loss was huge. The value of the accumulated gold, silver, and other plunder collected at Cajamarca alone was estimated at a staggering 1.5 million gold pesos (1.8 million ducats)—over three times the value of Moctezuma's treasures and thirty times the value of the entire silver output of New Spain. Twenty lucky Spaniards were permitted to return to Spain (with Hernán Pizarro) in 1533 knowing that Pizarro had broken his promise to free Atahualpa. The unfortunate Inca emperor was executed by a so-called Spanish tribunal in late July.[40]

Pizarro had improvised his way through to the largest treasure in American history. But his conquistadorial style was very different than that of the aristocratic, smooth-talking Cortés. Pizarro was a gruff, illiterate warrior who favored the use of brute force over diplomacy. It was the more diplomatic Hernán Pizarro who was sent to Spain to defend the family's claim to Peru via a binding royal cedula. Francisco resumed the offensive. He left Cajamarca in August 1533 with a six-hundred-man invasion force, supplemented by Almagro's men and reinforcements from Panama, and made sure to safeguard the Inca treasure at the highland village of Jauja in early October. Word was out about the Spaniards. As Pizarro's army headed south to Cuzco along the increasingly high-altitude Inca trails, de Soto's advance party was saved from near annihilation only because of the arrival of Almagro and his men. When Pizarro finally entered Cuzco on November 15, 1533, the one-year anniversary of his arrival at Cajamarca, local resistance was surprisingly minimal.[41]

Pizarro occupied Cuzco's palaces and facilities, discovered another large cache of treasure (mainly silver) in town, and awarded encomiendas to eighty Spaniards. The establishment of Spanish Cuzco on March 23, 1534, signaled that the conquest of Peru was essentially over. However, the adelantado had already made plans to build his capital city at coastal Lima. Construction started in January 1535. In the meantime, Francisco withdrew most of his men to the midway village of Jauja, left a mere forty Spaniards to defend Cuzco from a possible Inca counterattack, and issued additional encomiendas from Jauja during the summer. He even allowed sixty more Spaniards to return to Spain. Pizarro's choice of Lima was a fateful one. Cortés had selected imperial Tenochtitlan for his capital (Mexico City), but Cuzco and Jauja were considered too remote and Cajamarca too political. With this one stroke, Pizarro unwittingly abandoned the trappings of imperial Inca power that Cortés had skill-fully manipulated (via the Aztecs) in New Spain. The choice of Lima did not prevent a grateful Charles V from issuing two binding cedulas to Pizarro in the summer of 1534. The cedulas extended Pizarro's domains to as far south as the Inca border at Santiago, Chile.[42]

Prior to Cajamarca, the Spanish Crown had received American treasure shipments in fits and starts. Average annual treasure imports had fluctuated between 163,247 and 239,110 pesos during the 1506–1520 period before declining during the 1520s. Revenues from the sugar and cattle industries of the West Indies were approaching the levels realized from gold. Large portions of Moctezuma's treasures had been lost or pirated away. Mining results from New Spain were disappointing. An annual royal quinto of around thirty-nine thousand ducats was insignificant.[43]

Pizarro's plunder of Inca Peru was on a completely different scale. The total value of the precious metals that were collected, plundered, or ransomed between 1531 and 1535, highlighted by the plunder of Cajamarca and Cuzco, has been estimated at a staggering 2.3 million gold pesos (2.8 million ducats). The modest amounts of gold and silver collected en route to Piura in 1531–1532 were dwarfed by the smelting of the Cajamarca metals in July 1533 and the smelting of the Cuzco plunder in February 1534. The numbers were huge. The smelted plunder yielded roughly 12 tons of Inca gold, or twelve times the annual Caribbean volume at its peak, and 105 tons of Inca silver. The silver take was equivalent to the entire annual silver output from Central Europe. Together, the Inca gold and silver was worth nearly three times the value of the otherwise huge ransom payment received from François I of France in 1531.[44]

The spectacular plunder of Inca Peru allowed Charles V to reconsider his geopolitical options. This included some second thoughts on the Crown's adherence to the traditional quinto. While Suleyman had been thwarted in Austria in 1532, Hayreddin Barbarossa's Ottoman–backed corsairs were wreaking havoc in the Mediterranean. In addition, the French army was still at large and bent on reversing the humiliations suffered with the Treaty of Cambrai in 1529. Yet the pending receipt of a huge royal quinto from the Inca treasure would not be enough to cover the cost of either a naval campaign against Barbarossa at Tunis or a possible land campaign against France. This meant that most of the Inca treasure would have to be confiscated outright.

Pizarro's first treasure ship to Spain had delivered a royal quinto of around four hundred thousand ducats to the imperial vault at Medina del Campo in late December 1533. The delivery triggered a 1 percent transfer fee to Cobos for his efforts and prompted an astonished emperor to grasp for more. He confiscated 600,000 ducats in privately owned treasure in December 1534 and followed up with an even larger confiscation of 800,000 ducats on March 4, 1535. The unlucky confiscatees apparently received juros at 3 percent in exchange for their Inca bullion. By then, the emperor was focused solely on Tunis. The

Castilian treasury authorized the transfer of around one million ducats of American treasure by July 1535, nearly half of which was used to hire mercenaries and outfit warships for the planned invasion of Tunis. The remainder was applied either to some of the Fuggers' outstanding debts in Antwerp or to unidentified military needs.[45]

Cobos was in the middle of these and other transactions. Rumor had it that the less-than-sentimental Cobos had melted down Cortés's silver cannon gift to the emperor to recover its twenty-thousand-ducat silver value. Cobos was an entrepreneur at heart. He negotiated the Venezuela contract with the House of Welser in 1528, earned nearly eighteen thousand ducats from a dubious Venezuelan salt mining venture in 1529, and contracted with Pedro Alvarado to export West African slaves to Guatemala at a cost of ten pesos each. This was just after Charles had officially rejected Cortés's claim to the governorship of New Spain. Cobos even protected his interest by having Alvarado knighted, salaried at fifteen hundred ducats, and appointed adelantado and governor of Guatemala. While there is no evidence that Cobos conspired against Cortés, his political influence in New Spain was enhanced by the official demotion of Cortés and the official promotion of Alvarado.[46]

The Pizarros had to contend with Cobos as well. It had been Cobos, after all, who had secured Francisco Pizarro's coveted capitulación in August 1529. When Francisco dutifully delivered his very modest pre-Cajamarca plunder to Toledo in April 1530, he had been forced to wait in line with Cortés and Alvarado—all three conquistadors happened to be in Spain at the same time in search of royal favors. Cobos received initial word of the spectacular plunder of Cajamarca in December 1531 from Gaspar de Espinosa, Pizarro's Panama-based partner, along with an estimate that the treasure was worth a jaw-dropping 2.3 million gold pesos. Cobos's 1 percent share gave him a stake of over twenty-three thousand pesos. Reality intervened when Hernán Pizarro sailed into Seville with plunder, gold nuggets, and exotic animals in mid-January 1534. Cobos confiscated Hernán's entire cargo on behalf of the Crown and then persuaded the emperor to extend his *fundidor mayor* to all of South America.[47]

The emperor's military campaigns against Barbarossa and France would squander over three million ducats in windfalls (from France and Peru) between 1531 and 1536—with yet another round of military initiatives on the drawing board. But if the House of Habsburg was banking on Peru to finance the ongoing campaigns against France, Istanbul, and even German Protestantism, it would be disappointed. The former Inca Empire would be engulfed in civil war through 1548 and most of the fighting would involve Spaniards against Spaniards, Spaniards against Incas, and even Incas against Incas. Manco Inca remained a hunted fugitive until his death in 1545 and withdrew to a temporary Inca refuge-state at Vilcabamba in the jungles west of Cuzco. Yet the internal struggles in Peru failed to dampen a host of prospecting expeditions. Most everyone, especially the Spanish invaders of Inca Peru, was engaged in a feverish quest to discover the underlying mineral deposits from which the treasures of Cajamarca and Cuzco had been fabricated.[48]

One of the invaders, the thirty-two-year-old Hernán Pizarro, would shortly uncover the first meaningful silver deposits in the Inca Empire. In the meantime, Hernán had been sent back to Spain with the royal quinto in late 1533 to obtain a binding royal cedula and, in part, to relieve his endless bickering with the non-Pizarros. Hernán has been described as well educated, arrogant, self-centered, mean-spirited, hotheaded, money grubbing, reviled—and very talented. He had served with his father Gonzalo in Navarre in 1521 and, unlike Francisco, had received a royal appointment as an army captain and the rights to his father's estate. If Francisco was jealous of his much younger brother, he trusted him completely. He awarded Hernán a generous seven shares of the Cajamarca plunder and recognized that his negotiating skills made him the most logical envoy to the royal court. Hernán was skilled with licenses, tax exemptions, grants, mining equipment deals, and removing would-be rivals. The large capitulación that he won for Almagro in May 1534, giving Almagro the rights to two hundred leagues of territory south of the Pizarros' domain, was intended to remove Almagro from Peruvian affairs completely. He may not have known that

Anton Fugger, of all people, had obtained the right in 1532 to colonize Chile over an eight-year period. Anton's Chilean project never got off the ground, but Almagro had a good excuse to be absent from Cuzco when the city and her 180 Spanish defenders were besieged by an Inca army in 1536. Hernán returned just in time to be attacked by Manco Inca and thousands of Inca warriors.[49]

## BOLIVIA, COLOMBIA, AND MORE

If Diego Almagro, Hernán Pizarro, or any Spaniard knew that the Inca Empire stretched as far south as Santiago, Chile, it is unlikely that they knew in 1534 that Kollasuyo (Bolivia) contained massive deposits of silver. So if Almagro's capitulación for western Bolivia and Chile was a belated reward for his loyalty and competency, he was in for another unpleasant surprise after he left Cuzco in 1535 for a two-year expedition to the southern wilds. The ambitious expedition party comprised over three hundred men, including a few Inca officials, a large cadre of Inca porters, and pack trains of llamas. From Cuzco, Almagro followed the Inca highway across southwestern Bolivia—via Tiahuanaco, the Desaguardero River, Lake Poopó, and Tupiza—to a 13,776-foot-high pass through the Andean Cordillera. While the party managed to travel through Bolivia without incident, they had traveled right past the primary deposits of silver. It got worse after Amalgro had crossed into the Copaiapo River valley in Chile. His native bearers decided to abandon their Spanish friends to the cold, snow, diminishing food supplies, and dangerous Araucania tribes. Almagro struggled on but found no evidence of precious metals between the Copaiapo and Valparaiso. He had the good sense to return to Cuzco via the safe Atacama Desert route along the Chilean coast.[50]

Almagro arrived in Cuzco just in time to recapture the city from Manco Inca. Juan Pizarro, the third brother, had already died from a single stone-shot to the head during an assault against a nearby Inca fortress. Almagro's timely appearance in the spring of 1537 prevented Manco Inca from destroying the remnants of the Spanish army—they had been cut off from the coast for over a year. In addition, Almagro's personal valor won the respect of the rank and file. Flush with victory, Almagro dared to claim Cuzco and southern Peru as his own to compensate for the forgone treasure of Cajamarca and an apparently useless capitulación in Bolivia and Chile. It was a big mistake. The Pizarros had discovered intriguing deposits of minerals in Bolivia in Almagro's absence and were not about to part with them. Francisco instructed Hernán's militia to attack Almagro's men at Salinas (near Cuzco) in 1538, capturing Almagro during the action, and then had him strangled to death. The subsequent transfer of Almagro's capitulación to Pedro de Valdivia in 1539 created an instant enemy in the person of Almagro's mestizo son.[51]

The youngest Pizarro, Gonzalo, had been somewhat anonymous. Like his two-year-older brother Juan, he was a recognized illegitimate son of Captain Gonzalo Pizarro and regarded as an excellent horsemen and outdoorsman—if somewhat of a loudmouth. Gonzalo bragged that he and his fellow Spaniards had killed over one-third of the 200,000 Indians who besieged Cuzco in 1536–1537. He followed up his narrow (Almagro-assisted) escape by killing a number of his fellow Spaniards during the Battle of Salinas and receiving a commission to occupy the mineral-rich province of Bolivia in 1539. He was also instructed to convert the Indian village of Chuquisaca to La Plata (the future Sucre). From there, Gonzalo was sent north to Tomebamba (Cuenca, Ecuador) in 1540 to monitor Belalcázar's prospecting activities and to explore the unclaimed eastern wilds on his own.[52]

The threat to royal authority posed by the Pizarro brothers escalated when it was learned that valuable mining properties had been discovered in Bolivia. Unwilling to believe Almagro's disappointing accounts of the region, the Pizarros had investigated Bolivia for themselves in 1538 and established a staging area at Chuquisaca (La Plata) to catalyze prospecting activities throughout the altiplano. Of course, the Aymaras were less than happy about it. They had been shocked by the Inca conquest, a process through which private property was abolished, taxes were raised to two-thirds, and thousands of Aymara workers were drafted (via mitas) to construct Inca highways, warehouses, fortresses, and agricultural terraces. Yet they still preferred the Incas to the reviled Spaniards. Most Aymaras joined Manca Inca's general rebellion and participated in the siege of Cuzco. The one Aymara group who sided with the Spaniards, the Collas, were rewarded with military support from the Pizarros in late 1538. It was partly on their behalf that the Pizarros marched into the altiplano to crush Inca resistance at Chuquito and in the Desaguardero River valley.[53]

The unfortunate Aymaras had a new landlord when Hernán and Gonzalo Pizarro remained in Bolivia to continue their prospecting activities. Even earlier, in 1535, the forward-looking Pizarros had sent Hernán Sanchez de Pineda to the altiplano to scout for precious metals. It was Sanchez who apparently discovered a nearly abandoned Inca mining site at Porco and encouraged Hernán to import European mining engineers, mining equipment, and African slaves to exploit the site more fully. While Porco's silver output during the 1537–1540 period is unclear, Hernán's investments eventually raised Porco's production total to as much as 200,000 gold pesos during the early 1540s. This was greater than anything uncovered (to date) in New Spain and surpassed the modest quantities of Bolivian gold discovered near Chuquiago in 1538–1539. These strikes encouraged the Pizarros and scores of other Spaniards to continue their search for the geological source of the Inca treasures that had been plundered at Cajamarca and Cuzco. Fearing their success, the emperor was already scheming to remove the Pizarros from power—the energetic family had to be prevented from controlling the mineral wealth of an entire continent.[54]

In the meantime, Hernán Pizarro's investments at Porco were consistent with his transition from warrior to self-styled business tycoon. Hernán realized significant profits from the sale of imported European goods to Spaniards and native chiefs and became skilled at circumventing the Casa de Contratación in Seville. He was also willing to apply extortion to further his personal ends. The royal authorities eventually caught up with him. After returning to Spain (unwittingly for good) in 1540 to explain the circumstances of Almagro's death, Hernán was imprisoned immediately and deprived of any opportunity to assist his brothers further in the New World. An incarcerated Hernán learned that his brother Francisco, the wealthiest man in South America, was assassinated in Lima in 1541—either by supporters of Almagro's mestizo son Diego or by one of the long arms of Charles V—while Gonzalo was on an expedition in eastern Ecuador. Hernán was eventually released and was even allowed to build an impressive estate on the site of his father's homestead in Trujillo. However, it is unclear whether the last surviving Pizarro brother managed to retain any of his American properties before his death in 1578.[55]

The assassination of Francisco Pizarro in 1541, following the death of Juan (1536) and the removal of Hernán (1540), left the fate of the fledgling Pizarro empire in the hands of Gonzalo. His dispatch to inland Ecuador in 1540 had been fueled by exciting reports of gold from Colombia's upper Cauca and Magdalena rivers and rumors that an El Dorado was waiting in the uncharted wilds to the east. The thirty-year-old Gonzalo left Quito's capital of Tomebamba (Cuenca) in February 1541 with 220 armed Spaniards and as many as 4,000 Indian allies for an inland trek into uncharted Amazonia. Joined en route by his lieutenant, Francisco de Orellana, and twenty-three other Spaniards, the expanded expedition party marched two hundred miles across the rugged Andes to a point along the Coca River. At that point, the local Omaguas encouraged the Spaniards to construct a single brigantine, which they did, to continue their journey down the Coca into the jungle-infested Napo River. Subsisting on horsemeat, dogs, roots, and irregular fruits, Gonzalo and his barely fed men reached the junction of the Amazon River sometime in late December.[56]

If the murky story is true, Orellana was given twelve days to take sixty men and the brigantine down the otherwise unknown Amazon River in search of food. Not only did Orellana fail to return, he traveled down the Amazon's entire length and reached the Atlantic

Ocean. He also named the mighty river to honor the female natives who were encountered en route—these native Brazilian women were more real than the Greek myth. Orellana defended his unauthorized voyage by claiming, quite logically, that the strong downstream current had prevented him from rejoining the main group. But he had left Gonzalo with the task of constructing a second brigantine and spending the next six months finding a way back to civilization. Gonzalo somehow struggled his way back to Tomebamba from where he filed a report on September 3, 1542. The report accused Orellana of betrayal at the junction of the Amazon River some thirteen months earlier. However, by the time that the case was heard (and dismissed) by the Council of the Indies in Valladolid in early June 1543, Gonzalo Pizarro was a wanted man.[57]

The few survivors who returned to Quito in the late summer of 1542 carried great respect for Gonzalo's capabilities and courage. It was then that Gonzalo learned for the first time that Hernán had been imprisoned in Spain since 1540 and Francisco had been murdered in Lima in 1541. Since Francisco's will had specified Gonzalo as next in line for the governorship of Peru in the event of Hernán's death or absence, the recent events should have transferred the governorship of Peru to Gonzalo Pizarro then and there. Gonzalo had already built a sizable fortune from his large encomiendas in Bolivia, including silver interests, plus holdings in Cuzco and Arequipa.[58]

Unfortunately for Gonzalo, Charles V's cash flow difficulties had traveled across the Atlantic in the person of Licenciado Cristóbal Vaca de Castro. Castro had arrived in Peru in 1540 to preside over a new audiencia of Lima and to stir things up following Hernán's imprisonment, Francisco's death, and Gonzalo's prolonged absence. Castro was authorized to renounce Gonzalo's rightful claim to Peru and to redistribute the large encomiendas that had been issued to and by the Pizarro family during Francisco's nine-year governorship (1532–1541). Some of these encomiendas just happened to include potentially lucrative mining properties in Bolivia. As long as the royal quinto and confiscation opportunities had been dispatched to Spain, the Crown had been a passive partner in Peruvian affairs. But

Charles V was badly in debt in 1540 and the underexploited mineral wealth of Bolivia (and Colombia) appeared to be too significant to be left to an independent-minded family. Cortés had been handled in New Spain and so could the Pizarros. So with royal backing, Francisco dead, Hernán behind bars, and Gonzalo conveniently away (and probably dead) in Amazonia, Castro seized the opportunity by adding the governorship of Peru to his titles and raiding the prolific silver mines at Porco. When Gonzalo returned unexpectedly in the late summer of 1542, he was apparently too exhausted to challenge Castro's coup d'état. He even offered to assist Castro's move against Diego Almagro Jr.'s rebel army at Chupas (near Cuzco). Gonzalo's loyalty to the Crown would be tested very shortly.[59]

Gonzalo Pizarro's harrowing trip down the Napo River had come up empty in 1541–1542, but prospecting activities were exploding in uncharted Colombia. The initial objectives were to find the geological source of the Inca treasures and identify a possible diagonal river passage between the Caribbean and the Pacific Ocean. Between 1537 and 1543, over six major expeditions—led by Gonzalo Jiménez de Quesada, Nikolaus Federmann, Sebastián de Belalcázar, Jerónimo Lebrón, and Alonso Luis de Lugo—were directed at Colombia's upper Magdalena and Cauca rivers. The emperor had helped to catalyze the Colombian gold rush by leaving open the Indies trade to all subjects of the Holy Roman Empire. The objective was to encourage non-Castilian merchant bankers and entrepreneurs to participate in a marketplace that was short of capital and business skills. Many of the Colombian expedition parties included Germans, Conversos, and Moriscos who had been otherwise excluded from the transatlantic trade. No matter that many of the competing territorial claims overlapped and upper Colombia was the traditional homeland of the Muisca civilization.[60]

The Spaniards sought to find and exploit ceremonial treasures and gold deposits that had been worked in the lower Cauca and Magdalena valleys for two thousand years. As early as 200 BCE, the lowland peoples of Colombia were collecting placer gold from rivers and streams, hammering gold and gold-copper alloys into thin metallic sheets (tumbaga), and ornamenting (often embossing) the sheets in the Peruvian style. Colombian tumbaga was typically reserved for ornaments, sacred vessels, and nose-rings. The availability of alloys of any kind suggest that Peruvian metallurgical practices like roasting, smelting, gilding, and embossing had reached Colombia over many centuries. If Colombian metallurgists failed to match the skill levels of the Peruvians, they borrowed many of the techniques developed by the Moches, Sicans, and Chimus and even added a few of their own—most notably, a "lost-wax" casting method to produce hollow-cast objects from various combinations of copper-gold alloys. The Colombians also helped to spread the Peruvian style northward to the backwater chiefs of Central America and the Caribbean.[61]

This metallurgical tradition was unknown to the few Spaniards who had visited Colombia's shores since 1499. Most European explorers, starting with Columbus, had determined that the northern reaches of the uncharted South American continent, Tierra Firme, were not worth the effort. But Cortés's plunder of inland Tenochtitlan had piqued everyone's interest in the American interior. As early as 1525, one year after Charles V had opened up the Indies trade to all of his subjects, Rodrigo de Bastidas had received the right to colonize the eastern region between the lower Magdalena River and Cabo de la Vela. Another royal contract was issued to Pedro Gonzalo Fernandez de Oviedo to colonize the western region between the lower Magdalena and Panama. When a third contract, covering Venezuela, was issued to Ambrosio Alfinger and Georg Hohenmuth von Speyer of the House of Welser on March 27, 1528, the northern Caribbean coast was split into three separate governorships. Exploration activities had been confined to the mineral-poor lower Magdalena until Pizarro shocked everyone with his conquest of Cajamarca in 1532. By 1535, Bastidas's contract had somehow been transferred to Pedro Fernandez de Lugo, the adelantado of the Canary Islands, and Oviedo's contract was in the hands of Pedro de Heredia. The geological source of the Inca treasures (or other bonanzas) was assumed to be in Colombia or even Venezuela.[62]

The business experience that the Welsers had gained from Joachimstahl and a recently opened office in Santo Domingo under Ambrosio Alfinger was of little use in Venezuela. Pearls had been discovered in eastern Venezuela as early as 1499, but the Welsers' thirteen-year contract to explore, mine, and colonize the wild interior was something else. Pushed by Francisco de los Cobos, who had a piece of the action, the Welsers were prepared to import an entire crew of professional miners and metallurgists to Venezuela from their works at Joachimstahl in return for the mineral rights and an authorization to establish their own mint. With high hopes, Alfinger arrived at the port of Coro in February 1529 with Nikolaus Federmann and 180 men to assume the first governorship of Venezuela. This was the peak. When seventy prospectors died from local disease during a ten-month exploration campaign, Alfinger decided to return to the comfort of Santo Domingo and leave Federmann in charge of the project. There was little extractable gold or silver to be had in the wilds of Venezuela. No more than one-half a ton of alluvial gold (worth around ninety thousand pesos) had been excavated when the ambitious thirteen-year project expired in 1541. The ordeal left the Welsers with a 100,000-ducat write-off and a series of lawsuits that would drag through the courts until 1550.[63]

Bartholemew Welser's attempt to colonize Venezuela turned out to be far more costly than Jacob Fugger's modest exposure to the failed expeditions of Loaisa (to the Moluccas) and Cabot (to Rio de la Plata) in 1525. Charles V softened the financial blow by selling off some of his unpledged properties in the Netherlands and Naples to repay most of the Welsers' outstanding claims. Since Charles still owed substantial sums to the House of Fugger, he recognized that he needed the creative Welsers as an alternative source of funds for his next round of military campaigns. In addition, the emperor persuaded the Welsers to divert some of their professional mining talent from Venezuela to the more promising mines of New Spain.[64]

Venezuela was a bust. Federmann failed to discover any minerals during an unauthorized expedition to the Acarigua River region east of the Venezuelan Andes in early 1531 and returned to Ulm, Germany in disgrace. Unfortunately for Alfinger, he and one hundred Spaniards perished during a southwesterly trek from Coro to Colombia's Magdalena River in 1531–1533. It was Alfinger's death that prompted the Welsers to recall Federmann from Germany to continue their quest for the fabled riches of "Xerira"—even if these riches were not necessarily located in Venezuela. However, the Welsers stationed the independent-minded Federmann at Coro for the time being and transferred the title of governor to the more acceptable Georg von Speyer. History repeated itself after Speyer arrived at Coro in February 1535. Another one hundred men died during Speyer's expedition to Cabo de la Vela, and Federmann, stuck at Coro, decided to disobey orders once again. In December 1536, he led an unauthorized 140-man expedition back to the Acarigua River.[65]

Federmann must have heard about the promising developments next door. Colombia rivaled newly discovered Peru as the hottest prospecting region in the hemisphere after Jerónimo de Melo had traveled eighty miles up the Magdalena in early 1532 and Pedro de Heredia had founded the port of Cartagena (1533). Pedro Fernandez de Lugo was next. Having grown wealthy from a portfolio of sugar mills and limestone quarries on the Canary Islands, the sixty-year-old Lugo organized a massive twelve-ship, one-thousand-man fleet under the command of Licentiate Gonzalo Jiménez de Quesada after signing a royal contract in January 1535. Since each expedition member was required to invest as much as twenty ducats of his own money for the right to participate in the project, the incentives were all in the right place. Lugo also knew that Melo had received reports that the Magdalena flowed nearly all the way into Ecuador and that the governor of Santa Marta, Garcia de Lerna, had followed up with a 140-man expedition up the Magdalena in 1533. This suggested that the massive expeditionary force that arrived at Santa Marta in January 1536 could narrow its focus on the upper Magdalena River. Hopes were high after Lugo's own son, Alonso, helped to recover an estimated three thousand pesos of gold during a series of preliminary expeditions.[66]

On April 5, 1536, Quesada left Santa Marta with six hundred men for an expedition up the Magdalena River. While as many as four hundred succumbed to the same deadly diseases that had killed off hundreds of earlier Spaniards, Quesada pushed forward after a series of salt storage depots were discovered in early 1537. The surviving Spaniards had been in the Colombian wild for nearly one year when they approached the densely populated lands of the Muisca tribe. However, Quesada's decimated army was still powerful enough to defeat the Muiscas in battle and followed up this triumph by stumbling into more salt mines, emerald deposits, and more local resistance near the future site of Santa Fé de Bogotá. The overmatched Chief Bogotá Zupa and his successors were captured and executed between August 1537 and June 1538, leaving the plunder of highland Bogotá to Quesada and his 172 confederates.[67]

Quesada failed to find any gold until he traveled south to the Muisca workings at Neiva along the upper Magdalena. He was not alone. The energetic (but unauthorized) Federmann had penetrated through Venezuela's southern llano region in early 1538 and followed his Indian guides through the Rio Ariari region to the eastern banks of the Magdalena. When Federmann reached the Muisca village of Pasca in February 1539, he had already lost one hundred men—some during the frigid crossing of the Cordillera Oriental and one at the hands of a local jaguar—and was in no position to challenge Quesada's claim. Federmann and the other survivors wandered into Bogotá in March only to discover that there were other Spanish adventurers in the vicinity. A southward-moving expedition under Juan de Vadillo had followed the Rio Sinú, crossed the Cordillera Occidental, and uncovered a possibly rich gold field at Buriticá near the Cauca River. The unexpected arrival of the Vadillo party at Cali created a jurisdictional dispute with the esteemed Sebastián de Belalcázar.[68]

Belalcázar had also followed the rumors into upper Colombia. Sent to defend Piura in 1533 with a decent (if disappointing) share of the Cajamarca plunder, he had organized his own expedition (on behalf of Pizarro) to conquer the Quito region of Ecuador in 1534.

It was at that point that Pizarro summoned Pedro de Alvarado (from Guatemala) to ensure that the loyal Belalcázar remained loyal. Belalcázar didn't enjoy being monitored. He defied Pizarro by heading farther north and establishing the upper Cauca River towns of Popayán and Cali by 1536. He would have had competition if Pedro de Andagoya had managed to trace an overland route between Cali and the newly established Pacific port of Buenaventura—the proposed route was blocked by the dense tropical Choco and groups of gnarly Indians. When Belalcázar left Quito again in early March 1538 with two hundred Spaniards and five thousand allied Indians, he discovered significant deposits of gold en route to Cali and unexpected competition from the southward-moving Quesada and Valdillo parties. But Belalcázar turned out to be a winner. His claim to the adelantado and governorship of Popayán was eventually affirmed in Santo Domingo in 1540, and he was wise to side with the Crown against Gonzalo Pizarro in 1546. By then, Belalcázar had managed to outlive three Pizarros and three of his peers—Almagro (d. 1538), Alvarado (d. 1541), and de Soto (d. 1542).[69]

Quesada had been aware of Belalcázar's exploits in the upper Cauca River and invited his formidable competitor to work out a deal. After the two adventurers met in April 1539, it was settled that Belalcázar would control Popayán and Quesada would be free to establish Santa Fé de Bogotá as the provincial capital of New Granada. Quesada also agreed to join a thirty-man mission down the Magdalena (to Cartagena)—comprising some of his men, representatives of Belalcázar, and even Federmann—to present their competing claims to the Spanish courts in Santo Domingo. When the missionaries arrived in June 1539, they were surprised to discover that the governorship of Santa Marta had been transferred temporarily to Jerónimo Lebrón of Santo Domingo (since May 1537) until the raft of competing claims could be sorted out. Lebrón muddied the waters with a claim that his governorship extended to New Granada and possibly even Popayán.[70]

Lebrón must have been impressed by the thirty-man mission that arrived at Cartagena in June 1539—Quesada, Federmann, and Belalcázar were each carrying between eight thousand and twenty thousand pesos worth of gold and emeralds. Lebrón had done pretty well for himself as well. Between January and August 1538, he had supplemented his monthly salary of 120 pesos by forming a compania with two merchants and a royal treasurer. The compania earned fifteen thousand gold pesos simply by sending a brigantine loaded with food, clothing, equipment, and other trade goods to Spaniards located on the San Jorge and lower Magdalena rivers. The arrival of the Quesada mission encouraged Lebrón to forward his own claim to upper Colombia and to arrange for the importation of additional provisions from Santo Domingo.[71]

On January 10, 1540, the intrepid Lebrón ignored the significant risks by leading a party of 6 supply-laden brigantines, 180 horses, 300 men, 10 priests, and a number of black slaves on a trading and political mission to Bogotá. The expedition party fared just as poorly as its predecessors. It was decimated by hostile native attacks en route as well as from diseases and food shortages. When Lebrón, the other survivors, and the remaining trade goods finally reached Bogotá in November, the local Spaniards declined to recognize Lebrón's dubious claim to the governorship of New Granada. No one knew that the Council of the Indies had already transferred the coveted title to the rightful heir, Alonso Luis de Lugo, after the competing claims had been sorted out during the summer of 1538. Lebrón and his business partners had to settle for over twenty-two thousand gold pesos in trading revenues.[72]

The governorship of New Granada was handed over to one of Lugo's representatives in Santa Marta in November 1541 while Lugo was making his way across the Atlantic from Tenerife. When he finally arrived at Cabo de la Vela in April 1542, he impressed his rivals by leading his own expedition party of five supply-laden brigantines and as many as two hundred men to the lower Magdalena River. Lugo fared little better than Lebrón—over half of his men and most of his horses died en route to Velez, a river outpost located to the north of Bogotá—but he was at least recognized as the official adelantado and governor of New Granada on May 3, 1543. This completed the official carve-up of Colombian territory.[73]

## RIO DE LA PLATA

The wild free-for-alls in Colombia, Peru, and Bolivia were in marked contrast to the relatively limited competition experienced in the West Indies and even Mexico. Gold and silver did that to people. But if the southern activities held intriguing potential, there was no indication that the minerals extracted to date represented the primary geological sources of the Inca treasures or a possible El Dorado or Xerira. This quandary left the cash-strapped Crown to conduct a multifront war against Gonzalo Pizarro, the clock, and even the Portuguese in South America. The Crown was even preparing to attack the geological problem from the vast, underexplored territories to the southeast of Bolivia. No matter that a remarkable series of northwesterly expeditions launched from the Rio de la Plata region, most recently in 1537, had all failed—the emperor's financial clock was ticking.

King João III of Portugal (1521–1557) had become engaged in South America as well. If Portuguese Brazil appeared to contain even less precious metals (for now) than Venezuela, João had instructed incoming governor Martim Afonso de Sousa in 1531 to follow up on some of Cabot's recent expeditions in the Paraná River region. Cabot's adventures, the brazilwood trade, and a fledgling sugar industry had attracted Spanish and French privateers to Brazil during the late 1520s. Coupled with the Welser efforts in Venezuela and the early Pizarro expeditions to Ecuador, João was forced to dispatch an armed patrol fleet to Brazil on a regular basis. Otherwise, there had been no serious Portuguese attempt to explore the Brazilian interior until Governor Sousa arrived at São Vicente in February 1531 with five ships and five hundred men. The plan was to evict the pesky French, prospect for precious metals, determine a formal boundary line for Portuguese Brazil, and colonize the coastline with a generous land grant (captaincy) system borrowed from Madeira. In short order, Brazil was divided into eleven feudal-style captaincies to the east of an estimated "Tordesillas" meridian line.[74]

Sousa immediately sent Pero Lobo and eighty men to explore the uncharted Paraná River and to obtain some information about Garcia's earlier efforts to find a "mountain of silver." It was another big mistake. News of the Lobo party's slaughter at the hands of local natives persuaded Sousa to leave the upper Paraná and Paraguay rivers to native Brazilians and suicidal Spaniards. Transplanted sugar from Madeira and the Cape Verdes, not even brazilwood, would catalyze the future development of Bahia, Pernambuco, and the other nine captaincies that were in place by 1534. While entrepreneurs like Duarte Coelho, the son of Goncalo, and Pereira Coutinho helped to maintain the Portuguese presence in Brazil after Sousa's departure in 1533, João III eventually confiscated Bahia in 1548 and appointed Tome de Sousa to rule as the first governor-general of Brazil (1548–1551). An ambitious three-year effort to colonize the northern Maranhão territory was also a bust, but sugar flourished and Jesuits founded São Paulo upriver from São Vicente in 1554.[75]

Sousa's departure in 1533, coupled with the Pizarros' spectacular success in Inca Peru, encouraged Charles V to promote a Spanish expedition to Rio de la Plata in 1535. In that year, Pedro de Mendoza was commissioned to lead another northwesterly invasion into Paraguay and points north. To date, the Pizarros had not penetrated Bolivia and no one had done as well as the Garcia expedition of 1524. Garcia had managed to cross the dangerous Chaco Boreal west of the Paraguay River and may have traveled up the Pilcomayo River to as far north as Chuquisaca. Mendoza arrived as the first Spanish governor of Rio de la Plata province with an impressive eleven-ship fleet. He promptly established Nuestra Señora Santa Maria de Buenos Aires near the mouth of the fabled river but had to abandon the tiny post in 1537. The local Guaranis refused to cooperate. However, one of his lieutenants, Juan de Ayolas, was so intrigued by Garcia's earlier expedition of 1524 (and the Pizarros' more recent success at Cuzco), that he sailed up the Paraguay River in 1537 and reestablished Asunción as an inland staging area at the junction of the Pilcomayo. The surviving members of Mendoza's ill-fated Buenos Aires colony were eventually transferred to Asunción, but Ayolas wasn't so lucky. If the murky story is true, Ayolas marched northward toward the Bolivian frontier, stole a large quantity of silver from the local Paraguay Indians, and was subsequently killed by avenging natives during his hasty return.[76]

Although Mendoza's campaign was yet another failure, Ayolas had at least established Asunción as way station. By then, the Pizarros were prospecting for metals in Bolivia, Colombia was a work in progress, and Charles V had been devastated financially by the military setbacks in Provence (France) and in Greece. The emperor needed new sources of mineral wealth to finance his next military campaigns, and the Pizarros seemed to be everywhere. It was the inadvertent arrival of Hernán Pizarro in 1540, the developer of Porco and the foremost Spanish authority on the Inca Empire, that persuaded the emperor to try again in that very same year. After having Hernán imprisoned and interrogated, Charles turned to the legendary Alvar Nuñez Cabeza de Vaca to uncover the inevitable Inca mines that had evaded the Pizarros to date and most probably lay between Porco and the uncharted region north of Asunción. The objective was silver. More specifically, the objective was to beat the Pizarros to the geological source of the silver that had been plundered in Cajamarca and Cuzco and had prompted an Argentine river to be named Rio de la Plata (River of Silver).[77]

Idled unhappily in his hometown of Jerez de la Frontera when the invitation arrived, the forty-year-old de Vaca apparently hadn't had enough of the Americas. Between 1528 and 1536, he had been engaged in a surreal form of adventure with three other members of Pánfilo de Narváez's ill-fated expedition to Florida. The foursome was subsequently shipwrecked off the coast of Galveston, Texas, captured by a series of less-than-civilized Indian tribes, and then forced to wander the vast wastelands of south Texas and northern Mexico as traders, slaves, and medicine men. The four had plenty of tales to tell when they emerged at the Pacific outpost of San Miguel de Culiacán in early 1536. De Vaca even recorded their often hair-raising adventures after returning to Spain in August 1537. No specific mention was made of fantastic treasures or El Dorados, but the observations published in *Relación*, released in 1542 while de Vaca was attending to matters in Paraguay, helped to catalyze a series of treasure hunts throughout unexplored North America.[78]

Cabeza de Vaca landed on Santa Catarina Island in March 1541 as the second governor of Rio de la Plata with 4 ships, 400 men, 9 priests, and 28 horses. Ayolas's whereabouts were still unknown, and legal ownership of what would be the southeastern coast of Portuguese Brazil was still unclear. A massive swath of territory, covering southeastern Brazil, Paraguay, southern Bolivia, and most of present-day Argentina, was inhabited by roughly sixty-five thousand native Guaranis in the coastal areas and thousands of less-civilized natives in the interior. De Vaca had paid for the rights to Rio de la Plata. He contributed eight thousand ducats in return for an annual salary of two thousand ducats and the rights to share in the region's unknown commercial potential. It was only when Ayolas's death was confirmed in May 1541 that de Vaca was able to assume undisputed provincial power at the relocated capital at Asunción.[79]

After determining that the northern route from the abandoned post at Buenos Aires was too dangerous, de Vaca left Santa Catarina in early November 1541 and followed Aleixo Garcia's diagonal route of 1524–1525. The route took the Spaniards past the spectacular Iguazu Falls at the intersection of the Iguazu and Paraná rivers and across the plains of southern Paraguay. When the party arrived at the four-year-old town of Nuestra Señora de la Asunción in mid-March 1542, they joined a rowdy frontier village of 150 Spaniards and possibly hundreds of Guaranis. Another seventy Spaniards were on their way from Buenos Aires. The local Guaranis were so friendly that Asunción had become a melting pot of Spanish, Indian, and mestizo cultures in four short years. Of course, the melting pot was rife with competing factions and rivalries that were beyond the control of the relatively mild-mannered de Vaca. If the Pizarros were ruthless killers, de Vaca was an enlightened liberal.[80]

Unfortunately for the colonists, the territory to the north of Asunción was inhabited by tribes that were downright hostile. This

helps to explain why de Vaca declined to explore the Pilcomayo
River and chose to place his bets on the Paraguay. The upper
Paraguay was no bargain. It was not until a scouting party returned
to Asunción in February 1543 that an overland route to Bolivia was
identified through the river post of Puerto de los Reyes in southeastern Bolivia. This appeared to be a possible pathway to the Inca
mines for which de Vaca, Charles V, and everybody else in Asunción
had been waiting. No matter that Los Reyes was over five hundred
miles due east of Chuquisaca. De Vaca waited until early September
to lead an expedition force of 10 *bergantinas*, 120 canoes, 400
Spaniards, and over 1,000 Guaranis up the Paraguay River to Los
Reyes. After learning en route that Ayolas and eighty Spaniards had
in fact been massacred in 1538, the party reached Los Reyes in late
October. Los Reyes was an agricultural village of possibly two thousand Indians that sat on the eastern edge of the densely forested
Chaco Boreal. De Vaca's Bolivian invasion ended almost as soon as
it began. In late November, after wandering through the wooded
maze for nearly a week, de Vaca decided to return to Los Reyes for
additional supplies. Uncharacteristically, he procured food supplies
by violent means and became stricken with a spreading disease. In
late March, a weakened de Vaca turned his woebegone expedition
party back to Asunción only to be arrested there in late April under
a contrived charge of treason. Many of his colleagues had grown
tired of his humane treatment of the Indians and preferred the more
conventional (ruthless) approach that had served other Spaniards so
well. De Vaca was sent back to Seville shortly thereafter.[81]


## NEW SPAIN

De Vaca would be forced to overcome a raft of legal proceedings,
but he had at least escaped with his life. Four other regional
claimants—Solis, Garcia, Ayolas, and Almagro—had been slaughtered by unfriendly natives or by their fellow Spaniards. Similar
fates had been realized by Alfinger in Venezuela, Lobo in Brazil,

Narváez in Florida, and de Soto in Mississippi. However, these and other disappointments were overwhelmed by the positive news coming out of Colombia and Bolivia. Even New Spain was looking up. The initial silver strike at Zumpango in 1531 had shifted prospecting activity to other exposed (oxidized) outcrops along the southwestern escarpment of Mexico's central plateau and to uncharted regions in the north. The northern territories had been fair game since Cortés's claims to the Panuco territory, stretching between the Gulf of Mexico and the Pacific Ocean, had been transferred to Nuño Beltrán de Guzmán in 1525. Guzmán and ten thousand Indian allies had set off in 1526 to assert his gubernatorial authority over a string of undefined territories that would later become the provinces of Jalisco, Zacatecas, Aguascalientes, Nayarit, and Sinaloa.[82]

Like Cortés, Guzmán made alliances with local Indian chiefs as he moved northward and impressed almost everyone with his extreme cruelty. After establishing the northwestern towns of Santiago de Compostela (1529), San Miguel de Culiacán (1529), and Guadalajara (1532) as staging areas in New Galicia, Guzmán encouraged prospecting activities in all directions. However, mineral discoveries were insignificant and Guzmán was even less capable of following royal orders than Cortés. The governor was stripped of his title (and domains) in 1533, then arrested, imprisoned, and exiled. Guzmán was out of power when the four-man Cabeza de Vaca party arrived miraculously at Culiacán in 1536 and he missed out on the subsequent establishment of San Luis Potosí (1533), Querétaro (1537), and Guanajuato (1542). Guzmán apparently died destitute in 1544.[83]

Guzmán had also overshot discoveries made along Mexico's southwestern escarpment. Cortés had stumbled into Taxco's modest copper and tin deposits back in 1524, in an attempt to manufacture his own bronze cannons, and had advised Charles V in 1525 that the Zapotecs of Oaxaca occupied "the richest mining lands to be found in New Spain." This was not the case, but *asientos de minas* were established after silver was discovered at Zumpango (1531),

Amatepec (1531), Sultepec (1531), Taxco (1532), and Zacualpan (1540). Ever the entrepreneur, Cortés managed to acquire interests in twenty or so mines in Sultepec, twelve in Taxco, and three more in Zacualpan prior to his departure to Spain. It was Cristóbal de Oñate, one of Guzmán's lieutenants, who made a minor silver strike in Guadalajara and established the town as a northern staging area. An east–west line between Guadalajara and Panuco had served as the northern limit of Aztec settlement. While none of these early mining sites matched the scale of Porco in Bolivia, a royal accountant claimed in 1539 that he was personally receiving a half ton of Mexican silver per year.[84]

Even after sizable silver deposits were unearthed at Sultepec and Taxco, it still took a few years to construct processing facilities, organize work crews, and arrange for the transportation of the silver back to Seville. Cortés did his part. Having built foundries in Veracruz, Tlaxcala, Puebla, and Taxco during the early years, he added a water-pumping system and stamping mills at Taxco and an ambitious three-hundred-foot-long adit at nearby Tehuilotepec. Unfortunately, the adit and the "Cortés Tunnel" at Sultepec were poorly engineered and subject to cave-ins. Cortés also missed a rich Taxco vein that was discovered in 1542. But exploration activities continued. Scores of frontiersmen, traders, soldiers, and lower-level aristocrats invaded the wild Chichimeca territories to the north of the east–west line. Cortés had warned Charles V in 1525 that the Chichimecas were "a very barbarian people and not as intelligent as those of the other provinces," but added that some Spaniards had learned that the northern region was "very rich in silver." The task of removing the powerful Caxcanes Indians was left to a combined force of Spanish soldiers and thirty thousand Indian allies in 1541–1542. The subsequent defeat of the Caxcanes in the so-called Mixton War finally opened the Chichimeca frontier to Spanish exploration and development.[85]

The next phase of Mexico's treasure-hunting activities was fueled by hyperbolic reports of vast mineral riches—Seven Cities of Cibola, Quivira, and El Dorado—that had been catalyzed, in part, by Cabeza de Vaca's miraculous return to Mexico City in 1536. The list of volunteers included Cortés, Alvarado, and a few of Guzmán's ex-lieutenants. Cortés was the headliner. After receiving his Pacific capitulación in October 1529, he had constructed four more ships at Tehuantepec and Acapulco by May 1532 and schemed of ways to participate in the South American treasure hunts. Two of his ships eventually found their way to Peru with Alvarado in 1534, but the Pizarros' receipt of a royal cedula had made Peru off limits to the conqueror of Tenochtitlan. Cortés shifted his attention to the unexplored coastline north of Compostela, Mexico. He left Tehuantepec with three ships and 320 recruits in April 1535 only to be wracked by a storm off the southern tip of Baja California. His two supply ships were scattered during the landing. Cortés and his men were stranded there for months and survived only after locating and repairing the two heavily damaged vessels.[86]

Prior to his landing on Baja California, Cortés had somehow learned that António de Mendoza, a member of one of Spain's wealthiest families, had been appointed as the first viceroy of New Spain. Not only was Cortés crushed by the appointment, he was instructed to return immediately to his feudal estate at Cuernavaca and to abandon his northern campaign. Cortés returned to Mexico but refused to abandon his Pacific quests. He wrote the Council of the Indies on September 20, 1538, that he had constructed nine new ships in preparation of a second northern expedition. He also claimed the right, under his royal capitulación of 1529, to discover the fabled riches of "Quivira" that had been fueled by Cabeza de Vaca's cross-continental adventures of 1528–1536. When Cortés discovered in 1539 that the viceroy had already handed the Quivira expedition to the unknown Francisco Vásquez de Coronado, he immediately dispatched one of his lieutenants (Francisco de Ulloa) to the uncharted Gulf of California and left for Spain in the spring of 1540. He would have to assert his claims in person once more.[87]

Coronado was chosen to lead the Quivira expedition because Cortés was still too powerful, too talented, and the Crown had learned to spread its power around to encourage competition. Coronado received the governorship of Nueva Galicia in 1539 and selected the northwestern outpost of Miguel de Culiacán—founded by Guzmán and visited by de Vaca in April 1536—as his staging area. He also sent an advance party to the north under Franciscan Fray Marcos de Niza, a Frenchman with previous adventures in Peru, with de Vaca's famous colleague Estevanico, a slave, serving as a guide. The Niza party may have followed sections of the ancient Turquoise Trail as it traveled north to Hawikuh, a village located fifteen miles southwest of Zuni Pueblo in northwestern New Mexico, before it encountered some hostile tribes. Many were killed during the retreat, including the unfortunate Estevanico, but Niza managed to return safely to Compostela. There, he compounded his failure by fueling exaggerated (Indian-inspired) accounts of at least one of the fabled Seven Cities of Cibola. Mendoza ordered everyone familiar with Hawikuh to remain silent, so as not to attract the attention of other would-be conquistadors.[88]

Following a viceregal send-off, Coronado left Compostela in February 1540 with 292 soldiers, 5 Franciscan friars (including Niza), 1,100 friendly Indians, 1,000 horses, and 600 pack animals. Traveling north and then northeast, Coronado followed a route between the Pacific Coast and the Sierra Madre Occidental to Arizpe, Sonora, crossing the Gila River in eastern Arizona, and then due north to Hawikuh. Upon reaching the first of the fabled Seven Cities in July, an irate Coronado reported back to Mendoza that Niza had lied about almost everything he had said except for the name of the village and the fact that it comprised mud-plastered stone dwellings. Undeterred, Coronado pushed on to Zuni Pueblo and sent a humiliated Niza back to Mexico to explain his misinformation.[89]

A victorious skirmish with the Zunis enabled Coronado to head east through El Morro, Acoma, and Bernalillo (north of Albuquerque) on the upper Rio Grande. From El Morro, a separate party under Garcia López de Cárdenas was sent west and discovered the Grand Canyon in the fall of 1540. From Bernalillo, another side trip was made to Taos. Coronado's main party continued through eastern New Mexico, bridging the Pecos River en route, and meandered

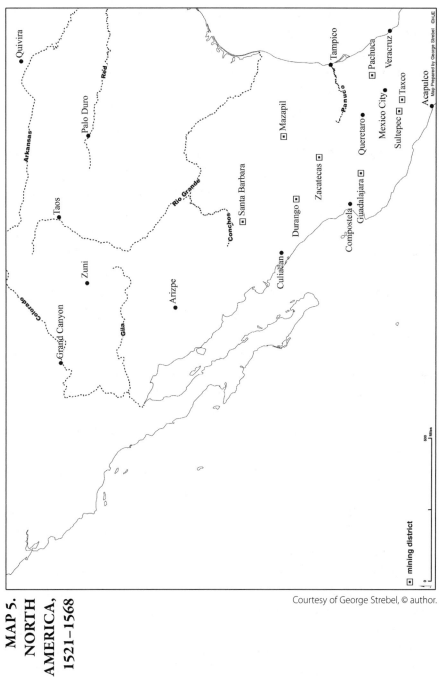

MAP 5.
NORTH
AMERICA,
1521–1568

Courtesy of George Strebel, © author.

through the Llano Estacado and Palo Duro Canyon in the Texas Pan-handle. Finding nothing of value, Coronado split his forces in June 1541 and headed north with a smaller party of thirty to forty men. After reaching the Arkansas River in southwestern Kansas, the party passed to the east of future Dodge City, stopped at the Great Bend of the Arkansas River, and concluded their fruitless quest among the Wichita villages of Central Kansas in the spring of 1542. There was no Quivira. Coronado and his lieutenants struggled back to Mexico City empty-handed and humiliated.[90]

Coupled with de Soto's concurrent failures between Florida and Mississippi, all funded by the plundered treasures of Peru, the consensus was that the action was in Mexico, Colombia, Bolivia, and possibly uncharted Amazonia. This left it to Francisco de Orellana, the Pizzaro lieutenant who had traveled down the Amazon River by accident in 1541–1542, to launch one of the last hair-raising expeditions of the period. Gonzalo Pizarro had returned to Quito almost miraculously in the summer of 1542 while the independent-minded Orellana was leading sixty men down the mighty Amazon to its mouth. No matter that the eastern Amazon region was as flat as a pancake, devoid of precious metals, and located in the Portuguese zone. On February 13, 1544, with Gonzalo engaged in a full-scale rebellion against the Spanish Crown, Orellana managed to secure a capitulación from Charles V to conquer the Amazonian province of New Andalusia from the Atlantic side. Orellana was instructed to find precious metals and establish two Spanish towns—one at the river's mouth and the other in the interior—in return for 5,000 ducats in anticipated profits and one-twelfth of royal revenues. Of course, Orellana would have to finance the expedition himself.[91]

Orellana was unable to raise the funds necessary to equip a three-hundred-man invasion force and a fleet of riverboats. The Crown was even more heavily in debt in 1545, and the Sevillian investment community was divided—while a few Genoese invested in the project, others recalled the fate of the Cabot expedition of 1525. The undaunted Orellana decided to sneak out of Spain with four undersupplied ships on May 11, 1545. He should have stayed at home. Two ships were lost before they had even reached the mouth of the Amazon in late December 1545, and Orellana struggled hundreds of leagues upriver into Amazonia. Traveling downriver in 1541–1542 had been a much easier task. The party was subsequently lost in a watery maze, split up into smaller groups, and attacked regularly by Indians. Most of the expeditionary force, including Orellana, were either killed or died from hunger or disease. One of the few survivors managed to reach Margarita Island (off Venezuela) to report these unhappy events to a group of Spaniards, including Orellana's own wife, in November 1546. By then, the single greatest silver bonanza in American history had been discovered at Cerro de Potosí.[92]

# CHAPTER FIVE

# AN UNLIKELY BANKRUPTCY

In early 1541, Charles V was struggling to finance his next campaigns against France, Istanbul, and German Protestantism. He had already squandered most of the 2.3 million pesos in Inca treasure—obtained mainly from outright confiscations of private holdings rather than from the traditional quinto—on inconsequential campaigns in France and the Mediterranean. The Habsburg treasuries of Castile, Brussels, Austria, and Naples were tapped out, and even Anton Fugger was reluctant to lend him a single ducat. Six years of feverish prospecting activity in the New World had yielded only modest deposits of silver at Sultepec, Taxco, and Porco and intriguing, but still modest, deposits of gold in upper Colombia. There had not been a single El Dorado, not even a Joachimstahl, and the emperor was becoming concerned that Hernán Cortés and the formidable Pizarros were capable of beating him to the geological sources of the native treasures.

Pizarro's spectacular triumphs at Cajamarca and Cuzco had provided the means for Charles V to respond aggressively to Hayreddin Barbarossa's capture of Tunis in August 1534 and a brewing alliance between Suleyman and François I. Humiliated by his forced retreat from Vienna in September 1532, the sultan had shifted his attention back to the east. He earmarked fifty new war galleys for Barbarossa's Mediterranean fleet and invested over thirty thousand ducats in a new arsenal and shipyard at Suez. Coupled with the eighteen warships and 106 large guns that were positioned at Jeddah, Suleyman's plan was to defend the recovering spice trade, threaten the Portuguese bases at Hormuz and Muscat (Oman), and march his

Ottoman army into Safavid Persia. The plan worked. The Ottomans captured Azerbaijan, Tabriz, and many of the prime silk-producing towns of Persia in 1534 and then most of Iraq—Baghdad was taken without a shot and Basra became an Ottoman naval base in 1538.[1]

Charles V had attempted to counter the Ottoman advance at sea. With his Danube flotilla in progress, he had dispatched Andrea Doria's seventy-eight-ship fleet from Sicily in August 1532 to find Barbarossa's elusive navy. Doria managed to capture the Greek port of Coron but no more. The Ottoman fleet was still in a rebuilding mode and was in no hurry to engage the heavily armed Christians. Finally, in May 1534, Barbarossa left Istanbul with seventy galleys, ninety thousand ducats worth of gold, silver, and precious stones, and instructions to restore a friendly regime at Tunis. The fading caravan gold depot had changed hands regularly in recent years and was incapable of stopping Barbarossa. Tunis was not Constantinople, but, located a mere one hundred miles south of Sicily, it moved the Ottomans that much closer to Italy. The talented Barbarossa was subsequently appointed grand admiral (*kapudan pasa*) of the entire Ottoman navy.[2]

The emperor was outraged by the loss of Tunis and an apparent naval alliance between France and Istanbul—Barbarossa was provided access to French harbors in February 1535. Flush with a spectacular windfall from Inca Peru, the emperor decided to personally lead a Christian armada against Tunis in 1535 as a prelude to invading France in due course. Charles, Doria, and Pope Paul I viewed a retaliatory campaign against Tunis as a stepping stone to a possible invasion of Istanbul herself. By June, Peruvian treasure receipts had helped to assemble a multinational Christian armada of over seventy ships—from Castile, Aragon, Portugal, Naples, Sicily, Genoa, and the Knights of Malta (but not Venice)—at a rendezvous point off Sardinia. The Christian armada and twenty-six thousand troops sailed to Goleta, the peninsular fortress that protected Tunis, and routed the Ottoman defenders completely on July 21. Scores of Barbary ships were captured, and as many as twenty thousand Christian prisoners were freed. While the victors returned the favor by selling ten thousand Muslim captives into slavery, Barbarossa and twelve Ottoman galleys managed to escape to Algiers.[3]

The triumph at Tunis was financed by the confiscated Inca treasures (860,000 ducats), the royal quinto (400,000), and the postwar plunder of Tunis (300,000). These sums more than covered the one million–ducat cost of the entire campaign—Charles's personal expenses were contained to around fifty thousand only because there was little room for lavish royal ceremonies at sea. Cobos had skillfully managed the confiscations, the treasure-for-juros exchanges, and the storage of the Inca treasure at La Mota. He even witnessed the treaty signing on August 6, 1535. But even Cobos struggled to finance the emperor's pending invasion of France. The plan was to punish François for his negotiations with Suleyman and to eliminate the French claim to the Duchy of Milan. The problem was that most of the Inca treasure had already been spoken for, all of the Habsburg treasuries were tapped out, and land armies were extremely expensive. By the time that Charles's fifty-thousand-plus infantrymen had reached Savoy in March 1536, Cobos was warning the emperor that he was on the verge of bankruptcy. Charles even challenged François I to a personal duel in an attempt to conserve cash—he declared from Rome on April 17 that the victor would receive Habsburg Milan and French Burgundy. François declined the invitation.[4]

Charles V was undaunted. Backed by additional confiscations and an upturn in silver shipments from New Spain, the emperor felt strong enough to launch a full-scale invasion of France in July 1536. The campaign was a total disaster. Disease proved to be more fatal than military engagements after the vaunted Habsburg army had reached as far as Aix. The emperor was forced to retreat from Provence in early September. The campaign did nothing to alter the status of the Duchy of Milan, and it squandered an estimated 1.5 million ducats on a huge mercenary army and associated provisions. The funds were somehow raised from a variety of sources. Cobos arranged nine separate loans totaling 915,198 ducats between April and September 1536, secured mainly by confiscated American treasure, a donativo from the Kingdom of Naples, and the prepledging of future American treasure shipments. Adam Centurione and Ansaldo Grimaldi reluctantly agreed to exchange 280,000 ducats for 6 percent juros.[5]

The House of Fugger advanced 100,000 ducats (at 14 percent) only on the condition that they receive the first American treasure shipment prior to January 1537. The transaction was repeated in 1537 with a change in the security—the Fuggers were released from their annual advance payment on the maestrazgo contract. Charles's financial position deteriorated so badly in 1537 that he even resorted to a currency devaluation. The Castilian treasury was ordered to widen the official gold-to-silver ratio from 10.1:1 to 10.6:1 to lighten the royal debt burden. But Cobos knew what he was doing. These and other machinations pulled Charles through. The Fuggers' 780,000 ducats in imperial claims were secured by an unexpected upturn in American silver receipts, thanks to the recent discoveries at Sultepec and Taxco, and their willingness to exchange 14 percent debt for 6 percent juros—in those days, juros were viewed as relatively risk-free investments.[6]

Suleyman responded to Charles's financial pressures by raising the ante in the Mediterranean. He dispatched Barbarossa and a massive fleet of 320 ships to Corfu (Greece) in the summer of 1537. Corfu was controlled by Venice, an erstwhile ally, but the Venetians trusted no one. They had recently doubled their own navy to sixty galleys and fortified Corfu with a state-of the-art *trace italienne*. In the tit-for-tat world of the sixteenth century, Suleyman hoped to avenge the loss of Tunis with any victory that would move the Ottoman navy closer to Italy. Corfu was held, but Venice was outraged by this unprovoked assault on her Greek interests. The Ottoman advance pushed Venice into a papally organized Holy Christian League with the Habsburgs in February 1538. France remained on the sidelines. Fortunately for the imperial treasuries of Castile and Istanbul and the thousands of poor souls who would have lost their lives, a looming naval battle between the league's 130-galley fleet and the Ottoman navy never materialized. The Ottomans won a mainly psychological victory in the Gulf of Prevesa (Greece) in late September only because Doria declined to attack.[7]

The losses in Greece, on the heels of the financial squander in Provence, reversed much of the momentum that Charles had gained in Austria in 1532, in Peru in 1533–1534, and in Tunis in 1535. While the emperor had almost single-handedly prevented a possible Ottoman invasion of Habsburg Europe, he had failed to curtail Ottoman activities in the Mediterranean. Barbarossa simply relocated to Algiers, and Suleyman returned his attention to Safavid Persia. Doria's caution at Prevesa left Castile with more debts to pay, Portugal with a Barbary navy that threatened her tenuous positions in Morocco, and Venice with a need to cut one of her typical side deals with Istanbul. The Venetians, true to form, dropped out of the Holy Christian League and sent Suleyman a huge tribute payment of 300,000 ducats in October 1540 to restore peaceful relations.[8]

In early 1541, a discouraged Charles V would not have predicted that a staggering volume of American treasure would shortly fall into his hands. In fact, the emperor had been underwhelmed by the explosion of prospecting activity that followed the Pizarros' triumphs at Cajamarca and Cuzco. While the recent discoveries at Sultepec and Taxco had raised silver production in New Spain, these contributions from Porco (Bolivia) and the streams of upper Colombia were quite modest compared with the one-shot plunders of Inca Peru and the last phase of the Central European mining boom. The setbacks in Provence and Greece had weakened the emperor's financial position severely, and yet he was already planning his next campaign against Istanbul.

An invasion of Hungary, proposed by Ferdinand of Habsburg after the death of the Ottoman puppet ruler John Zapolya and the loss of Buda and Pest in August 1541, was out of the question. The Battle of Mohacs in 1526 had reduced the confines of Habsburg Hungary to the mineral-rich western and northern regions of the former kingdom. This slice of territory was sliced again in 1541 while Ottoman Hungary was converted from a tributary vassal to a formal province. Removing the Ottomans from lower Hungary would require another expensive Danube-like campaign, and the funds for such an initiative were simply not there. Another idea,

Andrea Doria's audacious plan to launch a full-scale naval assault against Istanbul herself, was rejected as well. Instead, the emperor decided to focus on easier (and cheaper) prey—Barbarossa's naval base at Algiers. The opportunity to repeat the glory of the Tunis campaign of 1535 would attract a diverse group of nobles and adventurers, including the irrepressible Hernán Cortés and his two sons. Those who managed to survive the ordeal, including the Cortéses, would be very disappointed.[9]

Charles and Cobos scrambled to finance the cost of what would be a massive 155-ship armada against Algiers. Some 842,000 ducats were required for the ships, the twenty thousand or so troops, supplies, and royal entourage for a five-month campaign. This was less severe than the costs of the aborted campaign in Provence in 1536 but a sizable addition to the emperor's existing financial burden. Most of the debts from Provence and Prevesa were still outstanding in 1541, and raising the funds for the Algiers campaign was not easy. A Fugger commitment of around 250,000 ducats left the bulk of the campaign costs to be borne by Habsburg Naples and Sicily. The southern Italians came through in July 1541 with a three-year donativo of 772,000 ducats that, unfortunately, was whittled down to around 465,000 after interest and transfer charges. It is unclear how the shortfall was bridged, but Cobos usually found a way.[10]

If the emperor intended to make a major statement at Algiers, he chose to ignore the advice of one of his notable passengers—Hernán Cortés. The fifty-five-year-old conquistador had left New Spain in a huff in the spring of 1540 after learning that the unknown Francisco de Coronado had been commissioned (over Cortés's head) to find the fabled riches of Quivira in northern Mexico. Cortés's major problem since 1525 had been that his undeniable talent had interfered with royal interests. He returned to Spain in 1540 in the hopes of persuading Charles V to restore his territorial claims to past and prospective conquests in North America. He had no chance. The great conquistador was welcomed as a hero in the streets of Seville, but he was out of royal favor. Charles V, focused on the upcoming campaign against Algiers, was in no position (or mood) to receive him in early 1541. So when the emperor left to join Doria's invasion fleet in Italy, Cortés, being Cortés, decided to join the Algiers campaign himself. It was his only chance to argue his case directly with the emperor. Cortés left Spain with two of his three sons, some one hundred thousand ducats in emeralds, and a motley crew of glory-seeking nobles, adventurers, and mercenaries.[11]

Like most of his fellow crusaders, Cortés assumed that the Algiers campaign would be a repeat of Charles V's epic conquest of Tunis in 1535. He was wrong. Two separate Christian fleets under the emperor and Doria gathered at a rendezvous point off Algiers on October 20, 1541, and landed the bulk of the 21,300-man army seven miles east of port. It was then that an unusually fierce storm blew in, scattering most of the Christian fleet on October 24 and destroying supply ships that carried the artillery and provisions. Horses had to be slaughtered to feed the landing party. Charles's fourth personal crusade ended in a humiliating retreat. The emperor ignored Cortés's military advice throughout and essentially laughed off the conquistador's offer to remain onshore and lead the capture of Algiers himself. While Cortés and his two sons survived the ordeal, he lost one hundred thousand ducats of emeralds at sea and any opportunity to win the emperor's favor.[12]

This last royal snub must have been painful to Cortés. The greatest conquistador of a conquistadorial age spent the last years of his life dealing with the ramifications of his brash circumvention of royal authority back in 1519. No matter that he had conquered the Aztec Empire, created Mexico City, co-founded the Mexican mining industry, financed a number of Pacific expeditions, and built sugar mills, cattle ranches, shipyards, hospitals, monasteries, and schools. Cortés was burdened by the loss (theft) of his conquests and an awareness that his brutality against native Americans had helped to institutionalize the practice. He never saw Mexico again. In early 1547, he provided a one-hundred-thousand-ducat dowry for the eldest of his four daughters (Maria) and specified in his will that he be buried in Mexico within ten years of his death. He died near Seville on December 2, 1547, at age sixty-two. The status of his remains were just as controversial as the man. They were finally transferred to Texcoco in 1566, reinterred inside a series of Franciscan chapels in Mexico City between 1629 and 1794, lost within one of the chapel walls for over a century, and rediscovered only in November 1946.[13]

## FINANCIAL MACHINATIONS

Mother Nature had refused to cooperate at Algiers in October 1541. Not only was the retreat from the North African coast a humiliation to Charles V, the emperor had added another 842,000 ducats in debt to the imperial balance sheet. He had some serious fundraising work to do if he wanted to launch another campaign against France. Much greater attention would have to be paid to the Spanish Americas, and ways found to extract more funds from the Castilian treasury. When times were tight, the emperor had had no qualms about confiscating private treasure shipments from the Americas and exchanging the bullion for Castilian juros. However, the associated expansion in juro debt had shifted the royal creditor base from the friendly confines of Castile to the less-tolerant money markets of Antwerp. Spanish capital (and Spanish American capital) was flowing into the hands of the Fuggers, Welsers, and Genoese. The added problem in the early 1540s was a lack of collateral—the Peruvian plunder had been spent, Central European silver was on a downward slope, and conventional American mineral output was disappointing. As military-related debts mounted and went unpaid, outstanding loans were consolidated and converted into even larger amounts of juros.[14]

The emperor left Spain for good in 1543, leaving behind his sixteen-year-old son Felipe as regent, after having revised the method through which the 10 percent alcabala (and *tercia*) were collected in Castile. These taxes accounted for over four-fifths of Castile's 1.1 million ducats in total ordinary revenues and were the principal means with which to service the outstanding juros. Charles had needed a series of extraordinary servicios so badly starting in 1537 that he agreed to Castile's introduction of a fixed head tax for the 1537–1546 collection period. The deal made each of Castile's municipalities liable for a predetermined quota of alcabala (and tercia) revenues, plus a pro-rated share of juro interest charges, and shifted the servicing of juros from the Castilian treasury to the full faith and credit of the municipalities. While the concept sounded good in theory, the renewal of the fixed quota program (through 1566) deprived the Castilian treasury (and the emperor) of incremental revenues driven by inflation. These revenues flowed, instead, into the coffers of the Castilian municipalities. Charles may have needed the extraordinary servicio funds during the 1537–1546 period, but the other side of the deal, a thirty-year head tax quota (1537–1566), came back to haunt him and Felipe II in the decades ahead.[15]

Charles's financial position was deteriorating. Even before Algiers, he had had to crush a tax revolt in his hometown of Ghent in February 1540. François added to the emperor's woes by declaring war in July 1542 and invading the Habsburg Netherlands in conjunction with a rebel army under Maartin van Rossum. Faced with the prospect of a three-front war against France (with Van Rossum), the Schmalkaldean League of Protestant German princes, and the Ottomans, Charles chose to invade France in the spring of 1543. He was backed by the recent servicios from Castile and a superior military machine. The Habsburg army captured a string of French border towns by late November and prevented Van Rossum's rag-tag militia from taking Antwerp.[16]

When François permitted Barbarossa and 110 Ottoman war galleys to winter in Toulon harbor in 1543, he had gone too far. The implied naval alliance outraged the Habsburgs, the German Protestants, and even the English. At the Diet of Speyer, the Protestant princes agreed to raise an army of twenty-eight thousand men and join their arch rivals, the Catholic Habsburgs, in a campaign against France (and possibly the Ottomans). Charles also arranged an alliance with Henry VIII of England. A huge Habsburg army reached the Marne River by August 1544 and would have advanced on Paris if an equally huge English invasion force of forty-four thousand had not been halted between Calais and Boulogne—Henry VIII was more careful with his resources than Charles V. If Henry's caution saved France, François got the message. The resulting Peace of Crépy of September 19, 1544, ended hostilities between the Habsburgs and France and lasted through the death of the French monarch on March 31, 1547.[17]

Of course, the invasion of France was extremely expensive—a staggering 2.8 million ducats were required to march a Habsburg army of 38,000 infantry, 9,000 cavalry, 1,400 trench diggers, 63 cannons, wagons, horses, provisions, and the royal train between Speyer and the western front. This was on top of the 820,000 ducats incurred to defend the Habsburg Netherlands in 1542. Nearly all of the costs were financed (or refinanced) by the Fuggers, Welsers, and the Castilian treasury. Some of the loans were secured by confiscated American treasure, the royal quinto from Peru, a so-called imperial states grant, and the early receipt of the servicio for 1546–1548. Others were secured by Duke Cosimo de Medici's repurchase of two Habsburg-occupied fortresses in Florence, another donativo from Naples, a contribution from the Diet of Speyer, and a dowry from João III for the pending marriage of his daughter Maria to Felipe.[18]

The financial machinations of 1542–1544 left a very large mark on the Castilian treasury and help to explain why Charles's presence at the Cortes of 1542 was his last. In the two-year 1543–1544 period, annual short-term borrowings by the Castilian treasury averaged a jaw-dropping 1.6 million ducats—nearly four times the average level incurred between 1520 and 1542. Since the Crown was paying as much as 13 percent interest on its short-term debt and another 12 percent (or more) in cases involving international fund transfers, Cobos was faced with a severe cash flow problem. His solution was to convert as much of this short-term debt as possible into lower-coupon (3–6 percent) juros. Coupled with the increasingly regular confiscated-treasure-for-juro exchanges, Castile's juro exposure was beginning to get out of hand.[19]

Having tapped out the Castilian treasury, Charles relocated

("fled") to Brussels in 1543 to tap into the strongest financial system in Europe. The emperor intended to finance his escalating military requirements (from Brussels) with 6.25 percent revenue bonds (*renten*) backed by the tax-raising capabilities of the individual provinces. In the same way that juros were now backed by the individual municipalities of Spain, renten had been issued by the municipal governments of the Netherlands since the 1200s. The bonds were backed by predictable revenue streams, like excise taxes on beer and wine sales, and were regarded as relatively risk free. Renten had gained such a high degree of credibility with investors that Margaret of Austria continued the practice during her regency. Brussels borrowed only rarely from the gnomes of Antwerp. Even Charles V had been careful with the system. Although he had received subsidies totaling 762,000 ducats over a seven-year period (1529–1536), he had left Brussels pretty much alone. The emperor knew that Brussels, the provinces, and their renten investors would not subsidize military investments outside of the Netherlands.[20]

The regents, Charles's aunt Margaret of Austria (1506–1514, 1519–1530) and his sister Mary of Hungary (1531–1555), had skillfully combined the traditional renten system with some of the financial innovations developed in Antwerp. They also presided over a booming regional economy. The royal domain revenues in the Habsburg Netherlands had more than tripled to 552,000 ducats between 1522 and 1536. While much of this growth was debt financed, it was manageable. The City of Antwerp issued nearly one million gulden in loans in one three-year period, backed by revenues covering a fifteen- to twenty-year period, and the so-called Council of Finance in Brussels managed to sell 755,541 gulden in renten to backwater investors in Holland and Zeeland between 1534 and 1540. Unfortunately for all concerned, these funds proved to be insufficient during the crisis years of 1542–1544. The emperor's never-ending military campaigns would shortly transform the Habsburg treasury at Brussels into the same tenuous version that had been created in Castile.[21]

It was Charles V who pushed the States General into approving three emergency measures in December 1542. The emperor needed help in financing a projected 5.4-million-gulden increase in military spending for 1543–1544 now that he had essentially tapped out the existing renten system. He called for a 10 percent tax on income from real property, a second 10 percent tax (the "Tenth Penny") on so-called commercial profits, and a third 1 percent tax on the value of export sales. The new taxes were extremely controversial. While the two 10 percent taxes raised over 2.6 million gulden in 1543 and 1544, revenues from the 1 percent export tax were disappointing. Even Gaspar Ducci, the principal (if roguish) adviser to the Brussels Court, failed to recoup the 200,000 gulden that he had paid upfront for the tax-farming rights. The States General reluctantly approved an extraordinary subsidy of 677,851 gulden in 1544 to bridge the budgetary shortfall. Even worse, the credibility of the Receivers General itself was challenged when it was discovered that Ducci had allowed the receivers to guarantee various loans (via letters obligatory) on the basis of ordinary revenue streams that had already been pledged. Needless to say, the seventeen provinces of the Habsburg Netherlands were in a state of shock.[22]

Charles added to the confusion by approving "lending at interest" in 1543, a financial bombshell that defied a still-binding papal decree of 1312. Rome had allowed annuities, instruments through which a creditor sold a fixed annual sum to a third party and regained his initial purchase price with an embedded interest component, only because the return component was viewed (unlike interest) as a perpetual feature. But credit had replaced annuities as the principal method of public finance—aided by the application of security-enhancing "payable to bearer" clauses, assignments, endorsements to bills of exchange—and trading in interest-bearing short-term and long-term bonds had exploded on the Antwerp Bourse. The emperor was only catching up with reality. However, borrowing rates did not decline with the Peace of Crépy with France in September 1544, the death of Hayreddin Barbarossa in July 1545, and Suleyman's willingness to sign a peace treaty with Habsburg Austria in 1547 (to conserve funds for his latest campaign against Persia). There was simply too much debt outstanding. The treaties and an apparent upturn in American mining revenues in 1546 only allowed Charles V to return to the battlefield. This time, the target was the Schmalkaldean League of Protestant princes. The heavily armed league threatened the Habsburg's Catholic dominions in Germany and the Netherlands.[23]

Protestantism was an increasingly powerful force in Germany. In 1530, Charles had attempted to make peace with the Protestant German princes at the Diet of Augsburg but was rebuffed. The application of an Inquisition against heretical Protestantism served only to harden attitudes even more. The independent-minded rulers of Saxony and Hesse, backed by mining revenues of their own, responded by organizing a military alliance at Schmalkalden in December 1530. They renounced Ferdinand's election as king of the Romans, a largely ceremonial title that was intended to convey Christian solidarity to Istanbul, and created a Schmalkaldean League. The general plan was to consolidate the Protestant position in Germany and possibly invade the Habsburg Netherlands spiritually. Although the league shifted temporarily to the Habsburg side in 1543–1544 to deal with the Franco-Ottoman alliance, the Protestant princes returned to the religious battle after the Treaty of Crépy was signed in September 1544.[24]

The Protestant threat was so severe that Pope Paul III offered to contribute 200,000 scudi, 12,000 infantrymen, and half of church revenues in Habsburg lands (for one year) if Charles would wage war against the Schmalkaldean League. The pope's timing was excellent. He knew that France was too broke to intervene—François had just paid England over two million crowns for the return of Boulogne in June 1546—and the Ottomans were preparing to invade Safavid Persia. Suleyman subsequently signed a five-year truce with Habsburg Austria in June 1547 in return for a modest annual tribute payment of thirty thousand ducats. If that were not enough, the pope and the emperor's always-anxious creditors had received news of a possibly huge silver strike in the wilds of Bolivia.[25]

The financial inducements from Rome and the prospect of

another windfall from Peru were enough for Charles V. In June 1546, he agreed to campaign against the two most powerful Protestant princes—John Frederick of Saxony and Philip of Hesse—and moved immediately to relieve the Protestant sieges of Leipzig and Dresden. The emperor assumed personal command of the fifty-one-thousand-man Habsburg army (from Ferdinand) by January 1547 and prepared to meet the princes head-on at a site fifty miles east of Leipzig on the Elbe River. A famous portrait depicted the dashing emperor crossing the river on horseback. When German Catholics met German Protestants at Muhlberg on April 24, 1547, it was the Catholic Habsburgs who emerged victorious. Charles still had the finest army in Europe. But even though John Frederick was captured during the action, the great triumph was as short-lived as most Habsburg victories. A whopping 3.1 million ducats had been required to check (but not defeat) the expansion of Protestantism in Germany. Matters were not helped by the recent death of Francisco de los Cobos (1547) and the abrupt suspension of Peruvian treasure shipments. But Charles found a way. Over half of the campaign costs were assumed by the papacy and the Spanish church—handled, of course, by the House of Fugger—and another 630,000 ducats in reparations were received from the Protestant princes. Anton Fugger, a devout Catholic, helped out with Protestant Augsburg's payment as a goodwill gesture.[26]

The remaining eight hundred thousand ducats in German campaign costs were assigned to the Castilian treasury. Try as it might, and despite the American silver bonanzas uncovered in 1545 and 1546, the Castilian treasury was fighting a losing battle. No wonder that Charles chose to live abroad between 1543 and 1556—he might have been lynched had he remained in Castile. According to historian Ramon Carande, the eight hundred thousand was simply folded into 1.9 million ducats in other loans issued between May 1546 and July 1547. Some of these loans were secured by future American silver shipments while others were used to restructure (or refinance) a mountain of debt that was approaching 3.3 million ducats. The Fuggers had advanced nearly five hundred thousand ducats themselves on the basis of the recent American silver bonanzas—even if Peru was engulfed in a full-blown civil war. By July 1548, all but 132,000 ducats in estimated Castilian revenues for 1549 (one million ducats) had already been prepledged to service existing loans. A cash flow shortfall of 868,000 ducats loomed, and no investor, including Anton Fugger, was likely to accept another round of juros without an identifiable stream of unpledged income. It was no accident that virtually all of the new loans of 1546–1547 had been transacted outside of Iberia. Antwerp lenders (like Anton Fugger) refused to be repaid anywhere except Antwerp.[27]

## BONANZA IN BOLIVIA

The mines of New Spain had delivered for Charles V since 1544. The annual volume of silver from Sultepec, Taxco, and the other western sites had soared from a very modest ten thousand marks in 1534 to a Joachimstahl-like eighty-five thousand marks (425,000 gold pesos) in 1544 and 1545. These impressive totals helped to secure some of the emperor's war debt, but they were nowhere near the scale of the Inca plunder or even the declining silver volumes being produced in Central Europe. The wild series of treasure hunts on both American continents had been disappointing as well. However, two unexpected silver strikes in 1545–1546—one at Zacatecas in the northern Chichimeca territory of Mexico and a second, even larger one at Cerro de Potosí in Bolivia—changed everything. These two strikes were instrumental to the financing of the Muhlberg campaign and provided a new source of collateral for the emperor's next campaigns—once the political situation in Peru was under control.[28]

Even before the strike at Cerro de Potosí in April 1545, the emperor's escalating debt burden had called for a military solution to the Gonzalo Pizarro problem in Peru. There was simply too much potential wealth at stake to do otherwise. A new Viceroyalty of Peru was established in Lima in 1542 to reassert royal authority following the murder of Francisco Pizarro, the imprisonment of

Hernán, and the assumption that Gonzalo had died in Amazonia. But Gonzalo's unexpected return muddied the waters. Incoming viceroy Blasco Nuñez Vela was authorized in 1544 to remove the usurping Castro from power, ignore Gonzalo's rightful claims to the governorship of Peru, and declare legal war on Peru's existing *encomenderos*. Vela and the newly established Audiencia of Lima were also authorized to implement Charles V's New Laws of 1542, well-intended ordinances designed, on one hand, to abolish the encomienda system (and slavery) in Spanish America and, on the other, to destroy the "Kingdom of Pizarro." Of course, the proposed destruction of the encomienda system was a call to arms for the endangered encomenderos of Peru and Bolivia. Gonzalo was recruited to lead a rebellion against royal authority in Lima and Cuzco. The rebellion succeeded. By October 1544, the under-manned Vela had evacuated Lima, and Gonzalo had assumed the governorship of Peru.[29]

As Gonzalo Pizarro prepared to take on Vela's royalist forces near Quito, he and everyone else in Peru were unaware of a dis-covery (or rediscovery) that had been made almost by accident in southwestern Bolivia. Cerro de Potosí's 15,700-foot-high peak, rounded and maroon-colored, was not hard to miss along the Inca highway running between Porco and Sucre. And inhabitants of the two Aymara villages below must have known that Aymara miners had extracted modest amounts of silver from this same mountain many decades earlier. Potosí's surface-level deposits needed only to be "rediscovered" in early 1545 by two natives, Gualpa and Guanca, to launch a globe-shattering boom that would make Villa Imperial de Potosí one of the richest cities on earth. Whether or not Gualpa and Guanca knew about the earlier Aymara efforts, they relayed their find almost immediately to a Spanish prospector named Villaroel. It was Guanca and Villaroel who jointly registered their claim at Porco on April 21, 1545.[30]

The initial quantities of surface-level ore yielded a jaw-dropping silver content of nearly one mark (8.0 ounces) per pound. Even Potosí's "impure" quantities rated at a record eighty marks per

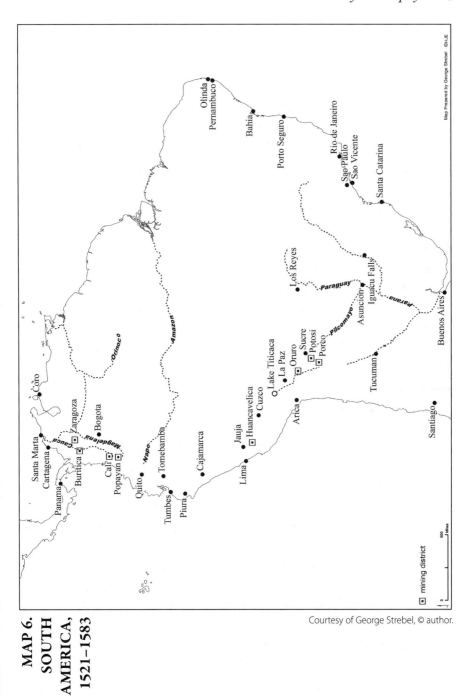

## MAP 6.
## SOUTH
## AMERICA,
## 1521–1583

□ mining district

0          500
        Miles

Courtesy of George Strebel, © author.

quintal. Better yet, the ore was easily extracted and could be smelted by encomienda natives in inexpensive clay furnaces (*huayras*). By year-end 1545, over six thousand huayras were operating along the slopes of Cerro de Potosí (Rich Hill) and launching the single largest bonanza in American history. Early production figures are unclear, but Potosí may have produced a staggering 195 tons of silver (800,000 marks) in 1548 alone. The associated six million silver pesos in value was nine times that of Joachimstahl at its peak (87,500 marks) and nearly four times that of New Spain—a recent strike at Zacatecas (1546) had even raised total Mexican production to an otherwise shocking 190,000 marks (1.6 million silver pesos) in that same year.[31]

The Viceroyalty of Peru, modeled on the same Viceroyalty of New Spain that had stripped Cortés of his political powers, held jurisdiction over Potosí, Bolivia (Charcas), and Peru whether Gonzalo Pizarro recognized it or not. No matter that the Pizarros had relaunched the Bolivian mining industry at Porco in 1537 and Guanca and Villaroel had filed their initial claim at the Porco registration office. Gonzalo and his rebel army were preoccupied with finding and defeating Vela's royalist forces near Quito in late 1545. By then, news of the great bonanza had reached Gonzalo, Vela, and even Charles V. The excitement in Brussels was tempered by alarming reports that Gonzalo was engaged in an open rebellion against the Crown, was directing brutal reprisals against his enemies, and had even suspended shipments of the royal quinto (and confiscation opportunities). With the emperor planning a major campaign against the Schmalkaldean League, the suspension of the treasure shipments was frightening. He made preparations to send an armed fleet to Peru under Licenciado Pedro de la Gasca to crush the last of the Pizarro brothers once and for all.[32]

When Gasca arrived in Lima in July 1547, he immediately repealed the New Laws of 1542 and offered pardons to the followers of Gonzalo Pizarro. Since the restoration of the encomienda system undercut the primary motivation for the rebellion, Gonzalo's support melted away. The last Pizarro surrendered to Gasca's royalist army at Jaquijahuana (near Cuzco) in April 1548 knowing that he had been sold out by his fellow encomenderos and had also been deprived of a major piece of Cerro de Potosí. Gonzalo was executed (beheaded) within days. Of course, there was a large pile of money involved. While Gonzalo had received no more than 2.5 shares of Cajamarca treasure, slightly less than his deceased brother Juan, the value of his estate had grown to as much as three times the two hundred thousand ducats left by Juan in 1536. Even without a stake in Cerro de Potosí, Gonzalo Pizarro may have controlled the largest private fortune in the Western Hemisphere.[33]

Gonzalo's estate and a large chunk of the Pizarro family's encomiendas, tributary streams, mining interests, and business properties were either confiscated by the Crown or redistributed to Gasca's troops and allies between 1548 and 1560. If the Pizarro enterprises were worth as much as one million pesos in 1548, they compared favorably with the private fortunes of Charles V, Anton Fugger, Hernán Cortés, and Francisco de los Cobos. The roughly 125,000 pesos generated by the family's remaining encomiendas in 1556 were more than double Cobos's annual income in 1546, nearly double the tributary revenues received by the Cortés estate in 1567, and one-third of the annual tribute realized by all of Mexico's 480 encomenderos in 1560. If one added in the prospective value of the Pizarros' confiscated mining interests, the family fortune would probably have been much greater than one million pesos. An imprisoned Hernán Pizarro (d. 1578) was allowed to retain some of his holdings until around 1560—mining interests, a tribute-paying population of nearly seven thousand Indians, and a few coca leaf plantations—but his effort to recover the family's Peruvian estate was dealt a fatal blow in 1570. In that year, incoming viceroy Francisco de Toledo terminated all of the family's outstanding lawsuits. The family title of marques was transferred to Hernán's heirs following the death of Francisco's last descendant in 1756, disputed for decades, and then retired for good in 1924.[34]

Of course, the confiscation of the Pizarro assets between 1548 and 1560 deprived the family of an opportunity to cash in on one of the greatest bonanzas of all time. While the Crown decided to halve the Mexican quinto to a *diezmo* (10 percent) in 1548 to encourage exploration activity, the full quinto (20 percent) was maintained in viceregal Peru. This tax mattered greatly to local mining magnates because Potosí's richest surface-level deposits were depleted rather quickly—a royal quinto of 668,000 silver pesos in 1550 meant that silver output in that year had fallen to 379,244 marks or 3.3 million silver pesos. Costly mining shafts had to be sunk to extract Potosí's lesser-quality ores. However, the magnates (and the Crown) would be rewarded when massive quantities of silver sulfides and chlorides were uncovered at the intermediate depths. Five principal silver lodes—Centenos, Rich, Tin, Mendieta, and Onate—were unearthed in subsequent decades and each lode, in turn, was subdivided into a series of mines. Between 1545 and 1610, the silver produced from Cerro de Potosí alone would outstrip the entire mining output of New Spain.[35]

Charles V should have been sitting pretty in 1550. He had invaded France successfully (1544), outlived his arch rival François I (1547), halted the territorial expansion of the Schmalkaldean League (1547), crushed the Gonzalo Pizarro rebellion (1548), and was in the middle of a five-year truce with Istanbul (1548–1553). If that were not enough, his huge debt burden had been eased by the spectacular returns from Cerro de Potosí, Zacatecas, and the fledgling gold fields of upper Colombia. An annual contribution of nearly two tons of Colombian gold, nearly one-fifth of global supply, had helped to triple the value of American treasure imports to around one million gold pesos annually between 1541 and 1550. Additional gold strikes at Buriticá and Popayán would help to double these annual totals to nearly two million gold pesos in 1551–1555. These volumes appeared to be sustainable, in contrast to the one-shot plunder deals from Cajamarca and Cuzco, but they were becoming increasingly difficult to predict.[36]

But Charles was stricken with the same illness that had afflicted his hapless grandfather Maximilian I—the more money he had, the more he spent. Virtually all of the emperor's American treasure shipments had already been pledged to securitize his escalating debts in Europe. Antwerp had become wise to the emperor's financial machinations even before his relocation to Brussels in 1543. The Fuggers had refused to advance more than token amounts until their 710,000-ducat claim was repaid and sat mainly on the sidelines until news was received from Peru. The Catholic Welsers, with a net worth of only 243,000 florins in 1543, attempted to stay out of the German Wars of Religion entirely. Charles responded to their disloyalty by appointing Gaspar Ducci to serve as his principal financial adviser in Brussels between 1542 and 1549. A notorious arbitrageur and occasional crook, Ducci had been suspended in 1540 for creating an artificial scarcity of gold that nearly wiped out the royal Portuguese factor in Antwerp (among other financial schemes). He was unable to travel through town without a group of bodyguards. But Ducci helped the Habsburg cause by raising a whopping five hundred thousand ducats in bonds guaranteed by the receiver general of the Netherlands—even if the guarantee turned out to be worthless and the powerful House of Fugger was among the unfortunate bond holders.[37]

Rogues like Gaspar Ducci were attractive to Charles V because they knew how to make money fast. Ducci had turned the ancient practice of foreign-exchange arbitrage into an art form. The practice of arbitraging bills of exchange had been advanced by the Fuggers, the Genoese, and the traders of Lyon before it became centered in freewheeling Antwerp in around 1540. The concept was simple. If a merchant banker could borrow in Venice at a low interest rate and in a relatively weak currency, a large profit could be realized by trading the same bill of exchange in a higher-rate, higher-valued market like Antwerp. Or vice versa. When the shifting exchange relationships between gold and silver were thrown into the mix, the opportunities expanded. Ducci's contribution was to extend a commodities-oriented practice derived from trade fairs to the new (and massive) public financial instruments that had been created by Charles V (in Antwerp) and by François I (in Lyon) to fund their never-ending military campaigns.[38]

## ANTWERP

Antwerp had handled a large volume of transactions generated by the Portuguese, the Habsburgs, and the Fuggers. When Spanish America and Charles V's increasingly large capital needs kicked in, Antwerp's economic power expanded even further. One might have expected that the influx of precious metals from Africa, Central Europe, and the Americas would have raised the money supply and reduced the cost of borrowing. This was not the case. In addition to escalating warfare, there was an associated explosion of business activity that stimulated even greater demands for investment capital. Funds became especially expensive after an inconsequential military campaign or two. The expansion of Charles V's revenue base in Spanish America merely generated fresh sources of unpledged cash flow to collateralize even more ambitious military campaigns and the assumption of even larger amounts of public debt. The cost of Charles's triumph at Muhlberg in 1547 was nearly four times the cost of his defeat at Algiers in 1541.[39]

Antwerp was the wealthiest and most powerful entrepôt in Europe. In addition to the capital markets, Antwerp was the principal European distribution center for Portugal, Spain, England, the Netherlands, and Germany. Portuguese and Spanish merchants could purchase American silver more cheaply in Seville than in Antwerp, but they relied on Antwerp almost completely for their stocks of American-bound trade goods. Although the port had been forced to share the European spice market with Marseilles and Venice after 1530, spices remained an important business in Antwerp. The Portuguese were still importing at least twenty thousand quintals of spices on an annual basis and relying on Antwerp for the distribution end. Declining profit contributions from the spice trade and the Central European silver industry had been more than offset by public finance, the transatlantic trade, the regional textile industries of Flanders and Brabant, and the importation of Baltic grain. Antwerp (and smaller-scale Amsterdam) captured most of the grain business from the Hanseatic

League during the 1540s and solidified this economic triumph after the Battle of Muhlberg.[40]

A host of new businesses were started in Antwerp: printing, brewing, goldsmithing, diamond cutting, and the manufacture of pottery, clocks, watches, leather, soap, and harpsichords. After a glassmaking business had been transferred from Venice in 1537, an estimated eleven thousand mirrors were manufactured in 1549. The local textile industry produced a variety of cloths: velvets, satins, threads, wools, linens, silks, carpets, fustians, embroideries, and tapestries. Then there were tanneries, shipbuilding, carriage making, the warehousing of grain, arms and munitions, fish salting, and the refining of metals, wax, and sugar. The twenty-six local guilds included the century-old Guild of St. Luke to represent Antwerp's world-renowned printers, publishers, and book sellers. There were 169 bakers, 78 butchers, over 90 fish sellers, 110 barber-surgeons, 594 tailors, 124 goldsmiths, and purveyors of milk, butter, cheese, and beer. Beer revenues surged whenever the King of Denmark closed the Danish Sound to Baltic shipping—in those periods, German malts were replaced by local varieties. While these and other businesses were dwarfed by Antwerp's core positions in metals, spices, textiles, and finance, they diversified the economy and raised traffic volume. The port area had to paved and supplied with a new crane to handle the five hundred cargo ships, one thousand wagons, and ten thousand carts that arrived daily at mid-century. All of this activity made Antwerp an expensive place to live. Thomas Gresham, England's royal agent during the early 1550s, estimated that the local cost of living was twice that of France. A five- or six-room house could be rented for no less than two hundred crowns per year.[41]

Antwerp was also a melting pot of cultures. In addition to the well-established English presence, the town attracted scores of displaced Hansa merchants (from Bruges), Iberian Conversos, Genoese merchant bankers, and even an occasional Venetian. While Portuguese Converso spice merchants like Francis and Diego Mendes were caught up in the wider campaign against alleged Protestant heretics, attempts to halt Converso emigration into Antwerp was often circumvented by fees, bribes, and assistance from allied merchants. The one thousand-plus foreign merchant companies operating in Antwerp at midcentury conducted their daily affairs with as many as five thousand merchant representatives, ship captains, and middlemen. One Florentine visitor, Lodovico Guicciardini, estimated that Antwerp was receiving three million crowns worth of Italian cloth and silk every year. Imported English cloth was estimated at five million crowns annually—five times greater than the value of the Portuguese spice trade. Gresham was so impressed with Antwerp's New Bourse, constructed for three hundred thousand crowns in 1531 and enclosed by thirty-eight sculpted columns, that he used it as a model for the later London Royal Exchange.[42]

Antwerp and the prosperous southern provinces (Flanders and Brabant) dominated the Habsburg Netherlands. The tax-conscious Habsburg Court at Brussels valued the Flemish economy at 2.5 times that of Brabant and five times that of Holland, leaving a federation of seventeen provinces that was unbalanced—held together since 1493 mainly because Regent Maximilian of Habsburg had somehow gained the allegiance of the regional nobility in that year. Otherwise, the seventeen *Stadholders* (governors) and the primary textile towns of Ghent, Bruges, Brussels, and Antwerp competed aggressively against one another for economic gain. Dutch freighters were doing well against the Hanseatic League when it came to fish, timber, grain, and salt, but the backwater northern provinces (Holland and Zeeland) were largely ignored by a succession of Brussels-based regents—starting with Charles's grandfather Maximilian, his father Philip I (through 1506), his aunt Margaret of Austria, himself (1515–1519), his sister Mary of Hungary, and eventually his son Felipe of Spain (1555–1557).[43]

While Antwerp was not immune to the religious wars, few Antwerpers complained when the Habsburg army was called upon to crush an occasional Protestant uprising. Law and order was good for business. The Habsburg military machine also needed vast quantities of food, clothing, material, arms, and munitions that

Antwerp's merchants were only happy to supply. The surrounding Netherlands, viewed as a buffer zone against Protestant Germany and France, was bolstered with scores of innovative fortress designs, inland riverway defenses, and armament configurations. While these, too, were good for business, they required large amounts of capital that could be financed only by controversial tax hikes and an expansion of the renten system. In 1548, when the Habsburg Netherlands was reestablished officially at Brussels, nearly two-thirds of Brussels's tax receipts were being provided by Antwerp, Flanders, and Brabant, and they weren't very happy about it. Holland, accounting for less than 13 percent of total tax receipts, was required mainly to provide loyalty to Charles V—when neighboring Utrecht and Gelderland revolted against Habsburg rule in 1543, Holland did nothing.[44]

The House of Fugger, despite Charles's financial machinations, was doing very well. In fact, if Antwerp had its own throne, it would have been occupied by Anton Fugger at midcentury. Jacob's nephew had diversified into the Antwerp money market just in time to reap spectacular profits from the transatlantic trade and the massive borrowing needs of Crown and country. In the 1540s, the Fuggers expanded the scale of their public finance transactions with the Habsburg Court at Brussels, the City of Antwerp, and even the English Crown. But even with the family's near-monopoly on the European metals trade, most observers were astounded by the profits revealed in their financial statements of December 31, 1546—the firm's net worth had soared to 3.6 million ducats, or more than double the value reported back in 1527. And although reported profits from Castile and Hungary accounted for 90 percent of the total earned during the 1539–1546 period, virtually all of these profits flowed through the Antwerp office. Antwerp's actual profit results remained a well-kept secret.[45]

The Fuggers had some issues. Some 2.7 million of their 5 million ducats in total assets were "book debts" (claims) that were less than certain. Over half of these claims were due in Spain (mainly to Charles V) and another 800,000 or so were split among the City of Antwerp,

Gaspar Ducci, Ferdinand of Habsburg, and the Crowns of England and Portugal. The Fuggers' 2.3 million ducats in tangible assets were more solid. These included copper and other inventory held mainly in Antwerp (840,000), mines and real estate (560,000), and assorted private partner accounts, cash, and "other" assets (805,000).[46]

The Fuggers borrowed money to make money. They owed nearly seven hundred thousand ducats to creditors in Spain and Antwerp, but their pristine creditworthiness allowed them to borrow on the Antwerp Bourse at around 9 percent and relend the same funds (so-called Fugger Bills) at 12 percent. Even Charles V had to pay up. Fugger loans issued to consolidate the emperor's accumulating unpaid debts carried an interest rate of 12 percent—even with security. When the imperial credit rating had been strong, Charles had found it relatively easy to borrow short-term bills of exchange that were redeemable at Antwerp or at the fairs at Medina del Campo, Lyon, and Besançon. The fairs rarely handled coinage, given the transportation costs and the risk, and had followed Antwerp's lead into bills and royal contracts (*asientos*). But the emperor's borrowing costs had risen with his debt burden, and he was relying increasingly on large asientos to finance his imperial campaigns. Handled mainly by the Fuggers and the Genoese, the asientos were initially less costly than short-term bills, and they limited coinage disbursements to troop wages and balances between nations.[47]

The issuance of asientos had been standard practice in the mining and wool trades for centuries. Royal asientos were eventually used for tax collection ("tax farming"), property development, financial dealings with the military and religious orders, and the provisioning of armies and fleets—often including bullion transfers, currency exchanges, and even conventional loans. In return for issuing a supply contract, the Crown would receive a large advance from the designated contractor at a reasonable rate of interest. Of course, the level of interest became less reasonable when the royal credit rating was under pressure. This would require either issuing new debt to cover the cost of the unpaid accruals, forcibly converting existing high-interest loans into lower-cost juros, or exchanging confiscated private treasure shipments into juros. As the Fuggers knew all too well, Charles was making a regular habit of these eyebrow-raising practices.[48]

Charles's shaky financial position should have been relieved by the resumption of Bolivian silver shipments in 1549, the confiscation of Gonzalo Pizarro's massive estate, and the receipt of record volumes from Zacatecas and Colombia. But when the emperor learned in December 1548 that record quantities of American treasure were finally on their way, he began to plot his next series of campaigns. According to Ramon Carande, the emperor borrowed a whopping 1.2 million ducats per year between 1543 and 1551— nearly twice the level borrowed annually in 1533–1542—before royal borrowings began to spiral out of control. A staggering two million ducats of American treasure was confiscated upon its arrival at Seville in October 1550 and exchanged for juros or applied to imperial debts in Italy and Spain. The moves freed up 951,000 ducats in fresh borrowings for new campaigns against France and Protestantism, as well as support for a papal siege of Parma.[49]

The two-million-ducat windfall from Spanish America was still exceeded by the emperor's 3.5-million-ducat debt position. Charles declined to repay much (if any) of his 1.5 million in unsecured floating-rate debt, a ticking time bomb, and operated as if money was no object. Even prior to receiving news about Gonzalo's execution, Charles had allocated an estimated 364,000 ducats for Felipe's "morale-building" tour through Antwerp and Brussels. The Fuggers, still stuck with huge unpaid claims, must have been concerned about the emperor's intensified campaign against heretical Protestants in Antwerp and the gathering military threat from France. A royal edict of June 1551 banned anyone suspected of heresy from even setting foot in Antwerp. Scores of convicted heretics, mainly Anabaptists, were burned at the stake in Antwerp's market square between 1551 and 1555. If that were not enough, the French threat was so real in 1551 that the City of Antwerp approved a twenty-five-year fortification plan at a cost of one million gold crowns. Since local, state, and imperial budgets were already under considerable pressure, the funding scheme for the new defense program was somewhat vague.[50]

## MILITARY ESCALATIONS

In the meantime, the bonanzas uncovered at Potosí, Zacatecas, and Colombia helped to pay off the arrears on existing Castilian debt and provided security for the next round of royal borrowings. Annual treasure shipments would average two million gold pesos between 1551 and 1555, double the level in 1546–1550, thanks mainly to the gold fields of upper Colombia. Otherwise, American treasure volumes were becoming unpredictable. The highest quality, surface-level silver ores had already been picked off at Potosí and Zacatecas, leaving silver output to decline steadily after 1551. Few magnates were willing to invest in costly mining shafts, drainage systems, and smelting facilities. Coupled with the continuing decline in Central European silver production, the Habsburgs (and their creditors) were being deprived of a large chunk of their projected collateral. Charles V had even lowered the royal quinto in New Spain to a tenth (a diezmo) in 1548 to encourage prospecting activities and to speed up the deployment of productivity-raising mining and smelting techniques.

The metallic windfalls of 1550 and 1551 secured financing for new fortifications in the Netherlands and the recruitment of a massive 148,000-man Habsburg army. Three-fourths of these troops were stationed in the Netherlands and Germany and the remainder in Lombardy, Naples, Sicily, North Africa, and Spain. Warfare in the mid-sixteenth century had changed dramatically from prior periods. Large armies became even larger, fortifications became far more complex, and many of the medieval specialties—archers, crossbowmen, broadswordsmen, and halberdsmen—had nearly disappeared. The change in direction had been signaled by the sixty thousand foot soldiers deployed in Castile's conquest of Granada in 1492—twenty-one years earlier, a Burgundian army under Duke

Charles (the Bold) had relied more heavily on its five thousand archers than its infantrymen. Of course, the survival rate of infantrymen was not necessarily dependent on warfare. Armies in the sixteenth century were as likely to be reduced by disease and desertion as by enemy fire.[51]

Military strategy had also been affected by pikes, muskets, and new artillery. Swiss military strategists had discovered that soldiers equipped with fifteen-foot-long iron pikes (pikemen) could repulse a traditional cavalry attack and protect an emerging class of muske-teers—musket-bearing Aragonese troops (*arcabuceros*) were capable of firing a twenty-ounce lead shot at three hundred feet. Coupled with the introduction of powerful artillery, like the forty siege cannons carried to the Italian front by French king Charles VIII in 1494–1495, cavalry charges were replaced by siegecraft (and their defense). The ratio of cavalry to infantry flip-flopped from a traditional 2:1 to 1:3. Italian engineers responded to the larger armies and the improved artillery with the trace italienne. The trace was an extremely expensive antisiege design that comprised earth-supported fortifications, triangular bastions, a surrounding moat, and wall-mounted artillery. It worked. Henri II captured Metz with thirty-six thousand infantrymen in pre-trace 1552, but Charles V would discover shortly that a Habsburg army of fifty-seven thou-sand was incapable of taking post-trace Metz in 1553.[52]

Traces were extremely expensive. A papal plan to refortify Rome with eighteen bastions was scrapped in 1542 because of the cost. But money was no object to Charles V. He hired a team of Italian fortifi-cation engineers in 1544 to construct a network of fifteen fortresses, backed by 1,012 artillery pieces, to secure the Netherlands' border line. He also planned to surround Antwerp with nine bastions and five massive gates. The cost of these two projects was 660,000 ducats. The fortification of the Netherlands would turn out to be a work-in-progress for the next sixty years. By 1572, a staggering 4.4 million ducats had been spent on nearly twenty-seven miles of modern forti-fications, four large citadels, and twelve new circuits. Annual expen-ditures had come in at roughly one hundred thousand.[53]

Charles V also upgraded the Habsburg army. No one in Europe fielded a regular standing army, so the emperor was forced to compete for the best mercenaries. While the French continued to pay up for skilled Swiss pikemen, the Habsburgs favored battle-tested Spaniards and *Landsknechte* troops from Bavaria. Infantries were expensive. The monthly cost of a typical Habsburg company of 279 men had increased from around three ducats per man in 1536 to at least four ducats by 1553. Brussels was paying up to six ducats for foot soldiers and as much as thirteen for cavalrymen. Then there were the expenses for officer pay, transportation, food, provisions, baggage, training, and the deployment of artillery—comprising over one-third of a campaign's costs—and the added expense to equip (and lavish) a royal entourage when Charles commanded a campaign in person. In general, the cost of cannon, muskets, pikes, and swords were huge. So were the charges to transfer funds from one domain to another—exchange fees and currency charges were at least 15 percent of the principal amount.[54]

The fifty-one-year-old emperor had additional military concerns in early 1552. Germans of all political stripes were outraged by his succession plan, approved by Pope Julius III, to transfer Habsburg Germany, Habsburg Austria, and the Holy Roman Emperorship to his son Felipe. The Germans refused to accept a Spanish monarch, and Ferdinand was apparently happy where he was. Outrage turned to warfare in January 1552 when Duke Moritz of Saxony, an erstwhile ally of Charles V, struck a deal with Henri II. Henri agreed to contribute seventy thousand gold crowns monthly to Moritz and his German allies in return for the rights to French-speaking Cambrai, Metz, and Verdun. Henri immediately led an army of thirty-six thousand down the Moselle River to seize pre-trace Metz in April 1552. He also raised the stakes by providing military support to Suleyman's assault on Transylvania (Hungary). Ferdinand had attempted to take the neutral province in 1550.[55]

These tit-for-tat actions set off the mutually destructive Habsburg-Valois Wars of 1552–1559. If the French army was manageable on the battlefield, Charles V had a real problem on his hands now that the formidable Moritz had joined the Schmalkaldean League. The emperor relocated to Innsbruck to organize a new army knowing that the recent American windfalls were long gone and that his principal lenders were tapped out. Moritz attacked Habsburg forces at Innsbruck in May 1552 and forced a reeling Charles V to withdraw to the Fugger-friendly town of Villach. Moritz was willing to deal from strength. He invited Ferdinand to broker the face-saving Treaty of Passau on August 15, 1552. The treaty paid Moritz to go away, released John Frederick of Saxony from his five years of captivity, and, most importantly, recognized Protestantism in the Holy Roman Empire. The last point was huge. It meant that the emperor's thirty-year war against Protestantism had squandered millions of ducats and tens of thousands of lives for nothing. The recognition of Protestantism was reconfirmed by the Peace of Augsburg on September 21, 1555.[56]

Prior to his retreat to Villach, Charles had met privately with Anton Fugger to convince the wealthiest man in Europe that the fate of the Habsburg Empire hung in the balance. A sizable amount of capital was required to bribe Duke Moritz of Saxony back to the Habsburg fold, respond to the French threat at Metz, and support a failing siege at Parma. Fortunately for Charles, the arrival in Antwerp of a large (if disappointing) American treasure shipment— well below the windfalls received in 1550 and 1551—triggered a 400,000-ducat loan (at 12 percent) from the Fuggers in November 1552 and a modest advance from the cautious Genoese. Most of these funds were delivered to Villach (via Brussels) and made payable in Spain. At this point, the Fuggers preferred to be repaid in Spain to reduce the likelihood that their share of the American treasure was not confiscated en route to Antwerp. They also expected to realize a handsome profit on the currency transfers.[57]

Moritz of Saxony was one thing; France was another. If the Treaty of Passau and the failed siege at Parma were blows to Charles's prestige, the French capture of Metz threatened the entire Netherlands. True to form, the emperor responded to these humiliations with an offensive of his own in September 1552. The financing would be sorted out at a later date. No matter that the citizenry of Metz had surrendered willingly to the French in April and that six thousand French troops were fortifying the city with an Italian-style trace. The new trace was capable of neutralizing the fifty-seven thousand Habsburg troops and scores of cannons that Charles V sent to Metz in early November. Metz was one of the costliest sieges in European history. It was ultimately abandoned on January 1, 1554, after it had saddled the already-broke emperor with another staggering bill for 3.3 million ducats. The attempted power play to promote Felipe as Habsburg ruler of Germany, Austria, and the Holy Roman Empire had been a very big mistake. It would get even worse.[58]

Metz was Charles's downfall. His leadership there was the last of nine personal commands that he assumed between 1529 and 1552. Prior to Metz, the emperor had led Habsburg armies in Italy (1529–1530), along the Danube (1532), Tunis (1535), Provence (1536), Algiers (1541), Germany and France (1543–1544), and Germany (1546–1547). Only the first three campaigns could be categorized as successes. Three were great failures (Provence, Algiers, and Metz). Success or failure, campaigns commanded personally by Charles V were twice as expensive as those in which the emperor stayed at home. His personal campaigns against France and German Protestantism totaled 13.6 million ducats over a ten-year period.[59]

The failed Siege of Metz, coupled with the 700,000 ducats squandered at Villach and at Parma, added four million ducats to the emperor's preexisting debt load. The Castilian treasury was forced to deal with 3.7 million ducats in new, restructured, or refinanced debt in 1552. Nearly half of this total was restructured by the Fuggers, Schetzes, Dorias, and Spinolas on the condition that there would be no treasure confiscations for one year. Cash proceeds from the disappointing quinto of 1552 (158,000), the sale of public offices, and a donation by fifteen Spanish grandees totaled less than 500,000. In addition, Charles's short-sighted tax collection deals with the Castilian Cortes back in 1537 had come back to haunt him. The municipalities of Castile had patriotically supported Charles's foreign adventures when they had many chances to do otherwise.

But the fixed quota system, backed theoretically by alcabala and tercia taxes collected at the local level, was incapable of servicing the rising tide of juros. Prior to Metz, half of Castile's one million ducats in projected revenues for 1552 and 1553 had been prepledged to secure the state debt. After Metz, all of Castile's projected revenues were prepledged through 1555.[60]

The news wasn't all bad. The prepledging of Castilian revenues for 1553–1555 had assumed that only one million ducats in American treasure would be delivered to Seville during the entire three-year period. When treasure imports came in at an unexpected two million ducats in 1552–1553, thanks again to the volatile gold fields of Colombia, the emperor had a new source of unpledged assets to confiscate. He applied 600,000 ducats in confiscated treasure and a higher-than-expected quinto to retire a substantial amount of debt issued in 1552. However, the Genoese, relative newcomers when it came to Habsburgian finance, were outraged by the forced exchange of confiscated treasure for increasingly suspect Castilian juros. António Maria Grimaldi and Domenico Grillo refused to give up their 250,000 ducats in American treasure and may have gotten away with it.[61]

The volume of treasure imports ebbed and flowed unpredictably between 1554 and 1559. This made financial planning nearly impossible. The 4.3 million ducats in debt due in Castile in 1554 included a large chunk of the sums squandered at Metz and 878,210 ducats in accrued (unpaid) interest expense. This was on top of a jaw-dropping 17.6 million ducats in accumulating juro debt. The pattern of confiscation, conversion, and prepledging continued. In Maximilian fashion, Charles prepledged all of Castile's projected revenues (including American treasure) through 1559 and forcibly converted huge amounts of high-coupon debt into 5–7 percent juros. The newly issued juros were increasingly suspect. After Erasmus Schetz had been forced to convert half of his 521,760-ducat claim into nearly worthless juros, the tapped-out Fuggers and Genoese declared that they would accept only (nonconfiscatable) American treasure as security.[62]

The accumulating avalanche of imperial debt prompted the

emperor to prepare for an early retirement. Of course, he planned to go out in style. In January 1555, his twenty-eight-year-old son Felipe was dispatched secretly from Spain with two hundred thousand ducats in silver bullion to arrange a typically lavish (and debt-defying) abdication ceremony in Brussels. Felipe, the heir apparent to the largest empire in Europe, had been married to the thirty-eight-year-old Catholic queen of England, Mary Tudor, in July 1554. The typically arranged marriage carried the potential to pay huge dividends politically, financially, and religiously. England had been a Habsburg ally against France since the Treaty of Canterbury in 1520, her annual trade with Antwerp was pegged at nearly twelve million gold crowns, and Thomas Gresham had helped to solidify England's shaky royal finances. Secured by English tin, lead, and textiles, Gresham had actually repaid 300,000 ducats in loans from the Fuggers and Genoese in 1553. He went on to reform England's chaotic currency system, establish the London Exchange based on the Antwerp model, and earn a knighthood—all after Mary had died unexpectedly in 1558 and the English throne was passed to the Protestant Elizabeth.[63]

Felipe was married to the queen of England when he assumed sovereignty over the Habsburg Netherlands on October 25, 1555. Of course, he was overwhelmed by a financial situation that had become increasingly dire. The ongoing war with France was sinking a never-ending stream of funds into the Habsburgs' armed forces and raising fears of a widespread mutiny. The financial condition of the Habsburg Empire was so bleak (and so challenging to decipher) that an unexpected silver strike made at Guadalcanal (Spain) in the summer of 1555 was hushed up. When Felipe II assumed the Spanish throne on January 15, 1556—giving him sovereignty over Spain, Spanish America, the Netherlands, Naples, Sicily, and Milan—he discovered that all of Spain's revenue streams had pre-pledged through 1561 and annual short-term borrowings were exceeding four million. Some of these loans carried interest rates as high as 49 percent. In the Netherlands, short-term loans charged against the Brussels treasury in 1555 reached 2.9 million ducats on top of a growing mountain of long-term debt. An escalation in the French campaign and another decline in American treasure shipments had combined to induce a severe financial crisis.[64]

Attempts were made to transfer some of the emperor's stupendous debts in Castile to the taxpayers of the Netherlands. The crisis years of 1542–1544 had been bridged with extraordinary subsidies and the sale of renten. The process was repeated in Brussels during the 1552–1555 period. The issuance of 847,000 ducats in renten backed by excise and property taxes might even have succeeded if Charles and Felipe had not sought to stick Brussels with some of Castile's high-coupon debt and juros. The outstanding debt in Brussels had already skyrocketed to nearly five million ducats, nearly five times the level in 1545, when Felipe assumed power in October 1555. Then it got worse. As much as 8.5 million ducats in Castilian debt—mainly juros, but also high-coupon (12 percent) loans that had been targeted for forced conversion—was sent north to Brussels between 1556 and 1559. Most of these loans were as unserviceable in the Netherlands as they were in Castile.[65]

The mounting chaos put the Fuggers to the test. Anton Fugger was getting along in years, and his continuing loyal support of Felipe II would have shocked his tight-fisted Uncle Jacob. He and his principal factor in Antwerp, Mathias Oertel, kept the Crown afloat by advancing nearly 1.5 million ducats in loans between February 1556 and January 1557. Some 652,000 ducats (at 12 percent) in loans, guaranteed in theory by the States General of the Netherlands and a group of wealthy state officials (as individuals), were allocated to the nearly mutinous Spanish troops fighting in the Netherlands. Another 600,000 crowns were advanced (at 23 percent) to cover unpaid salaries at the royal court in Brussels and at the front. The crowns were secured by vague streams of personal income from Felipe II and other state officials. Yet another loan of 430,000 ducats was secured by the next shipment of silver from the Americas—prepledged or not.[66]

Anton Fugger was supporting an untenable financial position. If Oertel was the catalyst for most of these transactions (or just the scapegoat), he had thrown a lot of good money after bad. Felipe's credit was finally tapped out in the Netherlands when the Rentenmasters halted all interest payments on floating-rate debt obligations in January 1557. Within three months, he was forced to suspend all quarterly repayments on outstanding debt obligations in Castile. Then he shocked everyone by declaring a state bankruptcy in June. Since Brussels owed massive amounts of floating-rate debt in Antwerp, the impact was devastating to the Fuggers and scores of lesser-heeled merchant bankers. Felipe immediately confiscated two more shipments of American treasure (worth 570,000 ducats) that had been bound for the Fuggers' warehouse in Antwerp. The Fuggers were promised that they would be repaid (at 14 percent) by June 1559. When the promise was broken, Oertel was sacked.[67]

The Spanish state bankruptcy of June 1557 called for a massive work-out plan. Royal creditors were forced to write down the carrying value of their outstanding claims and accept repayment in low-coupon (5 percent) juros or juros that sold at a discount from their face value. The conversion of an avalanche of high-interest, short-term debt into long-term *juros al quitar* hit unsecured short-term debt holders especially hard. And the Cortes would still have to approve a tax hike to service the new juros. Royal creditors had no choice but to submit. The shell-shocked Fuggers and a Genoese syndicate had each advanced a total of seven million ducats to the Habsburg Empire between 1521 and 1557, followed by the Welsers (4.2 million) and Schetzes (1.3). The financial chaos continued through the Habsburg triumph at the Battle of St. Quentin in August 1557 and the signing of the Treaty of Cateau-Cambresis in April 1559. Henri II had already declared a state bankruptcy of his own, wiping out thousands of patriotic French investors who had subscribed to the *Grand Parti*. Felipe II would have enjoyed his rival's financial pain if was he was not forced to declare a second Spanish state bankruptcy in 1560 to work out the remaining hangover debts.[68]

The House of Fugger barely survived the Spanish state bankruptcies of 1557–1560 and the associated financial carnage in Brussels and Antwerp. Their claim in the Netherlands—standing at three million ducats in 1558, with 2.2 million held against Felipe II and another 800,000 held against the States General—was matched by an equivalent three-million-ducat claim in Castile by December 1560. Assuming that the raft of postbankruptcy juros issued to the Fuggers, Welsers, Genoese et al. were worth half of their outstanding claims, the House of Fugger may have lost over three million ducats between 1557 and 1560. Scores of other Antwerp-based creditors, including Gaspar Ducci, were wiped out completely. But the Fuggers were not the wealthiest merchant bankers in Europe for nothing. Anton Fugger had directed a carefully planned distribution in 1553 to transfer half of the firm's net worth (1.4 million ducats) to himself and his familial partners as individuals. While the distribution left the corporate House of Fugger with a net worth of only 1.4 million ducats, a position that should have been overwhelmed by a three million write-down in 1557–1560, the position was supported by the partners' individual assets. The Fuggers could still borrow on the Antwerp Bourse at rates (9–10 percent) that were only modestly higher than the 7–8 percent rates received in happier days.[69]

Anton Fugger, the world's wealthiest nonmonarch, left a spectacular personal estate of 3.6 million ducats at his death on September 14, 1560. His series of partial (and secretive) distributions suggest that he had anticipated the financial collapse of 1557. But the merchant banking colossus that had been shaped by his uncle Jacob faced a very shaky future. Huge amounts of Spanish juros and bonds issued by the Netherlands receiver general had been written off, and Anton's heirs, nephews Max, Hans, and Jacob Fugger, were saddled with three million ducats in claims against Felipe II. The House of Fugger was essentially finished as a financial power. The firm would have to work its way through the current chaos as well as the Spanish state bankruptcies of 1575, 1596, and 1607—as much as eight million ducats in unpaid loans to the House of Habsburg were written off between 1519 and 1607.[70]

The primary beneficiary of the Fuggers' largesse, Charles V, had stuck his son Felipe II with a royal financial mess. This mess was many times worse than the one that nineteen-year-old Charles of

Habsburg had inherited from Maximilian in 1519. Charles did not go out on top. After transferring the Holy Roman Emperorship (reluctantly) to his brother Ferdinand in September 1556, the most powerful ex-ruler in Europe retired to a monastery in Yuste, in northwestern Spain, and died there on September 21, 1558. He was only fifty-eight. What should have been an extremely momentous death was forced to compete with the financial chaos of 1557–1560; the passing of Felipe's wife, Mary Tudor, on November 17, 1558; the signing of the Treaty of Cateau-Cambresis with France on April 3, 1559; and the fatal wounding of forty-year-old Henri II on July 10, 1559, during a joust. While Mary's death disengaged the Anglo-Spanish alliance against France, the alliance was no longer needed. By year-end, Felipe had married Elisabeth of Valois, a daughter of Henri II and Catherine de Medici, and was free to assume his duties as King Felipe II of Spain.[71]

# CHAPTER SIX

# MINING REVOLUTION IN SPANISH AMERICA

W hen Cabeza de Vaca passed away in Seville in the bankruptcy year of 1557, one of the most attractive characters in Spanish American history died knowing that his mercenary instincts, and those of some of his adventuresome predecessors, had been proven correct. There *had* been a "mountain of silver" to be found in Bolivia. Within months of de Vaca's departure from Rio de la Plata, two local natives uncovered one of the largest silver bonanzas in world history at Cerro de Potosí. The bonanza was in easy reach if only de Vaca had followed the Pilcomayo River up from Asunción. Yet the strike at Cerro de Potosí in April 1545 was not one of the fabled El Dorados—there were no El Dorados to be found in Spanish America—nor was it a collection of ancient ceremonial treasures. It was essentially a huge Joachimstahl, a massive conventional lode of silver that was very familiar to people like Anton Fugger, Bartholeméw Welser, and Charles of Habsburg.

It was small consolation to de Vaca that he had fared better than Ayolas in Paraguay, Pero in Brazil, Alfinger in Venezuela, Orellana in Amazonia, and de Soto in Florida. Faced with sickness, mutiny, imprisonment, and temporary banishment, de Vaca had at least escaped with his life. Treasure hunting on behalf of the Crown was a dangerous business. So was treasure finding. Cortés had also managed to survive, but the Pizarro brothers were either assassinated (Francisco), executed (Gonzalo), slain in battle (Juan), or imprisoned (Hernán). Yet these same conquistadors had provided the huge volumes of treasure that were squandered on the battlefields and sea lanes of Europe. The great disconnect was that borrowings secured by the plunder of Tenochtitlan, Cajamarca, and Cuzco between 1520 and 1534 and the development of genuine American mining industries in the late 1530s had resulted in a Spanish state bankruptcy in 1557. The more money Charles V received, the more he spent. It had also been Cortés and the Pizarros, in their attempts to make the transition from bloodthirsty plunderers to respectable mining magnates, who had launched genuine silver mining industries in New Spain and Bolivia. Their early efforts at Sultepec, Taxco, and Porco were linked critically to the unexpected discoveries at Zacatecas and Potosí in 1545–1546, the mining revolution that followed in 1555–1574, and the even larger volumes of silver that were squandered by Felipe II during the second half of the sixteenth century.

The Spanish state bankruptcy of 1557–1560 had been influenced in part by unpredictable treasure shipments and a steep decline in mining revenues from Central Europe. These regions had provided Charles V and his creditors with the hardest of hard assets to collateralize the imperial borrowings. Unfortunately for them, the mining expansion in New Spain, Bolivia, and Colombia was irregular and prone to the same type of boom-and-bust cycles that had characterized mining-centric empires since the days of ancient Sumer. In addition, the Colombian boom and the depletion of the rich surface-level deposits at Potosí and Zacatecas had tilted the value (not volume) of American metal production in gold's favor. The yellow metal was valued at at least eleven times silver, meaning that two tons of Colombian gold was worth more than the twenty-two tons of Spanish American silver exported annually during the 1540s. Spanish American mining output in 1560 was back to less than half of the annual volume produced in Habsburg Europe—and the great mines of Central Europe had even been on a downward slope since 1540. Charles V may have been reluctant to accelerate the transfer of European mining and refining technology across the Atlantic. Gold was the preferred metal, the Central European industry was in decline, and there was no need to pressure European silver prices any further.[1]

Charles V's failure to elect his son to the Holy Roman Emperorship spared Felipe II from the demise of the great Habsburg mines of Central Europe. The future was in the Spanish Americas, not Habsburg Austria, and even an unexpected silver strike at Guadalcanal in Spain—adding nearly nine tons per year between 1555 and 1563—was a short-term event. The influx of American silver pressured silver prices in Europe and forced a number of marginal works to be abandoned. While this raised the stakes in the volatile mining districts of Spanish America, Felipe II had little available funds to support them. The task was left essentially to private enterprise. Thus, the developments, improvements, and military defenses that did occur in Spanish America were typically structured as reward/risk incentives with interested parties. The Crown had little capital but a lot of power. The viceroys of New Spain and Peru could grant licenses, concessions, and land to entrepreneurial individuals who believed that the investment opportunity outweighed the short-term risk. A mere fifteen thousand Spaniards had immigrated officially to the Americas prior to 1550, but another eighty-five thousand or so unofficial Spaniards were participating in the epic gold, silver, and land rushes. Royal restrictions were unenforceable and had rarely been applied to associates of the Fuggers, Welsers, and Genoese.[2]

The Spanish American mining revolution that transpired between 1555 and 1574 was unexpected. It was also built atop a rickety foundation. Prior to 1546, the Mexican silver mining industry could only be described as primitive compared with the great mines of Central Europe and the Balkans. Sultepec and Taxco required little capital and technology during their early years. After the easily oxidized surface-level deposits had been excavated, Indians were hired to identify possible silver-bearing veins and to evaluate (assay) the quality of the deeper ores. The earliest mining shafts rarely exceeded eighty feet in depth, excluding the 213-foot (thirty-*estado*) shaft sunk at Taxco in 1548, and costly drainage tunnels were avoided in favor of primitive adits (*socabones*). Well-engineered tunnels, adits, and mining shafts would become mandatory in later centuries. One at La Valenciana (Guanajuato) would reach a depth of nearly 1,700 feet.[3]

Few of Mexico's deep-seated ores contained significant amounts of native silver, but there were large quantities of silver-bearing ores—silver chlorides (cerargyrites), lead carbonates (cerussite), and silver sulfides (argentite)—that were combined with sulfur and other metals. These base metals had to be separated and smelted. The Welsers had dispatched a team of German metallurgists to Sultepec and Taxco in the early 1530s to supervise the process through which lead or lead oxides (litharge) were added as flux during the smelting and refining stages. A single mine at Taxco consumed twenty-five quintals of litharge in order to refine a mere one hundred ounces of silver in 1539. The Germans at Sultepec, a colony that included a few ex-employees of the House of Fugger, helped to direct the construction of stamping mills and smelting furnaces starting in 1536. Otherwise, underground mining and smelting furnaces of the Central European variety were viewed as too costly and too complicated. There were few professionals (German or otherwise) on hand to supervise them.[4]

Mining magnates (*mineros*) like Hernán Cortés improvised with low-technology, labor-intensive techniques. Aside from a single Saxon-styled water-power system constructed at Sultepec in 1543, most mineros relied on hand-powered vertical winches (*malacates*) and mule-powered hoisting, bailing, and milling. Candle-guided mineworkers (*tenateros*) negotiated rickety, notched-log ladders with at least one hundred pounds of excavated ore strapped to their backs in rawhide bags (*tenates*). Mining supplies that were metal-based in Europe were fabricated from local materials in Mexico— these included wooden buckets, tubs (*canoas*), and bowls (*bateaus*), calabash tree gourds (*jicaras*), rawhide sacks, wicker baskets, cloth screens, and sieves (*tilmas*) made from agave fibers. In later years, after iron bars, tongs, and platters had been introduced, volatile mercury was still being carried in rawhide sacks and animal skins. Alexander von Humboldt categorized the Mexican mining system as a *sistema de la rata* ("rat-hole system") as late as 1800.[5]

It was the mineros who introduced cupellation to the New World. The ancient smelting process had been developed by the Sumerians in 3000 BCE and spread pretty much unchanged to succeeding generations of fellow Near Easterners, Muslim Caliphates, Saxons, and even the Chinese. The Near Easterners reduced the lead component of a smelted silver-lead alloy by blowing a stream of air into the molten mix. The temperature-raising air stream created a recoverable oxide of lead (litharge) and left behind a separated component of silver in the cupel. First-millennium writers like Al-Hamdani described silver smelting under the Muslim caliphates but made no mention of their need to add lead to the refining process. This omission implied that the silver-bearing ores of Turkestan, the primary source of Baghdad's silver for centuries, already contained sufficient amounts of lead. This was also the case at Sultepec and Taxco. Central Europe's silver-bearing ore deposits were apparently split between leaded and lead-free zones, a feature that helps to explain why the duke of Saxony and the Venetians had been interested in tweaking (not replacing) the ancient process of cupellation.[6]

The smelting process was relatively straightforward. After teams of Indians had washed, separated, and hammered the excavated ore, the broken pieces were crushed by a circular mule-drawn system borrowed from Europe's olive oil industry. The finely crushed ores were mixed with charcoal and litharge flux and heated in small clay (or stone) furnaces to produce a silver-lead alloy. Prior to the introduction of bellows (made from sheep and goat skins), American furnaces were either perforated to receive "katabatic" winds—the Bolivian word for furnace (*huayra*) meant "wind" in the native Quechua language—or heated by human blowers. The introduction of the cupellation process meant that the alloy would be cupelled in a second furnace to separate the silver from the litharge. Large numbers of native Mexicans (and Bolivians) opened backyard cupellation furnaces of their own in return for a profit-sharing agreement (a *pepena* or *partido*) with the mine owner. Unfortunately, they discovered also that litharge flux was costly, damaged the furnace walls, and emitted poisonous fumes.[7]

Mexico's silver-bearing ores were extracted, hauled, crushed, and smelted by a combination of Indian workers—free, forced, and enslaved. During the 1530s and 1540s, most Mexican mines were worked by encomienda Indians who were delivered to the mine owner as tribute. The practice was little changed from the Aztec system. Encomienda Indians were usually assigned to less-difficult jobs like chopping wood for charcoal, building facilities, or carrying food and supplies. They typically worked in two-week shifts. Slave labor was reserved for the more challenging work of excavating and smelting. Charles V's New Laws of 1542 attempted to abolish slavery in New Spain, but the laws were usually ignored and European diseases were devastating the native population. The prices paid for slave labor soared. A slave that could be had for seven pesos in 1529 might cost as much as fifty pesos in 1536. Cortés and his fellow mineowners in Sultepec and Taxco owned hundreds of slaves until disease and extremely harsh working conditions began to take their toll. After 1550, free (paid) native labor and imported African slaves comprised the bulk of Mexico's workforce. As many as eight hundred African slaves were working in Taxco in 1569.[8]

Although front-end capital costs were relatively low, the profit returns to early entrepreneurs like Cortés, de la Sosa, Castilla, and Oñate are unclear. The challenge to organize and import labor, food, clothing, fuel, smelting lead, mining supplies, and equipment was daunting. Food consisted mainly of maize, beans, and peppers until cattle ranches (haçiendas) were established in the north. Wood and charcoal fuel were delivered as tribute from local villages until regional timber supplies were depleted. Smaller-scale operators eked out a living by financing their supply needs with loans or participations. If a mine failed to produce as expected, a miner or a creditor could be wiped out entirely.[9]

The Mexican mining system was still in its early stages when Cristóbal de Oñate, the lieutenant governor of New Galicia, helped to uncover modest deposits of silver near Guadalajara in 1543. The discovery, in turn, stimulated exploration activities in the uncharted territory north of the Chichimeca line. Prospecting activities came up empty until 1546. In that year, a fellow Basque frontiersman named Juan de Tolosa led a small party of Spaniards and Indian allies on a 150-mile northeasterly trek from Guadalajara to the barren wilds of Zacatecas. Tolosa's expectations were limited, but he could not help noticing that the local Zacatecan Indians were trading silver ore in routine barter exchanges. When the ore was linked to deposits located near Tolosa's campsite below Cerro de la Bufa, Tolosa's discovery of September 8, 1546, unleashed the greatest silver rush in Mexican history. No matter that Zacatecas was 350 miles away from civilization (Mexico City), the quality of the initial ores was poor—nowhere near the purity of Potosí's early output—and the surrounding region was extremely dangerous. However, the silver strike at Zacatecas would dwarf anything found in Mexico during the sixteenth century and would catalyze an exploration boom in the Chichimeca territories to the north.[10]

Diego de Ibarra, who cofounded Zacatecas with Tolosa, Oñate, and Baltasar Temiño de Bañuelas, provided most of the leadership in the early years. The town of Zacatecas was established on January 20, 1548, followed by the discovery of rich (second-phase) ores in the surrounding hills of Serrania. These were the ores that would make the city's fortune. Zacatecas's three primary veins (*vetas*) were discovered among the exposed outcrops (*crestones*) of the Serrania de Zacatecas during the course of 1548—La Albarrada (within the larger Veta Grande vein), San Bernabe, and the first of the so-called Panuco mines. The latter were uncovered by Diego Ibarra himself. These three discoveries, located within an eight-mile radius of Zacatecas, helped to raise total silver output in New Spain from 112,000 marks in 1546 to 190,000 marks (1.6 million pesos) in 1548. These volumes were especially prized because Gonzalo Pizarro had suspended Bolivian treasure shipments to Charles V.[11]

With future confiscations in mind, Charles reduced the onerous quinto to a royal tenth (diezmo) in 1548 in order to catalyze exploration activities in Mexico. The full quinto remained in place in Colombia and Peru. The royal take had been a fact of economic life since Alfonso XI introduced the quinto in 1386 under the concept of "fundamental perpetual law." Any mine located within the Crown's dominions belonged exclusively to the Crown. On February 5, 1504, Isabella and Ferdinand instructed American miners that they, too, were obliged to contribute the fifth part of any metals extracted, free of all cost to the Crown. A long list of royal rules and regulations followed. In September 1519, Charles V decreed that pieces of gold (or silver) assayed by royal inspectors must be marked with a royal stamp to signify whether the metal was intended as barter or bullion. By May 1522, quinto rules and regulations had become so daunting that Hernán Cortés hired his own men to manage the royal quinto. A decree in July 1543 required that a "hall-mark" be stamped on metal bars to guarantee their value and to certify that the royal quinto had been paid.[12]

The royal diezmo and labor were only two of the many costly items at Zacatecas. While the silver-bearing galena ores of Sultepec and Taxco already contained ample amounts of lead, Zacatecas's ores were lead-free and required a lead additive. English lead had to be shipped to Seville, carried across the Atlantic to Veracruz as ballast, and hauled up to Zacatecas. At a certain temperature, the lead-additive combined with the ore's silver component to produce an intermediary silver-lead compound that could be separated by cupellation. Unfortunately for seller and buyer, the cost of shipping lead between England and Zacatecas was prohibitive from the start. English smelting lead would eventually be limited to the fading works in Central Europe and occasional opportunities like Guadalcanal in Spain.[13]

Zacatecas's reliance on English lead became more prohibitive after the surface-level ores were depleted in 1550. Based on the volumes registered officially in Mexico City, silver production in New Spain appears to have declined between 1549 and 1556. This after a steady fifteen-year expansion. As the royal share of Mexican silver output fell from 45,500 marks in 1549 to just 16,000 in 1553, a viceregal report of March 8, 1552, warned that mines that had produced four or more marks of silver per quintal were now yielding less than one mark per quintal. These deeper ores could be smelted only with costly lead and litharge (flux). As with Potosí, profit margins were being sandwiched between falling yields and the cost of imported lead, charcoal, wood, and labor. The hills of Zacatecas had already been stripped of their timber, and few of Zacatecas's mineros were prepared to invest in costly mining shafts and drainage systems. In addition, Zacatecas was affected by a growing labor shortage—the native American population was being decimated by disease, extreme working conditions, and occasional warfare. Viceregal officials were also complaining about the high cost of African slaves (180–200 pesos) and an ill-advised change in tribute payments from food to cash—most Indians were broke.[14]

This was cause for alarm. The bountiful treasure ships of 1552–1553 were bolstered by the recent gold strikes at Buriticá and Popayán, two Colombian gold strikes that masked the sharp declines in New Spain and Bolivia. Most of the high-quality surface-level ores had been excavated, the cost to excavate deeper-seated ores was prohibitive, and the addition of lead (or lead additives) to the smelting mix was either expensive or problematic. The basic problem, as poorly understood as it was, was that the silver-bearing ores of Spanish America varied greatly in their geological properties. Standard smelting processes that had worked well in Central Europe, the Balkans, and even Sultepec were not necessarily suited to the increasingly low-quality ores being mined at Zacataecas and Potosí. Since native American metallurgists had rarely mined or smelted ores deeper than one hundred feet, it appeared that new types of smelting processes were needed to better exploit the two largest silver works in the New World. One of these new processes, silver amalgamation, would make history.

Those who grappled with the metallurgical problem at least had the benefit of secretive Saxon know-how. The widespread introduction of printing presses had led to the publication of the first book on mining geology, *Eyn Nutzlich Bergbuchlin*, by Ulrich Ruhlein von Kalbe (Calbus of Freiburg) in Augsburg in 1505. This was followed by a series of influential mining pamphlets (*Probierbuchlin*). The real breakthroughs, however, were Vanoccio Biringuccio's *De*

*Re Pirotechnia* in 1540 and Georgius Agricola's *De Re Metallica* in 1550—these works were decisive to anyone involved with mysterious smelting processes like cupellation and even amalgamation. While Biringuccio was most interested in applying metalworking techniques to the manufacture of artillery and munitions, he provided the first (vague) comments on the application of mercury to silver amalgamation and liquation. He stated that washed (concentrated) silver-bearing ores may be reduced either by mercury or lead and even described an early version of the so-called patio process. Biringuccio also commented on reverberating furnaces. Agricola's *De Re Metallica*, completed in the Erzgebirge mining town of Chemnitz between 1533 and 1550, made no mention of silver amalgamation at all but it filled in many of the technical holes in Biringuccio's work. Sponsored by Duke Moritz of Saxony, the work was regarded as the definitive handbook on mining practices until the publication of Schluter's *Metallurgy* in 1738.[15]

## BARTOLOMÉ DE MEDINA

*De Re Metallica* was published just in time to educate a new generation of miners, metallurgists, and entrepreneurs in Spanish America. Although Pliny had claimed that the Romans had used mercury to separate gold by amalgamation, no one had mentioned the use of mercury to separate silver until Biringuccio's description in 1540 and Agricola's stamp of credibility. Biringuccio's work suggested that a rudimentary silver amalgamation process may have been practiced in Central Europe prior to 1540. This was apparently a cold process that added mercury, common salt, and copper-salt to the silver-bearing ores and only applied heat to remove the mercury from the pressed amalgam. It would be left to a Seville-based textile merchant named Bartolomé de Medina, a pure amateur in the fields of metallurgy and chemical engineering, to introduce a silver amalgamation process in the New World. Medina's trial-and-error experiments in Pachuca, Mexico, between 1554 and

1555 would uncover the missing link—the addition of copper pyrites (magistral) to the mix—to convert a murky textbook concept into arguably the most profound technological innovation of the sixteenth century.[16]

Few if any of Medina's predecessors were familiar with Biringuccio's thesis that silver-bearing ores could be amalgamated with mercury if they existed either in the metallic state or were found in combination with either sulfides or chlorides. The thesis was apparently unclear to Johann Enchel and the other German mining engineers who were recruited to New Spain in the 1530s under a royal decree. The Germans were most comfortable with lead-additive processes like cupellation and Thurzo's Saiger separation system. In Central Europe, roughly 2,400 tons of lead had been required to process 4,800 tons of Saiger copper annually during the 1530s. But when Enchel arrived at Sultepec in 1535, he found that the local silver deposits (and those at Taxco) were already combined with lead-rich galenas. They didn't need lead. This suggests that the lead-additive smelting process that found its way to Zacatecas in 1546—a process that consumed either English lead, local *greta* (litharge), or *centrada* ("hearth-lead")—may have actually been imported from Hernán Pizarro's earlier works at lead-free Porco.[17]

Bartolomé Medina's silver amalgamation process did not hinge on methods described in Biringuccio and Agricola, but Medina must have been aware of metallurgical concepts that had been transferred to the Americas by the Fuggers, Welsers, relocated mining engineers like Johann Enchel, and possibly even Hernán Pizarro. Medina's home town of Seville was a hotbed of transatlantic trading schemes at midcentury. If it was nowhere near Antwerp, Istanbul, or Venice in economic status, Seville had surpassed Toledo and fair-based Medina del Campo in Spain and rivaled Lisbon as the business capital of Iberia. The transatlantic trade had created a new class of wealthy merchants, including the Medinas, among an expanding population of around sixty thousand. No matter that Seville's business climate was clouded by Charles V's suspect creditworthiness, the risk of royal confiscations, and a Castilian treasury that was held hostage to the gnomes of Antwerp, Augsburg, and Genoa. An entrepreneur like Medina faced daunting business challenges in 1550.[18]

The forty-six-year-old Medina resided in Seville's fashionable Santa Maria district and prospered as a textile wholesaler. Like most of Seville's merchant class, he purchased most of his textile products on credit and discovered that a strong credit rating allowed one to apply the wholesaling concept to a variety of business lines. Medina diversified into the importation of North African goat skins for the Sevillian tanning industry and served as an import-export agent for a variety of foreign-based merchants. Medina was not always successful—he participated in a marine insurance syndicate that lost a valuable merchant ship to pirates off the Cape Verdes—but he managed to amass a sizable fortune by 1550.[19]

Medina had no direct experience with mining and metallurgy. But like most Sevillians, he was very familiar with the silver bonanzas that had been struck in the Spanish Americas and the challenges presented by the increasingly deeper, lower-quality ores— especially those located in the lead-free zones of northern Mexico. The great silver strike at Guadalcanal, located ninety miles due north of Seville, did not occur until 1555, so the inquisitive Medina was left to ponder the ancient piles of silver chlorides, weathered copper-bearing pyrites, and residual sulfides (gossans) that were located at nearby Rio Tinto and Tharsis. Massive piles of waste slag had been left by a series of Phoenicians, Carthaginians, and Romans between 1000 BCE and 100 CE. Many centuries of leaching activity had made some of the exposed gossans suitable for testing.[20]

Medina studied traditional smelting methods (like cupellation) and their associated labor costs before delving into possible alternatives to treat the deeper, lead-free ores of New Spain. Armed with this analysis and an economic "rule of thumb"—that one quintal of ore had to contain at least ten ounces of silver to be profitable— Medina constructed a rudimentary smelter in his backyard in 1552 and performed a series of experiments. The evidence is murky, but it appears that Medina's experiments caught the attention of a visiting German metallurgist, a so-called Maestro Lorenzo, who was familiar with amalgamation through Biringuccio, Agricola, or by word of mouth. Lorenzo advised the struggling Medina to steep grounded, silver-bearing ore in a salty brine solution, add mercury, and then mix the brew daily for several weeks—until the mercury had "darkened." The darkening process would signal that the silver had been decomposed by the salt and had formed an alloy (an amalgam paste) with the mercury. After washing out the waste ore, the next step was to "retort" (distill) the residual amalgam to separate the mercury component from the silver.[21]

The amalgamation method worked. Medina was so impressed with the concept that he agreed to commercialize the new system in partnership with Lorenzo. In fact, he feared that the process was so compelling that it ran the risk of being exploited by other interested parties in New Spain and Bolivia. The partnership would be brief. Lorenzo was somehow denied an entrance visa to New Spain, leaving Medina to commercialize the new system on his own. When he arrived in Mexico City in late 1553, without his technical partner, wife, or children, he was welcomed by Viceroy Luis de Velasco. The viceroy was delighted to support anyone with a credible proposal to reverse Mexico's declining silver output. So was Doña Juana, temporary regent of Castile during Felipe's brief residence in England. Doña Juana was especially pleased that mercury was an essential component of the silver amalgamation process. The Crown controlled the ancient mercury mines at Almadén (with the Fuggers) and would establish a royal monopoly on mercury in 1559.[22]

Medina promoted his new system in Mexico City for a full year before relocating to a test site at Pachuca in 1555. By then, at least two mining engineers, Gaspar Lohman of Sultepec and Miguel Perez of Taxco, had obtained permits to prospect for cinnabar deposits (mercury-bearing ores) in New Spain—if Medina's amalgamation process was going to succeed, huge supplies of mercury would be needed. No matter that Medina was foundering at Pachuca. Located within the audiencia of Mexico, sixty miles northeast of Mexico City along the future Pan-American Highway, Pachuca had been a smallish Indian village of 1,275 inhabitants when two speculative mining claims were filed there in April 1552. The resulting silver strike was modest compared with Zacatecas, but it converted humble Pachuca into a boomtown between 1552 and 1555. As the boom faded, Pachuca was left with the same increasingly poor, lead-free ores that plagued Zacatecas and other mining districts in New Spain. Medina established his own amalgamation facility and homestead, Haçienda de Beneficio Nuestra Señora de la Purísima Concepción, at the base of Pachuca's Magdalena Mountain and went to work.[23]

Assisted by a Taxco-based mining magnate, Juan de Plazencia, Medina followed Lorenzo's recipe precisely. He poured his Pachuca slurry into pancake-like *tortas* and placed them atop the haçienda's stone-floored patio. Then he waited patiently for the mercury to darken. But something went wrong. Lorenzo's recipe had worked fine in Medina's backyard in Seville, but the mercury was not darkening in the tortas of Pachuca. Most of the visiting mineros left the haçienda in disgust, leaving Medina, the loyal Plazencia, and a handful of Pachuca-based operators to contemplate their alternatives. At this low point, Medina's experience with the Sevillian tanning industry came in handy. Lorenzo's magic formula needed to be tweaked to treat the special characteristics of Pachuca's silver-bearing ores. If chemical additives could convert raw goat skins into finished leather products in Seville, then chemical additives could possibly enhance the performance of Lorenzo's mixture in Mexico.[24]

Medina experimented with various combinations of additives. Biringuccio had specified cleansing the silver with vinegar, verdigris, and a corrosive sublimate prior to amalgamation. These reagents may have worked well in Central Europe, but Biringuccio (and Lorenzo) was unfamiliar with the specific properties of New Spain's silver-bearing ores. Medina's trial-and-error experimentation finally produced a winner when magistral, a copper pyrite that was mined in the Rio Tinto region, was found to act as a reagent in the presence of Pachuca's silver-bearing ore, mercury, and salt. No one understood how and why magistral worked—even present-day chemical engineers wonder if the additive was part of a so-called hydrometallurgical chloride leaching process—but the missing ingredient to amalgamate certain types of Mexican ores had been revealed. Later practitioners would discover that magistral was required in amalgamation unless the silver-bearing ores were either silver chlorides or contained iron or copper sulphates.[25]

A jubilant Medina received a six-year royal patent after demonstrating the new process to Viceroy Velasco's satisfaction. Since amalgamation would require a larger-scale power capability, Gaspar Lohmann presented the viceroy with a proposal to construct a series of Saxon-styled water-powered grinding and hydraulic systems (ingenios). Water-powered ingenios had been deployed in Central Europe to amalgamate gold, and Lohmann feared that his more competent rival, Perez, would beat him to the punch in New Spain. Perez was already working on a mechanical system in Taxco. Velasco agreed to award an eight-year patent to Lohmann's ingenio enterprise in July 1556 only after Medina had been invited to join Lohmann as a partner. The royalty rate was set at 660 *pesos de ocho* per license on the condition that the new ingenio system pass a pilot test at Sultepec on August 31. However, the demonstration never took place. The leaded ores of Sultepec and Taxco were found to be unsuitable for amalgamation, a conclusion confirmed by a subsequent test at Guadalcanal in 1558, and northern districts like Zacatecas were too dry to benefit from water-powered ingenios. Medina was left with sole rights to the entire process, a mule-powered crushing system, and a focus on the unleaded ores of northern Mexico.[26]

Medina's multistep "patio" process could be completed within three to eight weeks. After silver-bearing ore had been excavated by crowbar-wielding *barretos* and lugged to the surface by *tenateros*, the ore was crushed with mallets, mortars, or iron tilt-hammers, and carried in broken pieces to a mule-powered stamping mill (*molino*)—all sixty-five of Zacatecas's stamping mills were still powered by animals in 1597. A team of harnessed mules, operating in a circular fashion, powered a toothed grinding wheel (or sledge) and an assembly of ten seventy-pound iron shoes (*mazos*). The ore was crushed into smaller pieces, passed through a perforated leather sieve—leaving lumpier pieces to be sent back for another session—and then grinded down in a stone-floored *arrastra*. There, an assembly comprising four heavy stones and a rotating beam was dragged by mule teams until the mix was pulverized into very fine particles. The pulverized ore (*harina*) was now ready for amalgamation.[27]

The pulverized ore was sifted once more and arranged into piles in a paved courtyard (patio). The piles were mixed with the magic formula of water, salt, copper pyrites (magistral), and mercury—ten pounds of mercury for each quintal of ore—and then trampled upon by Indians, horses, or mules. The trampled mix (*repasos*) was then spread into large pancakes (tortas) atop the stone floor and monitored until the mercury combined with the silver to form a spongy ("darkened") amalgam. After the spongy silver amalgam (*pella*) was washed and stirred in water-filled tubs (*tinas*) to separate the waste ore, it was retorted (distilled) in containers—the only heating step in an otherwise cold (charcoal-saving) process—to separate (vaporize) the mercury component. The remaining silver amalgam was melted, cast into sixty-five-pound silver bars, and stamped with a serial number.[28]

Mercury, the critical element in the entire process, just happened to be mined at Almadén in the upper Guadiana River valley. The ancient cinnabar (mercury) deposits at Sisapo (Almadén) had been worked by the Romans long before they were transferred to a series of Muslim rulers (*al-ma'dan* means "mine"), the military Order of Calatrava, and Charles V. Charles and the House of Fugger gained control of the deposits when Calatrava's mastership (maestrazgo) properties were assumed by the Crown in 1523. For centuries, Almadén had produced mercury and mercury chlorides for use in medicines, paints, and pigments. Over one hundred quintals of vermilion, a grounded form of mercury, were being produced annually when Almadén was ravaged by fire in 1550. The fire reduced mining capacity by more than 75 percent and prompted the Crown to assume direct ownership through 1562. Almadén was refurbished just in time to exploit the roll-out of Medina's bombshell discovery. The initial 265 quintals of Almadén mercury reached Veracruz in 1559 under a newly established royal monopoly. This was the first installment of the roughly two thousand quintals of Almadén mercury that would be sent to the Americas annually through 1600.[29]

The royal monopoly on Almadén mercury dovetailed nicely with Medina's magical process. When it came to smelting low-quality, unleaded ores, Medina's mercury-based technique created a new standard that would last for over three centuries. Mexican mineros were no longer limited to high-quality *fuego* (fired) ores that could only be cupelled. Huge quantities of low-quality *azogue* (mercury) ores could now be smelted profitably despite the fact that they contained only one-tenth of the silver content of typical fuegos. A Mexican silver rush ensued. Focused on abandoned mining properties, residual slag heaps, and new discoveries, the rush was extremely successful and should have made Medina a very wealthy man. He stood to realize royalties of as much as 496 pesos de ocho (eight reales) on any deployment of the amalgamation process within the Audiencia de Mexico. That was the problem.[30]

Medina failed to accumulate even a modest fortune from his history-making invention. His six-year patent was restricted to the relatively silver-poor jurisdiction of the Audiencia de Mexico when New Spain's primary silver districts, headlined by Zacatecas, were located either in the Audiencia de Guadalajara or in the fledgling northern provinces of New Viscaya and New Galicia. The courts ruled that mineros operating outside of the Audiencia of Mexico could deploy Medina's amalgamation process on a royalty-free basis. By 1562, when the patent expired and thirty-five amalgamation facilities were operating royalty-free in Zacatecas, Medina had received a mere 10,812 pesos in royalties. Nearly broke, the crestfallen inventor preferred to donate half of his meager royalty receipts to an orphanage and a church. Medina appealed to the Council of the Indies for a either a special royal dispensation or a royal pension of 2,000 pesos. His petitions were either denied or ignored.[31]

Medina's troubles continued. En route to Seville in 1563 to reassert his claim, the star-crossed Medina was shipwrecked off Hispaniola and returned to Seville without clothes, papers, or money. However, he somehow managed to scrape together enough funds to relocate his family to Pachuca. Petitions filed by Medina's heirs in 1578 and 1580 prove beyond any doubt that Bartolomé Medina was the first to amalgamate silver-bearing ores successfully. A *probanza* signed by the Council of the Indies and Viceroy Martin de Enriquez in 1580 included at least one valuable testimony. A minero confirmed that Medina's amalgamation process had raised his silver output from less than eighty quintals weekly to seven hundred despite the fact that the silver content of his amalgamated (azogue) ores was only one-tenth that of his smelted fuego ores (ten ounces per quintal).[32]

## ZACATECAS

Medina's process was transferred to Zacatecas by August 1557, within months of Felipe's bankruptcy declaration. The plan was to treat the red-colored silver oxide ores (*colorados*) located in the Serrania's upper veins and the huge deposits of lower-level silver sulfides (*negros*). The iron-rich colorados were reduced easily, but trial and error was required to amalgamate the blue-gray–colored negros—the iron-poor negros contained larger amounts of lead sulfide (galena) and required a European flotation technique to enhance the washing and stirring operations. The response was immediate. Zacatecas's minero community invested over eight hundred thousand pesos in thirty-five new amalgamation facilities by 1562 and held their collective breath. A consistent separation of the silver-bearing amalgam (pella) from the nonmetallic part of the ore-mud (torta) was achieved only after the flotation technique had been perfected. Of course, the predominance of negro ores left Zacatecas mineros to measure their silver content in ounces per quintal rather than in the eight-ounce marks that had been enjoyed in earlier days.[33]

The eight-thousand-foot-high town of Zacatecas boomed during the transition to amalgamation. With sixty mines, thirty-five amalgamation facilities, over thirty-four companies, scores of traditional smelters, and an expanding population of Indian workers and black slaves, the town emerged as the second-largest city in New Spain by

1562. Cristóbal de Oñate and his partners owned a number of mines, built thirteen stamping mills and imported over one hundred slaves. Unlike mining districts located south of the Querétaro line, where the encomienda and *requerimiento* systems were still entrenched, the mines of Zacatecas were worked mainly by imported (paid) Indian labor and, to a lesser extent, by slaves. The conditions were dreadful but the pay wasn't bad. Skilled foremen and carpenters earned around two pesos per day and most mine workers received one-half a peso plus food rations. This was two to three times the wages paid to future haçienda workers in northern Mexico. Zacatecas would employ over five thousand mineworkers at her peak, or over half of New Spain's mining workforce during the 1630s.[34]

Zacatecas eclipsed Guadalajara and Compostela as the commercial capital of New Galicia. The town had already received a Royal Treasury (*Real Caja*) in 1552, before the explosion in amalgamated silver production and seventeen years before one was placed in Guadalajara. However, Zacatecas did not look like the second-most important town in Mexico. It was poorly sited within a dry, narrow valley between two peaks, La Bufa and El Grillo, and was laid out in helter-skelter fashion. The founders had paid scant attention to the initial mining camp of 1546–1548. They assumed that the site would be abandoned in typical Spanish hit-and-run style after the surface-level deposits had been exhausted. By the time that Medina's amalgamation process had converted Zacatecas into a longer-term bonanza, it was too late to apply the royal grid pattern. Zacatecas grew on its own. Water was essentially rationed from local wells and most foodstuffs had to be imported.[35]

Poorly sited or not, Zacatecas's instant prosperity catalyzed a modest building boom in and around town. Fashionable two-story stone houses were constructed for as much as two thousand pesos around the central plaza. These were occupied by mineros, merchants, and government officials. The Franciscans, the first of five religious orders to establish themselves in town, eventually built a church and monastery (San Francisco) between the central plaza and the base of Cerro de la Bufa. Annual salaries in Zacatecas were much higher than average. A high government official received around 1,650 pesos, a master mason 450, and a construction worker 165. While Indian mineworkers were well paid by Indian standards, they were in a completely different (exploited) category than Spaniards or successful *mestizos*. Indians who managed to survive diseases and overwork were often forced to compete with African slaves. Imported slaves could be purchased for two hundred pesos each.[36]

The three primary founding fathers of Zacatecas, all Basques, fared very differently. Juan de Tolosa, the frontiersman who launched the whole enterprise in 1546, managed to marry a mestizo daughter of Hernán Cortés before squandering virtually all of his wealth on a series of failed treasure hunts in the north. Cristóbal de Oñate, who had arrived in Mexico as a nineteen-year-old in 1524 and assisted Nuño Beltrán de Guzmán's brutal conquests in the north, accumulated a fortune of 1.5 million pesos from the mines and smelters of Zacatecas. However, Oñate somehow lost this huge fortune through a combination of bad investments and ill-advised military campaigns. He died nearly broke. The third cofounder, Diego de Ibarra, did pretty well for himself. Ibarra was probably as wealthy as Oñate during the 1550s and even married the daughter of Viceroy Velasco in 1556. He helped to finance the northern expeditions of his nephew, Francisco de Ibarra, and claimed to have invested 200,000 pesos in the conquest of New Biscay.[37]

Zacatecas's founding fathers led the move to Medina's amalgamation process. While precise production figures for Zacatecas are unavailable (or unreliable) for the early boom-and-bust period between 1548 and 1558, the Real Caja de Zacatecas reported registered silver volumes of 103,006 marks (worth 875,551 pesos) in 1560. This was barely half of the 190,000 marks of silver produced in New Spain in heady 1548, but it exceeded the peak years at Guadalcanal and Joachimstahl. Production volumes rose steadily after the first raft of amalgamation facilities came on stream in 1562. Zacatecas was producing 148,519 marks by 1572—well above the 114,878 marks produced at Cerro de Potosí in that temporarily weak year—and would reach a century-high peak of

171,004 marks (1.4 million pesos) in 1575. Annual production would decline to around 120,000 marks between 1595 and 1602.[38]

Zacatecas's mineros were pleased that Almadén mercury was less expensive than English smelting lead, but they were less thrilled about a royal monopoly that kept mercury prices artificially high. Mercury obtained by the Crown for twenty-five ducats per quintal was resold in New Spain for over one hundred. Of course, the costs to refurbish Almadén after the devastating fire of 1550, install a water pumping system, and transport mercury to Zacatecas were not insignificant. A royal inspection tour of Almadén in November 1557 had noted poor drainage, extremely hard rock formations at six hundred feet, and limited timber supplies. Little mention was made of the horrific working conditions and the foul smell of sulfur that afflicted Almadén's four hundred households—even after the mercury roasting plant had been relocated eight miles away. The entire process was managed by the House of Fugger between 1562 and 1645. Mercury was poured into sheepskin containers, hauled ninety miles south to Seville by mule-trains, inspected, weighed, packed in sealed barrels, and placed in stamped wooden crates for shipment to Veracruz. From there, the cargo was carried by wagon train to Mexico City and Zacatecas. The Fuggers' contract of August 1562, the first of a series of renewable ten-year deals, called for a ramp-up in mercury production from six hundred quintals in 1563 to one thousand in 1565–1572. The quota eventually reached three thousand quintals during the 1590s.[39]

The caminos reales that carried precious cargo between Veracruz and Zacatecas—mercury up, silver down—had been jump-started by Hernán Cortés. However, it was Viceroy António de Mendoza (1535–1550) who completed a northern road between Mexico City and Querétaro during the late 1530s and then a western route to Acapulco to facilitate the Pacific trade. The boomtown of Zacatecas was supplied by Guadalajara until a direct camino real was constructed between Querétaro and Zacatecas in 1555. This and other caminos reales were planned by the viceroy, funded by the silver magnates and merchants, and built by inexpensive requerimiento labor and troops. While most of the highway engineering techniques were transferred from Europe, it is possible that some ideas were borrowed from the Aztecs and Incas. The Incas were renowned for their remarkable systems of retaining walls, drainage systems, and cantilevered bridges. Otherwise, the flat northern plateau was a much easier construction task than the challenging mountainous terrain between Mexico City and the two coasts. Northern extensions of the caminos reales followed the mining industry north during the next thirty-odd years.[40]

With so much silver at stake, Mexican roads, wagons, and freighting systems had to be superior to those operating in the home country. Barrels of mercury and a wide variety of European trade goods left Seville for the two- to three-month sail to Veracruz, including stops in the Canaries and the West Indies, followed by a one-month journey to Mexico City. Another month was required to reach Zacatecas. The cargo was typically carried in four-thousand-pound iron-wheeled carros—arranged into eighty-wagon trains, drawn by teams of oxen, mules, or horses and fortified to defend against Indian attacks. Most of the freighting businesses were owned by Indians and mestizos. The main routes were eventually lined by a chain of walled garrisons, regulated inns (*ventas*), and ranches (haçiendas). The Real Caja in Zacatecas coordinated the shipment of amalgamated silver down to the Royal Mint (*Casa de Moneda*) in Mexico City until Zacatecas received a Royal Mint of her own in 1606. Minted silver was hauled east to Veracruz, shipped to the Casa de la Contratación in Seville, and redistributed to royal and private accounts in Spain, Antwerp, and Genoa.[41]

New discoveries made at Durango (1563), Santa Bárbara (1567), Mazapil (1568), and San Luis Potosí (1592) helped to offset the increasingly lower depths of Zacatecas's silver-bearing ores. While some mineros bragged in August 1567 that a single quintal of Zacatecas ore was still generating one hundred marks of silver, this was certainly an exaggeration. Yields declined, excavation costs rose, and drainage became a severe problem as the shafts approached the water table. Most mineros preferred to abandon a

***Prince Henrique (Henry the Navigator):*** Promoted a series of Portuguese expeditions to uncharted Africa, reaching as far south as Sierra Leone at his death in 1460. *Painting by Nuno Gonçalves. From the Polytriptych of St. Vincent in the National Museum of Ancient Art, Lisbon, Portugal.*

***Columbus briefs Isabella and Ferdinand:*** After Dias had rounded the Cape of Good Hope in 1487, the challenge was to beat the Portuguese to the spice markets of South Asia. *Painting by V. Brozik, 1884. Courtesy of the Library of Congress.*

***Vasco da Gama before the Zamorin of Calicut (India) in 1498:*** The depiction is contrived because the Zamorin's intermediaries refused to present da Gama and his insulting offers of clothing, coral, sugar, butter, and honey. *Copyright by John D. Morris & Company, 1905. Courtesy of the Library of Congress.*

***Mr. and Mrs. Jacob Fugger in 1500:*** A budding copper monopoly in Central Europe would raise the family fortune to great heights and fuel Portugal's metal-for-spice transactions in the Indian Ocean.

***Calicut:*** One of the few Muslim sultanates to resist the Portuguese successfully. *Image from Georg Braun and Franz Hogenber's atlas* Civitates orbis terrarum, *1572.*

*Afonso de Albuquerque:* He devised a bold plan in 1509 to manage Portugal's fledgling Indian Ocean empire from a handful of naval bases at Mozambique, Hormuz, Goa, and Malacca. *Portrait from the National Museum of Ancient Art, Lisbon, Portugal.*

*Jacob Fugger at work in 1517:* Huge profits were being generated by the copper-for-spice deals with the Portuguese and the assumption of Habsburgian mining properties. *Office of Jacob Fugger, with accountant M. Schwarz, 1517. From a biography of M. Schwarz, Herzog-Anton Ulrich Museum, Braunschweig, Germany.*

***Hernán Cortés:*** One of the century's greatest entrepreneurs invaded Aztec Mexico in 1519 without authorization and helped to jump-start the Spanish-American mining industry. *Engraving by W. Holl, published by Charles Knight.*

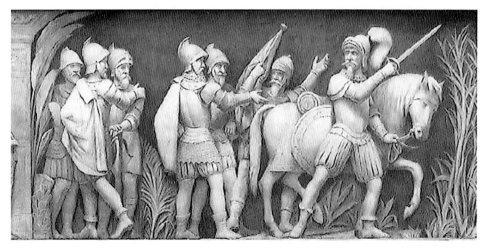

**The Pizarro brothers:** They plundered the Inca treasures of Cajamarca and Cuzco in 1533–1534, launched the Bolivian mining industry, and lost it all by 1548. *Courtesy of the Architect of the Capitol.*

**Inca stonework in Cuzco:** The cradle of the Inca Empire spawned some of the hemisphere's greatest feats of civil engineering and metallurgy. *Stereograph copyright by Underwood & Underwood, ca. 1906. Courtesy of the Library of Congress.*

***Antwerp:*** The world's most dynamic entrepot in the first half of the sixteenth century looks somewhat peaceful in this 1890s photograph. *Courtesy of the Library of Congress.*

***Holy Roman Emperor Charles V:*** Inconclusive military campaigns against the French, Ottomans, and German Protestants had soaked up huge amounts of Habsburgian capital when this portrait was completed in 1548. *Portrait by Lambert Sustris, 1548. Alte Pinakothek, Munich, Germany.*

*Cerro de Potosí:* The 15,700-foot-high peak in the Bolivian altiplano was the site of a spectacular silver bonanza in 1545 that was still going strong when this woodcut was published in 1553 by A. Skromnitsky.

*Zacatecas:* The silver uncovered near Cerro de la Bufa in 1546 led to much larger discoveries in the neighboring Serrania de Zacatecas in 1548. *Courtesy of the Library of Congress George Grantham Bain Collection.*

*Bartolomé de Medina's patio (amalgamation) process:* The revolutionary silver separation technique introduced at Pachuca, Mexico, in 1555 was deployed at the Comstock Lode, Nevada, works in 1866. *Image courtesy of the Library of Congress.*

*Anton Fugger:* Jacob's talented nephew remained loyal to Charles V, expanded the family fortune during the boom years in Antwerp, and survived the Spanish state bankruptcy of 1557 in reasonably good shape. *Portrait by Hans Maler, 1525. Staatliche Kunsthalle, Yorck Project, Karlsruhe, Germany.*

*Felipe II:* The formidable king of Spain squandered one of the greatest treasure troves in world history during an extremely eventful reign (1556–1598). *Portrait by Titian. Sammlung des Herzogs van Alba, Yorck Project, Madrid, Spain.*

**Francisco de Toledo:** The Viceroy of Peru raised silver output dramatically in the 1570s by investing heavily in the amalgamation process at Potosí, upgrading the mercury works at Huancavelica, and reorganizing the native labor force. *http://amerique-latine .com/ala/fr/inquisition.html.*

*Fernando Alvarez de Toledo, the third Duke of Alba:* He served as a virtual military dictator in the Spanish Netherlands between 1567 and 1574 but was undone by increasingly severe financial pressures. *Portrait by Tizian, 1551. Museo del Prado, Madrid, Spain.*

*William of Orange:* The Stadholder of Holland, Zeeland, and Utrecht was drawn reluctantly into the Dutch independence movement (1568–1581) only to be assassinated by an agent of Felipe II in 1584. *Portrait by Adriaen Thomasz Key, 1575. Rijksmuseum Amsterdam, Amsterdam, the Netherlands.*

***Oda Nobunaga:*** The energetic warlord conquered (and modernized) over half of Japan between 1568 and 1582 with Portuguese-assisted profits earned from the exchange of Japanese silver for Chinese gold and silk. *Woodcut by Yoshitoshi Taiso.*

***Battle of Sekigahara:*** Tokugawa Ieyasu completed the unification of Japan in October 1600 when his silver-financed army defeated Toyotomi Hideyoshi and other western daimyos in Central Honshu.

*Portuguese Macao:* The colony handled exchanges of Chinese silk-for-Japanese silver since its establishment in 1557, but suffered after the Portuguese were evicted from Japan in 1638. This map is from 1639.

*Malacca:* The crown jewel of Portuguese Asia had repulsed scores of naval attacks, including a siege by a VOC fleet in 1606, but eventually fell to the Dutch in 1641. *Etching by Wouter Schouten, 1676.*

flooded mine instead of investing in costly drainage adits (*socavones*). Zacatecas remained Mexico's leading mining district thanks to the wonders of amalgamation, but it was contributions from the northern strikes—supplied by an array of new haçiendas in the northern river valleys—that stabilized annual Mexican silver production during the 1560–1610 period. The 1.5–2.2 million pesos (forty-eight to seventy tons) produced during this period was less than one-third of Bolivia's annual output, but it exceeded that produced in the Ottoman Balkans.[42]

The new haçiendas were a response to the ongoing destruction of Mexico's native population, generous land grants, and the need to supply the mines with wheat, beef, rawhide, and tallow. The demographic carnage had weakened the traditional repartimiento, a native-based tributary system, and encouraged the establishment of huge ranches that required less labor per acre. Other responses included the recruitment of paid Indians (especially Tlaxcalans) from the south and the importation of black slaves. No matter that slavery had been outlawed by Charles V in 1542 and many of the twenty thousand black slaves working in New Spain by 1570 succumbed to the same deadly diseases that impacted the Indians. The better business deal was to pay Indian workers (*naboros*) when they were available. By 1600, Zacatecas was being worked by 4,606 naboros, 1,619 repartimientos, and 1,022 black slaves. The Crown may have cited religious reasons for its antislavery policies, but the truth was that Indian labor—when it was available—was far less costly than imported slaves.[43]

Colombia was a much different story than New Spain. The upper Magdalena and Cauca river systems failed to produce a single bonanza like a Zacatecas or a Potosí, but there was a steady stream of gold strikes after 1537—since gold carried at least eleven times the value of silver, almost any discovery was meaningful. One of Belalcázar's lieutenants, Jorge Robledo, had raised the stakes during his two-hundred-mile-plus expedition to the upper Cauca River north of Cali. Alluvial and vein gold were uncovered at Anserma, Cartago, and Buriticá between 1539 and 1541, followed by a host of mining works, and the introduction of black slaves to support the encomienda workforce. The work was labor intensive. The weathered gold had to be crushed by hand on stone mortars, sent downstream, and then recovered by panning. By 1550, Cartago and the more northerly staging area at Santa Fé de Antioquia were managing smelters (*fundiciones*), Royal Treasuries (*Cajas Reales*), and prospecting expeditions into the eastern flank of the Cordillera Occidental. Revenues from the Antioquia region boomed after even larger deposits of vein and placer gold were uncovered at Cerro de Buriticá in 1550. Buriticá had been worked by Indians long before Robledo's initial discovery in 1541.[44]

An estimated 1.5 million gold pesos of Colombian gold reached the port of Cartagena annually between 1538 and 1557—enriching the region's Spanish patriarch, Sebastián de Belalcázar, prior to his death in Cartagena in 1551 at age sixty-five, as well as Charles V and his royal creditors. The annual totals included some 240,000 gold pesos worth of ceremonial treasures, plundered from New Granada and the enlarged District of Santa Fé, but excluded 300,000 gold pesos from the neighboring province of Popayán prior to Popayán's transfer to the enlarged District of Quito in 1563. Altogether, Colombia may have generated as much as two million gold pesos annually between 1538 and 1557. This was equivalent to three million silver pesos and nearly matched the 3.3 million silver pesos produced at Cerro de Potosí in 1550. While the gold boom faded after 1560, Quito's annual output held up at around 300,000 gold pesos through 1580, and Buriticá was still being worked by 1,500 encomienda Indians, 300 African slaves, 200 transplanted Spaniards in 1582. Over two hundred years later, Von Humboldt rated the gold deposits of northwestern Colombia at an impressive twenty to twenty-one carats.[45]

## CERRO DE POTOSÍ

The looming problem for Felipe II was viceregal Peru. The returns from Cerro de Potosí were still much larger than those from

Zacatecas, but production fell gradually after 1550, local mineros were unwilling to invest in necessary infrastructure and processes, and the entire enterprise was dependent on native Bolivians. While it is unclear whether Potosí's founding fathers—Gualpa, Guanca, and Villaroel—had ever been regarded as mineros, the fact that at least one native Bolivian had been part owner in one of the greatest silver strikes of all time is of interest. It defies the notion that natives were excluded entirely from the reward component of American mining ventures. It was beneath the dignity of most Spaniards to perform manual labor, so someone had to do the dirty work of digging, hauling, smelting, and transporting the precious metals. Most of these tasks were performed by native Americans and most of these workers eventually succumbed to overwork, brutal conditions, and/or disease. However, Gualpa, Guanca, and hundreds of other native Americans defied the odds by accumulating wealth, or at least livelihoods, as owners or managers of mines, smelters, and freighting businesses. In the early days, Spanish settlers were few in number and business opportunities were presented to anyone with the ability to seize them.

Native Bolivians and Spaniards alike had to grapple with Potosí's challenging geology, topography, and geography. The top few hundred feet of Cerro de Potosí (Rich Hill) contained four extremely rich veins of silver that could not have been predicted by a professional geologist. Allowing for the bitter winter conditions associated with a 15,700-foot-high mining site and a 13,100-foot-high village below, these early high-altitude veins were relatively easy to excavate and eliminated the need for costly mine shafts and drainage works. Father Jose de Acosta's estimate that as much as twenty-eight million pesos of silver were excavated from Potosí during the spectacular 1548–1551 period was probably an exaggeration, but there is no doubt that production volumes, profit margins, and royal allocations declined after the easy surface-level ores had been depleted in 1550.[46]

In these early years, local mining magnates made do with excavation and smelting processes that were little changed from those deployed at Batán Grande in 1000. Capital investment needs were limited. High-grade ores yielding nearly one mark (eight ounces) of silver per pound were easily extracted and could be smelted by encomienda natives in old-fashioned clay furnaces (huayras). The native smelting process was simple. Silver oxide ores were collected from Potosí's exposed outcrops, loaded into wood-fired huayras, topped with perforated sections of damp clay, and then heated. The huayras were dug into the hillsides to catch the temperature-raising katabatic air flows at night. While thousands of these primitive furnaces served as Potosí's principal smelting works until around 1574, the deeper ores were "harder" than the Mexican variety and contained metals like antimony and tungsten that demanded higher heating temperatures to be separated properly. This placed additional demands on Potosí's limited supply of timber and charcoal. After the surrounding hills had been stripped of timber, wood fuel expenses climbed with the associated costs to find, cut, and transport timber (by llama trains) from distant locales.[47]

The cost equation changed after the surface-level ores had been picked off. The fact that Potosí's richest deposits had been no deeper than six hundred feet—somewhat below the depths of Mexico's highest-quality silver ores (rarely below four hundred feet)—was fortunate because most Spanish mineros had weak backgrounds in geology, mine engineering, metallurgy, and logistical planning. American mines, including Potosí, were typically managed "on the fly" without the benefit of sophisticated Central European engineering techniques. In the case of drainage, few mineros were willing to invest in costly hoists, pulleys, and pumps. New discoveries were almost always accidental and there was little concern for the unfortunate workforce. Indians and black slaves toiled away in harsh environments that were poorly ventilated, poorly drained, and prone to cave-ins below ground and characterized by an extremely high altitude and seasonal cold spells up above.[48]

Then there was Potosí's remote location on the outer fringe of Spanish civilization. Proximity to El Gran Camino Inca, the twisty north–south highway running through Sucre, La Paz, and Cuzco, was overwhelmed by the rugged, often-mountainous terrain lying between Potosí and Lima. Incoming viceroy António de Mendoza, a man who had guided the camino real system in New Spain, determined in 1551–1552 that Peru's challenging regional terrain made it impossible to construct a series of Mexican-styled wagon roads through the viceroyalty. His successor, the Marques de Canate, gave it a try in 1555. Canate organized a twenty-thousand-peso construction program, backed by a new one-half-percent tax on imported merchandise at Potosí, to upgrade bridges and roadways along the main Inca highway. He even recruited four civil engineers from Spain to supervise the effort. Unfortunately for Canate, Peru's topography was less than benign, and the reconstruction work was inconsequential. The Spaniards would have to make do with the same narrow trails that had served native pedestrians and llamas for centuries.[49]

The long stretch of highway between Potosí and Cuczo was broken up by Chuquisaca (Sucre), Cochabamba (1542), La Paz (1548), and the Inca motel system—inns (*tambos*) were sited every ten to twenty miles. The three Spanish towns solidified Spanish authority in the former center of the Aymara federation and presided over the twenty-three-carat gold strikes made in the highland streams of Carabaya, Oruro, Asillo, Asangara, and La Paz. These sites yielded 170,000 gold pesos annually between 1543 and 1553, highlighted by a one-hundred-pound, horsehead-shaped gold nugget that was sent to Charles V. Sucre served as a staging area for regional prospecting activities, the establishment of Santa Cruz in eastern Bolivia, and the subsequent founding of Tucumán (Argentina) in 1580. The future "back door" route between Potosí and Rio de la Plata was destined to run through Tucumán and Buenos Aires—not Asunción (or Santa Cruz). The dangerous Gran Chaco region of southeastern Bolivia remained in the unfriendly hands of the Chaco, Chiriguano, and Toba tribes for centuries.[50]

The new staging areas helped to manage the llama trains that operated between Potosí and the Pacific ports of Callao (Lima) and Arica (on the Chilean border). The grueling six-month journey to Arica required considerable planning. In one case, a single-file pack train of two thousand llamas and over one thousand Indians was enlisted in March 1549 to carry 7,771 bars of Potosí silver to the coast. Twelve days were required to start the tail-end of the train on its journey. After arriving at one of the two Pacific ports, the silver was shipped north to Panama over a wind-aided, fifteen-day period and then carried across the Isthmus of Panama to Nombre de Dios. The royal treasure fleet received the silver there and dropped off a variety of merchandise for the Potosí market. Unfavorable Pacific winds extended the return trip to Arica to around six weeks.[51]

While the boomtown marketplace at Potosí became one of the most diverse in Spanish America, the transportation costs were daunting. Fortune-hunters, merchants, and freighters raised product prices to off-the-charts levels—a single arroba of the lowest quality wine could cost as much as one hundred gold pesos—and ignored official attempts to limit trade goods to Spanish-only varieties. Although all merchants operating in viceregal Peru were required to register at the Casa de la Contratación in Seville, until a formal *consulado* was established in Lima in 1614, contraband trade was a regular feature of the Potosí economy. Slavery, abolished theoretically by the New Laws of 1542, was also pervasive. Native Indian merchants had no qualms about the slaving business and were happy to exchange fellow Indians and Africans to the highest bidder. Some slave traders earned thousands of gold pesos in a single day.[52]

Reincorporated as Villa Imperial de Potosí in 1561, the boomtown was critically dependent on local geology. The quality of Potosí's intermediate-depth ores began to decline in 1565, as it had at Zacatecas in the early 1550s, leading local mining magnates to study Medina's experiments at Pachuca and the roll-out of the amalgamation process at Zacatecas during the 1557–1562 period. Of course, the mineros recognized that a move to amalgamation would require large supplies of mercury—this was a problem because Almadén's capacity was committed to New Spain and Peruvian supplies had been discovered only recently. The search for local mercury deposits had actually begun in 1555, even before Medina's process had been demonstrated conclusively at Pachuca, thanks to reports in Lima that some Indian women painted their faces with cinnabar. Since the readily available red powder was usually found in combination with mercury, the task was to find the source of this material. One interested party, a Portuguese merchant named Henrique Garcés, even traveled to Mexico in 1558 to learn more about Medina's process. He returned the following year with an amalgamation expert, Pedro de Contreras, who compared Peruvian mercury samples favorably with those from Almadén. Garcés subsequently received a twelve-year royal monopoly to amalgamate locally produced silver with locally produced mercury.[53]

In the meantime, Garcés must have been thrilled by the discovery of a major source of mercury at Huancavelica by one Amador de Cabrera in December 1563—even if the deposits were confiscated immediately by the long arms of Felipe II. No matter that Huancavelica was located nearly one thousand miles northwest of Potosí and local metallurgists were struggling to adapt the Medina process to the special qualities of Potosí's ores. Garcés had tested the amalgamation process unsuccessfully at Potosí as early as 1558 and failed again with the Huancavelica samples in 1564. Different combinations of the magic recipe were mixed and matched with native smelting techniques throughout the 1560s. All failed. Huancavelica mercury found its way to Zacatecas before it was deployed at Potosí.[54]

The primary benefits of mercury-based amalgamation, that it could treat low-quality ores profitably and reduce wood and charcoal fuel costs, were clear to everyone. The offset was the costs associated with mercury, magistral, and salt and the trial-and-error experimentation required to find the right combination of ingredients. In fact, there would be many variations of the amalgamation process in Spanish America and many of them competed with traditional silver smelting techniques. In Mexico, traditional smelting methods continued to be applied to high-quality ores, leaded ores, and sites in which timber supplies were plentiful. The sixty-four operators at San Luis Potosí (Mexico) in 1598 got away with traditional methods even though amalgamation had been adopted by 296 operators in nearby

Zacatecas. Different processes had to be applied to different types of ores. This had been the case at Pachuca, Guadalcanal, and Central Europe, and it was repeating itself throughout Spanish America. The decision to deploy amalgamation at Potosí was not clear-cut. No one had recognized that Potosí's surface-level ores had been much richer in lead than those found at lower depths.[55]

Although annual output at Potosí had peaked at 379,244 marks (3.2 million silver pesos) in 1550, the local mineros got along quite nicely without amalgamation (and mercury) between 1550 and 1569. Production fluctuated between 207,776 marks and 284,443 marks (2.4 million pesos) during the 1554–1569 period. These volumes were still roughly three times greater than Joachimstahl's peak, and they delivered an annual royal quinto of between 360,000 and 480,000 pesos to the Spanish Crown—excluding confiscations. However, alarms were raised in Lima (and Madrid) when production dipped unexpectedly in 1570 and sunk to a record low of 114,878 marks (one million pesos) in 1572. This was only one-third of the level received in 1550. As we shall see, the local mineros were not the only injured parties. The production decline had challenged Felipe's ability to secure a flood of asientos for his military campaigns in the Spanish Netherlands.[56]

Incoming viceroy Francisco de Toledo would have to roll up his sleeves. After arriving in Lima in late 1569 as the fifth viceroy of Peru (1568–1580), Toledo learned that Potosí's four major veins had been worked out, her native silver and silver chlorides depleted, and tons of discarded tailings (*desmontes*) abandoned as slag. The remaining silver-bearing ores were located near or below the water table and were combined with a variety of metals—some with chlorine, bromine, or iodine (*pacos*), some with iron or copper pyrites (*mulattos*), and others with sulfides (*negrillos*). Excavation costs had soared, drainage had become a problem, ten small-scale stamping mills had reached their limits, and thousands of Indian miners (*baras* and *pongos*) were dying off from overwork, frigid winter weather, and disease. There was more. Smelting yields (and profits) had fallen with the deteriorating ore quality. Mulatto and negrillo ores now had to be roasted before they could be smelted. Potosí's six thousand *huayeros* could only dream of the five hundred pesos per quintal that they had earned in the early years.[57]

Prior to making an inspection tour in October 1570, Toledo had received advice from no less a mining authority than Felipe II. He was instructed to take royal possession of the desmontes and explore their possible treatment by amalgamation. Felipe and Toledo both knew that Garcés's twelve-year monopoly was set to expire in 1571. Garcés had produced small volumes of silver with Peruvian mercury in 1559, but his amalgamation trials at Potosí had been unsuccessful in 1564. While Bolivia's surface-level ores were essentially the same type of silver oxides that had been found in Zacatecas, the intermediate-depth ores were somewhat different. Toledo determined that the problem was more financial than it was geological or metallurgical. Potosí's lethargic mineros had been unwilling to support an aggressive trial-and-error R&D program, as Medina had at Pachuca, let alone invest money in a pilot amalgamation facility. As long as the low-cost smelting system had delivered the goods, the mineros had sat back and subcontracted the work to inefficient native Bolivians.[58]

Toledo decided to focus on R&D. In 1571, following the expiration of Garcés's twelve-year contract, he recruited Pedro Hernandez de Velasco from Mexico to develop an amalgamation process to treat the discarded tailings and the lower-quality ores. If the details of the R&D program are somewhat murky, we know that Velasco conducted experiments at Toledo's Cuzco residence for nearly three weeks in 1571 before making a visit to Potosí to collect sample tailings. Experiments involving the tailings were carried out secretly at Potosí, Cuzco, and even Lima during the course of 1571. They must have been successful. By March 1572, Velasco had been appointed "Chief Master of Mercury" at an annual salary of four hundred gold pesos, and two amalgamation facilities (ingenios) were placed under contract. One of the two was water-powered. It was the tailings enterprise, the desmontes, that provided a spectacular windfall to the Crown and the few fortunate mineros who had obtained licenses to amalgamate Potosí's mountain of discarded slag. A report of May 1572 indicated that many of the abandoned mines had been reopened and that Potosí was making the transition to Velasco's process. Toledo reported these developments to Felipe II by year-end without the knowledge that Potosí's average silver output was set to triple from 232,192 marks annually in 1551–1575 to 721,879 marks in 1576–1600.[59]

Toledo arrived at Potosí in triumph on December 23, 1572. The Audiencia of Charcas had just reported that Potosí's tailings, transferred entirely to the Crown in 1571, were somehow being amalgamated for the first time. Potosí's unlicensed mineros were outraged by the royal power play, but they had had their own chance. Velasco was already training a new generation of Spaniards and Indians to manage the new refining process. In a letter to Felipe II of March 20, 1573, Toledo estimated that over three hundred thousand pesos had been invested in local amalgamation facilities to date but that he had retained the existing six thousand huayras to ensure that smelting could continue during the dry winter months. Toledo added that over seven hundred Spaniards were residing in town and thousands of draft workers (*mitayos*) were on their way. The other "good news" was that Toledo had captured the Inca stronghold at Vilcabamba, established by Manco Inca in the jungles northwest of Machu Picchu, and had executed the last remaining Inca monarch, Tupac Amaru, in Cuzco.[60]

Toledo's reports were well received by Felipe II. The Crown was desperately short of precious metals to finance the Duke of Alba's mobilization of 1572 and to maintain the momentum in the Mediterranean after the Battle of Lepanto (1571). Hans Schedler, who had managed the works at Guadalcanal and Almadén for the House of Fugger, suggested that thousands of idle German mining professionals could be transferred to the primitive mines of Bolivia, New Spain, and Colombia to raise productivity below ground. Most of the Germans had been unemployed since the slowdown in Central Europe. The proposal was rejected. American mineros preferred the low-cost labor provided by Indians and slaves and sought to keep royal interference to a minimum. The mineros failed to stop Felipe's seizure of the desmontes in 1571, but they managed to block a number of royal schemes in the intermediate term. One such scheme, hatched by the Council of the Indies in August 1580, called for a doubling in the quinto to 40 percent and the placement of Peru's entire mining industry under a single mining magnate named Ventura Espino. The proposals were shelved.[61]

In the meantime, Velasco's amalgamation recipe worked spectacularly well. The chemistry was even less clear than the Medina formulation, but Velasco's magic combination of mercury, salt, magistral, temperature controls, and other measures somehow adapted the Medina process to Potosí's deep-seated ores and discarded tailings. The financial impact was huge. The silver output from Potosí's deep, low-quality ores was raised to roughly twenty-five pesos (five marks) per quintal. In addition, the very same process was applied to twenty-six-years' worth of smelted slag ores (tailings) that had been discarded by the primitive huayras. As much as ten pesos of silver (two marks) per quintal could be recovered simply by recycling the tailing ores. Further, the Velasco amalgamation process yielded much higher quality silver than the huayra method—fineness per mark was over 28 percent higher. By 1584, when virtually all of Potosí's silver output was being processed by amalgamation, most of the huayeros were toiling underground. There were 612 separately owned mines and mining shafts to be worked.[62]

Toledo accelerated the roll-out by extending low-cost credit to mine owners willing to invest in a Saxon-styled system of water-powered stamping mills. The Saxons had actually borrowed the concept from the Romans—Caius Plinius Secundus, otherwise known as Pliny (23–79 CE), described an integrated system of aqueducts, reservoirs, dams, and storage tanks that had been deployed at the gold mining works of northwestern Iberia. Water was also a scarce commodity in the Bolivian Altiplano, especially during the summer, but the mountains east of Potosí generated huge snowfalls that could be harnessed to support a Saxon (or Roman) water system. Toledo's long-term plan, requiring an investment in the millions of pesos between 1573 and 1621, was to link a high-altitude reservoir system—supported by thirty-two earthen and masonry dams, a connecting aqueduct, manmade central canal (*ribera*), and network of feeder canals—to seventy-two water-powered ingenios. The first phase of this plan was launched along the slopes of the Kari-Kari mountains in December 1574.[63]

The productivity benefits of water-powered amalgamation were clear to all. At thirty thousand pesos per unit, the front-end cost of a water mill was over three times higher than an animal or human-powered affair. However, it could process eight times as much ore per day (155 quintals) as animals; sixteen times as much as humans; and it carried a cost per quintal (170 pesos) that was less than half of that with animals and one-fifth of that with humans. By year-end 1576, nearly two million gold pesos had been invested in (or committed to) new stamping mill projects in and around Potosí—half allocated for 33 water-powered mills and half for 114 animal and human-powered mills. While a huge portion of Potosí's silver wealth had been confiscated by Felipe II in 1571, there was enough left over from the 4.6 million gold pesos in post-quinto profits generated in 1571–1576 to fund these and other projects. The huge cash flows generated by the tailings reprocessing work, which was performed mainly by animal- and man-powered mills, were reinvested in the water-powered mills. By the time that the tailings rework was completed in the late 1570s, a string of efficient water-powered mills were in place to offset the higher costs associated with Potosí's deeper, lower-quality ores.[64]

Toledo's yeoman efforts were reflected in the numbers. Measured in eight-ounce marks, registered production at Potosí expanded steadily from a trough of 114,878 marks in 1572 to 406,121 marks (3.5 million pesos) in 1577. Not only did this exceed Potosí's most recent peak in 1550, it was more than double the then-record 171,004 marks produced at Zacatecas in 1575, matched the one hundred tons of silver produced in Central Europe at its peak in 1540, and was nearly twice the annual output of the Ottoman Balkans. The dramatic recovery at Potosí was especially pleasing to

Felipe II and his Genoese creditors—and this was only the begin-ning. The conversion to water-powered amalgamation by 1584 and a series of technical improvements to the basic process itself would combine to produce an incredible run of success at Cerro de Potosí in the decades ahead.[65]

# CHAPTER SEVEN

# CASH FLOW SQUEEZE IN THE SPANISH NETHERLANDS

**W**ealth derived from an amalgamation-led mining revolution in Spanish America would be squandered in the Spanish Netherlands. Following the state bankruptcy of 1557–1560 and the peace treaty with France (1559), record volumes of American metals were dispatched to the fortunate few creditors who had not been entirely destroyed in the process. While the forced conversion of unpaid claims into half-worthless juros had nearly sunk the Spanish Crown, the House of Fugger, and the Genoese, these and other survivors recognized that genuine mining industries had taken hold in Spanish America. The fading mining industry of Habsburg Europe was now in the hands of Felipe II's uncle Ferdinand, leaving Felipe II himself to concentrate on a source of sustainable wealth that was proving to be far greater than anyone could have imagined. Whether the imported treasures were amalgamated or smelted, or received as a quinto, diezmo, or an outright confiscation, they provided Felipe of Spain with an opportunity to convert a royal financial mess into a cash flow machine of epic proportions.

Unlike his Burgundian father, Felipe II had been born in Spain in 1527 and was anxious to enhance the status of the Spanish Crown. He continued to appoint Spanish aristocrats to the positions of viceroy and governor, but he admired professional competence even if one's bloodlines were less than blue. This meant that positions of real power were open to men like Gonzalo Perez and his son António, two Aragonese Conversos who assumed the powerful secretarial role that Francisco de los Cobos had performed so well for

Charles V. Gonzalo Perez's primary challenge was to untangle the royal financial mess of 1560 in conjunction with two other senior advisers—the Portuguese Ruy Gomez da Silva and Fernando Alvarez de Toledo, the third Duke of Alba—and to position his son António as his successor. António Perez served Felipe ably between 1566 and 1579 and would have lasted even longer had his overambition not gotten him fired.[1]

The Perezes were incapable of reining in the free-spending habits of Felipe II, by now a clearly defined Habsburgian trait, even during the state bankruptcy proceedings of 1557–1560. The bare Spanish treasury failed to prevent the absolute ruler of Spain from squandering enormous sums in France through 1559 and initiating expensive new projects like a new capital city at Madrid, El Escorial, and a refurbished Mediterranean fleet. However, Felipe had assumed the Spanish throne in January 1556 with a morsel of good news. He knew that the largest European silver bonanza since Joachimstahl had just been struck on Spanish soil, ninety miles due north of Seville at the wine-producing town of Guadalcanal, during the summer of 1555. The unlikely discovery by Martin and Gonzalo Delgado was ironic. It suggested that the Phoenicians, Carthaginians, and even Romans had actually missed a major silver deposit in the foothills of the Sierra Morena. Within months of Felipe's accession, the first four thousand ducats of excavated Spanish silver was handed over to royal authorities. Guards were stationed at the mining site until the claims could be sorted out. There was no need to alert the royal creditors.[2]

The initial assays at Guadalcanal suggested a silver concentration of roughly 25 percent per quintal, many times higher than the silver found in most argentiferous lead ores. With much at stake, Agustin de Zarate was transferred from Peru to direct the mining operations, and one thousand Saxon and Tyrolian miners were transferred from the mercury works at nearby Almadén—idled in Central Europe, they had been summoned to Almadén by Anton Fugger in 1553. The timing was excellent. The Crown was on the verge of bankruptcy, the German miners were highly skilled, and

Guadalcanal's silver-bearing lead ores were nearly as rich as the early pickings at Cerro de Potosí. They could even be treated with standard Central European smelting processes. Zarate mixed thirty arrobas of ground silver ore with roughly half as much litharge flux (a lead additive), smelted the mix into silver-lead slabs in one of the five charcoal-and-wood-fired furnaces, and hauled the slabs by mule to one of his seven refineries. Labor accounted for over two-thirds of the operating costs at Guadalcanal, but the brutal summer smelting work and noxious fumes proved to be too much for some of the Germans. Nearly one hundred local Moriscos and African slaves were recruited to replace them. The remaining costs were assumed by oak obtained from nearby properties owned by the Order of Santiago, charcoal, litharge flux, equipment, utensils, animal feed, and lighting oil.[3]

If working conditions were nearly unbearable, Guadalcanal was a short-term bonanza for Felipe II. The 74,715 marks (eighteen tons) produced in 1556 alone contributed 457,200 badly needed ducats to be confiscated and squandered. This and subsequent volumes were well below those recorded at Potosí and Zacatecas, but the mines produced nearly nine tons of silver annually through 1563. Zarate and his successor, Francisco de Mendoza, were instructed to keep all information under wraps, especially from the Genoese. The Delgados and other claimants were slowed by the state bankruptcy declaration of June 1557. In addition, a royal pragmatic of January 1559 reminded everyone that the mines of Spain (and Spanish America) were the sole properties of the Spanish Crown. Although Martin Delgado's heirs eventually received 100,000 ducats or so in 1569, they missed the discovery of another important vein at Guadalcanal. An Inquisition-exempted Morisco named Francisco Blanco uncovered a new seam in 1571 that would contribute another thirty thousand ducats of silver annually until 1575.[4]

The unexpected bonanza at Guadalcanal was hidden from Felipe's many creditors. The funds helped to refurbish the Mediterranean fleet and supported some of Felipe's other pet projects. He selected backwater Madrid as his new imperial capital in 1561 and began construction on a new royal residence at nearby San Lorenzo de El Escorial to honor his father. While Felipe's principal advisers and the Spanish Cortes were incapable of stopping these and other extravagances, Rome was a big supporter. Felipe had led the Counter-Reformation in Spain by supporting humanistic religious orders—the Franciscans, Carmelites, Dominicans, and Jesuits—that had been promoted by the Councils of Trent (1545–1563). The most influential order, the Society of Jesus, had been cofounded by Ignatius Loyola during the 1530s and raised to great international heights by Francis Xavier. There was no Protestant Reformation in Spain. Less than thirty of eight hundred suspected heretics (mainly Lutherans) were burned at the stake in Spain during the 1559–1560 period. The heresy trials and the imprisonment of liberals like Bartholemé de Carranza, the Dominican archbishop of Toledo, were intended mainly to send a message to would-be reformers in the Spanish Empire.[5]

So if Felipe wanted to place his imperial capital at Madrid, it was all right with Rome. No matter that the economic cost of the move, in terms of infrastructure and opportunity costs, would be a further drag on the taxpayers of Castile. Madrid was a curious site for an imperial capital. Located in barren, inaccessible country between the barely navigable Manzanares and Jarama rivers, Madrid was not likely to stimulate business activity of any kind. The city's population had fallen to less than fourteen thousand in 1563 and would need a strong push from the lower Castilian nobility to reach sixty-five thousand by 1598. Madrid's future would be dependent on imperial administration and "white-collar" business opportunities in law, real estate, and investment management. Otherwise, Spain continued to be dominated commercially (and culturally) by Seville and Toledo.[6]

The focus on Madrid and El Escorial must have unnerved the overtaxed citizenry of Castile. The Spanish economy had been devastated by the state bankruptcy of 1557–1560, and the vast majority of Castilians received virtually no economic benefit from the plunder and bonanzas extracted from Mexico, Bolivia,

Colombia, and even Guadalcanal—these fortunes were squandered almost entirely on foreign military campaigns. In fact, New Spain was receiving more capital investments in highways, bridges, farms, haçiendas, and port facilities than the mother country. Most of Spain's major rivers were unnavigable, her highway system was mediocre, and enterprises like the Royal Association of Carters struggled to handle freighting business tied to the transatlantic trade. Industries dependent on internal transportation, like agriculture and wool, had been neglected for decades and continued to rely on medieval ox-carts (*carretas*) and 400,000 Spanish mules.[7]

American treasure not squandered on military campaigns was more likely to be spent on showpiece palaces, churches, and monasteries than on farms and business enterprises. Madrid and El Escorial were the most lavish examples of this trend. The home economy was starved of investment capital. Virtually all of the manufactured goods that were sent to the Americas were sourced from outside of Spain, marked up, and delivered to American customers at prices that were often double their levels in Europe. Meanwhile, a 20 percent decline in Spain's sheep population between 1552 and 1563 signaled that the Crown's array of protectionist measures had priced Spanish wool out of a number of markets. The looming crisis would cripple the Castilian Mesta, a powerful royal wool monopoly that had evolved from a thirteenth-century sheepherder guild, as well as Medina del Campo. The woebegone Spanish peasantry was already suffering from rising rents, taxes, and increasingly higher prices. Most of the nation's peasant income was soaked up by landed grandees and the Spanish Church.[8]

The state bankruptcy of 1557–1560 was a wake-up call to the Spanish nobility. In the early years of the transatlantic trade, credit was nearly nonexistent in Seville and the nobility had deferred to Seville's commercially oriented Genoese and Converso communities. These energetic groups were the principal investors in the transatlantic fleets until the Conversos were eased out of the picture. Spanish aristocrats, in turn, had preferred to sit back and collect untaxed agricultural revenues, grazing fees, and inflation-driven rents. But following the lead of the Duke of Medina Sidonia, one of few Spanish grandees with any interest in commerce during the early decades, a number of Castilian nobles stopped complaining about the Genoese and Conversos and entered the fray themselves. Spanish nobles launched as many as six companias based on Genoese (and Converso) business models.[9]

Enterprising Spanish nobles learned a lot from the Sevillian Genoese, the most powerful business community in Spain in the second half of the sixteenth century. Merchant bankers like Alberto Pinelli and Silvestro Catameo, having diversified from sea loans, sugar, and slaving, became Felipe II's most trusted financial advisers after 1560. If the Fuggers were tied to the Habsburg and Holy Roman past of Charles V, the Genoese were associated with the Spanish American future under Felipe II. The Genoese had been active in Seville and Medina del Campo since the Reconquest. They dominated Seville's slave market in partnership with Portuguese slave traders and were the primary investors in the West Indies sugar industry. Junior members of Seville's Genoese community had been sent to the Americas regularly to coordinate business with their extended families back home.[10]

By midcentury, the Genoese slaves-for-sugar business had become the second-largest transatlantic enterprise after precious metals. In addition to their huge sugar plantations in the West Indies, Genoese traders were shipping slaves to Hernán Cortés's sugar mills at Tlaltenango (Cuernavaca) and Tuxtla, penetrating the Brazilian market, and providing financial assistance to the Spanish Crown. Prince Felipe had received a 200,000-ducat loan from Alejandro Catano in 1550, secured by American treasure, in return for a royal slaving asiento. The Genoese also attempted to dominate the importation of Mexican cochineal, a plum-colored dye that was distributed (via Seville and Antwerp) to the textile industry in the Netherlands. Annual cochineal exports averaged around seven thousand arrobas between 1575 and 1600.[11]

The Genoese generated even larger profits from the transfer of private American funds to Seville. This was a Fugger type of enterprise. Accounts associated with the delivery of American bullion to the Royal Mints were so important (and so lucrative) that the bankers had to be bonded as individual guarantors, often to the tune of 200,000 ducats each. Since few Sevillians possessed that much capital, the Genoese served as the principal bondsmen for a banking industry that was dominated by themselves. The near-collapse of the Fuggers and Welsers in 1557–1560 and the failure of the Bank of Lizarragas, a native Sevillian enterprise that had followed the Genoese into ship owning, slaving, merchandizing, and merchant banking, made it inevitable that the Genoese would move into royal finance. Felipe II needed huge amounts of capital to address a resumption of the Ottoman threat in the Mediterranean as well as a brewing split between Catholics and Calvinists in the Spanish Netherlands—someone had to assume the financial role played by the Fuggers and Welsers.[12]

## OTTOMAN OFFENSIVE

The Ottoman Empire was not quite finished. Suleyman had failed to fully exploit the financial crises that befell Spain between 1550 and 1560, due mainly to the cost (and fallout) from his renewed campaign against Persia (through 1555). He had even signed a peace treaty with Charles V in 1548. But like most treaties in the sixteenth century, the 1548 document was merely a piece of paper. The sultan continued to send his war galleys into the western Mediterranean with some success. By 1552, he and his Barbary allies had taken Tripoli from the Knights of St. John, secured Algiers and Fez, and were essentially running free. If that were not enough, Felipe II was forced to dispatch his underpaid Habsburg army to Rome in late 1557 to reverse a recent pronouncement by Pope Paul IV—the pontiff had become so concerned about the Ottoman threat in a "post-Charles V world" that he supported the French claim to Naples and the transfer of Sicily to Venice. Felipe's army forced the pope to back down.[13]

The peace treaty signed with Persia in 1555 had also allowed

Suleyman to focus on his legacy. The sultan was so consumed with his massive public works program in Istanbul—highlighted by the spectacular Suleymanyye complex, the construction of six other large mosques, and a new water system—that he even ignored Henri II's intriguing request in September 1557 for a two-million-ducat loan to continue France's war against a mutual enemy. However, the sultan went back on the naval offensive in 1560 by raiding Christian shipping, sacking Sorrento, and recovering the island of Djerba. The island had been captured briefly in March 1560 by a refurbished eighty-ship Christian fleet under Gian Andrea Doria, Andrea's great-nephew, and the Knights of St. John. The Christian fleet had been rebuilt between 1556 and 1559 with some of the 200,000 ducats in Guadalcanal silver. Doria's triumph was temporary. A reinforced Barbary-Ottoman navy under Piali Pasha routed the Christians off Djerba on May 12, 1560, and sank or captured twenty-eight Christian galleys. Twenty-five more went down in a violent storm off Andalusia with an estimated eighteen thousand Christian sailors onboard. The losses were a devastating psychological blow to the king of Spain.[14]

Suleyman also benefited from the receipt of fifty or so tons of silver annually from the Ottoman Balkans (Rumelia) and an associated recovery in the spice trade with Venice. Istanbul was receiving eighteen tons annually from Serbia, highlighted by Zaplana (Kosovo) and more modest contributions from Srebrenica and old reliable Novo Brdo, twenty or so tons from the prolific Sidrekapsi District, including the mines of Sidrekapsi proper and those located further north at Demirhisari, and another thirteen tons combined from the mines surrounding Uskup (Skopje) in Macedonia and those at Lofca in north-central Bulgaria. Ottoman silver production helped to reinvigorate the Venetian spice trade. Annual pepper imports (via Ottoman spice entrepôts) averaged 650 tons between 1560 and 1564—levels that were even higher than those recorded prior to the arrival of the Portuguese. Venice should have been destroyed by the Portuguese advance, the astonishing rise of Antwerp, and Istanbul's consolidation of the Mamluk Empire. But it hadn't happened. The Venetians were not skilled merchant bankers like the Genoese and Florentines and they were not an industrial power of any distinction. But they knew how to build ships, including heavily armed carracks, and how to manage trading privileges. By 1566, the Portuguese viceroy in Goa, Antão de Noronha, was estimating that as much as twenty-five thousand quintals of pepper was being shipped to Ottoman-controlled ports on the Red Sea, twice the quantity being carried in Portuguese ships around the cape, and making their way west to Venice.[15]

The Ottoman navy maintained the offensive in the Mediterranean. Suleyman blockaded Naples with the help of thirty-five Barbary ships in the summer of 1561, captured an estimated twenty thousand quintals of Lisbon-bound pepper in that same year, and raided the coastline of Spanish Granada irregularly. Profits from the forty-thousand-quintal Muslim spice trade helped to formalize an alliance with the Barbary corsairs of Algiers and Tripoli. In May 1565, Piali Pasha directed an Ottoman armada of over two hundred ships and thirty-six thousand troops against the strategically important island of Malta. The 2,500 Knights of St. John who defended Malta, a gateway to Italy, would have been destroyed if not for the timely arrival of a Spanish war fleet from Sicily. The defense of Malta in 1565 was probably as important to Christian morale as the more celebrated triumph at Lepanto (Greece) in 1571. It occurred during an unusually lengthy truce between Istanbul and Persia (1555–1578) and a surge in Ottoman state revenues. In 1567, the year following Suleyman's death, Istanbul enjoyed a 2.5 million-ducat surplus on revenues of seven million. The huge surplus raised Felipe II's fears that Istanbul was capable of recruiting 150,000 disaffected Moriscos in Granada for a Fifth Column. The Moriscos had never been comfortable with Christianity and had plenty of scores to settle.[16]

Concerns about a possible Fifth Column intensified Felipe's focus on a Mediterranean navy that had been virtually ignored by Charles V after the debacle at Algiers. The king funded new and improved shipyards at Barcelona, Naples, and Sicily to produce a fleet of about one hundred war galleys by September 1564. The new fleet was useful to rescue the Knights of St. John on Malta in September 1565 but useless to deter thirty thousand Moriscos from revolting from the Aljapurras Mountains of Granada on Christmas Eve 1568. The Fifth Column had become a reality. A full-scale military response under the command of Felipe's half-brother, Don Juan of Austria, required nearly two years of effort (and expense) in concert with a royal decree that ordered 150,000 Moriscos to leave Granada for good. By then, Felipe had two new fires to put out. He was faced with a Protestant rebellion in the Netherlands and a new adversary in Istanbul. Recently crowned Selim II (1566–1574) hoped to place his own stamp on the over two-hundred-year-old Ottoman Empire.[17]

Flush with a huge budget surplus, Selim II recaptured Tunis in 1569, seized Cyprus from Venice in 1570, and organized a massive war fleet of two hundred galleys under Admiral Ali Pasha. The sultan also controlled his Hungarian flank by making peace with Holy Roman Emperor Maximilian II—Ferdinand's son and successor in 1564—and antagonized Venice some more by awarding trading privileges to Marseilles in October 1569. France was considered neutral, and Selim must have assumed that bitter commercial rivals like Spain, Portugal, and Venice were unlikely to join forces against the Ottomans. The latest Ottoman threat, coupled with the Morisco Revolt, had prompted Felipe to convene a so-called *Junta Magna* of seventeen Spanish government officials in 1568. A royal task force was asked to come up with new sources of revenue to combat Spain's enemies at home, in the Netherlands and in the Mediterranean. Consideration was given to raising the alcabala (sales tax), fine-tuning customs duties, adding more protectionist measures to further insulate the Castilian textile industry, and identifying new methods to raise mining productivity in Spanish America. Silver production from Potosí had turned down unexpectedly, and someone like Viceroy Francisco de Toledo was needed to introduce new technologies, recruit professional mining engineers, and make infrastructural investments.[18]

Selim had overreached with Cyprus. His decision to invade the Venetian-controlled island in July 1570 prompted Pope Pius V to immediately raise the annual papal cruzada to 400,000 ducats in the hopes of organizing a Holy Christian League. Alvise Mocenigo, the Venetian doge, signed on for a retaliatory campaign against his erstwhile business partners in Istanbul in alliance with Felipe of Spain. However, Felipe agreed to participate in May 1571 only after Fernando Alvarez de Toledo, the third Duke of Alba, had finally crushed the three-year-old Protestant rebellion in the Netherlands. The papal plan was to gather a massive Christian armada of two hundred galleys, one hundred galleons, and 54,500 troops at the Venice Arsenal for a great crusade against the forces of Islam. Although Felipe's financial position was woeful, he agreed to cover half the campaign costs with the other half shared by Venice, the papacy, and, to a lesser extent, Portugal.[19]

Under the command of twenty-four-year-old Don Juan of Austria, a mighty 208-ship Christian armada left Messina, Sicily, in mid-September 1571 in search of Selim's 230-ship navy. The destination was western Greece, a Venetian-controlled region that had been raided by Ottoman corsairs during that summer and appeared to be next on Selim's shopping list. The military key was that Selim had stuck with traditional war galleys while Spain, Venice, and Portugal had supplemented their own galley fleets with scores of heavily armed galleons and carracks. When the opposing armadas caught up with each other off Lepanto in the Gulf of Patras on October 7, 1571, one of the greatest sea battles in history ensued. An estimated 160,000 men took part in the action assuming that each galley was manned by 150 oarsmen and 250 troops. The battle, however, was determined less by manpower than by shipboard artillery. The Christians outnumbered the Ottomans in guns, 1,815 to 750, and the Ottomans ran out of ammunition. The Christians captured or sunk an estimated 195 Ottoman ships, induced Ottoman losses of thirty thousand dead and wounded, took three thousand Muslim prisoners, and freed an estimated fifteen thousand galley slaves. Ali Pasha was killed by a musket ball during some of the brutal hand-to-hand fighting. Christian casualties were limited to twelve galleys, nine thousand dead, twenty-one thousand wounded, and at least one injured hand—the twenty-four-year-old Miguel de Cervantes Saavedra would have to write *Don Quixote*, published in 1605, with his right one.[20]

The spectacular naval triumph at Lepanto confirmed a fact that the Portuguese had known for decades—galleys were no match for heavily armed sailing ships. But like the Battle of Tunis in 1535, Lepanto turned out to be somewhat inconsequential. The Ottoman treasury was so flush with silver-and-spice profits in 1571 that the shipyards at Istanbul and Sinop had two hundred new war galleys and five Venetian-style galeasses at the ready by May 1572—just in time to exploit the recent death of Pope Pius V, the principal motivating force for Christian solidarity. The rebuilt Ottoman navy evaded the Christian fleet off the Greek Peloponnesus in the summer of 1572, despite Felipe's orders to find it, and prevented the Christians from recapturing Cyprus on behalf of Venice. Although Don Juan and 137 Christian ships managed to regain Tunis in 1573, most observers viewed it correctly as a temporary event.[21]

If the Ottoman military threat remained very real to Spain (and to Portugal), war-weary Venice decided to withdraw from the Holy League in March 1573 and renounce all claims to Cyprus. This typically Venetian side-deal freed Admiral Kilic Ali Pasa and an expanded Ottoman fleet of 280 galleys and seventy thousand men to recapture the famous Spanish fortress at Goleta (Tunis) in the summer of 1574, retake Fez in 1576, and scheme to assert Ottoman authority in neutral Morocco. With Spain fully consumed with another revolt in the Netherlands, Istanbul raised the ante by installing Abd-al-Malik as a puppet ruler in Morocco, ousting the neutral Saadi dynasty in the process, and posing a direct threat to Portuguese interests in West Africa. The Ottomans hoped to solidify their western interests before resuming their hostilities against Persia in 1578. Twenty-four-year-old King Sebastião of Portugal, the brash grandson of João III, was left to ponder a retaliatory strike in Morocco in alliance with the deposed Saadis.[22]

Felipe II could not have known that the ages-old Ottoman advance against Christian Europe would end with Istanbul's invasion of Persia in 1578. Like his father, Felipe viewed himself as the primary defender of Christendom and was compelled to grapple with the powerful Ottoman threat in the Mediterranean. With France conveniently sidelined with her own Wars of Religion, Felipe may be excused for underestimating the potential of a brewing rebellion in the Spanish Netherlands. It is an accident of history that this otherwise insignificant territory took center stage during the final third of the sixteenth century. In fact, the financial relationship between Felipe's military campaigns in the Spanish Netherlands and the expanding mineral wealth of the Spanish Americas is about as crucial to the future course of world history as any single factor. While Portugal, China, Japan, and even the Philippines are important elements in this course, almost everything revolved around the otherwise insignificant Netherlands and the handful of America mining districts that financed Felipe's imperial war machine.

The seventeen provinces of the Spanish Netherlands had been a relatively problem-free pocket of the Habsburg Empire prior to Felipe's assumption of power in October 1555. The transfer in royal authority occurred within months of Felipe's accession to the Crown of Spain (January 1556) and the subsequent transfer of the Holy Roman Emperorship to Ferdinand (September 1556). Felipe's immediate problem in the Netherlands was that mighty Antwerp and the regional economy were being pressured by Spain's escalating war with France, an intensifying religious battle between Catholics and Protestants, and a deepening financial crisis. The States General in Brussels simply refused his request for an additional three million *guilders* in taxes until the French army had been defeated at St. Quentin in August 1557. But those who had hoped that the subsequent Treaty of Cateau-Cambresis and the state bankruptcy proceedings of 1557–1560 would restore Brussels (and Spain) to financial health would be disappointed—Felipe was already planning his next crusades against the Ottoman Empire and the latest outbreak of Protestantism.[23]

Felipe II began a process through which the otherwise autonomous Habsburg Netherlands was folded into Imperial Spain. Of course, trading activities among Spain, the Netherlands, and Spanish America had become so interdependent that one could argue that geographical boundaries were almost irrelevant. Since the Spanish economic system was driven almost entirely from Antwerp, it was logical to assume that Antwerp was even more integral to Spain than Seville or Medina del Campo. Other than wool, Spain relied on Antwerp for most of her internal needs as well as those of her American colonies. The usually vibrant economy in the Netherlands had also provided an alternative source of tax revenues to supplement the overstretched Castilian treasury since the 1540s.[24]

Unfortunately for Felipe, the federal estates of the Netherlands (the States General) had no interest in raising taxes unless there was a clear relationship between the hike and the needs of the local citizenry. Unlike in Castile, the provincial governments had refused to fund Habsburgian military campaigns in Italy or against the Ottoman Empire. While this refusal forced Felipe to rely on extraordinary grants to bridge the budget shortfalls at Brussels, he had his "own people" in positions of great power. He returned to Spain in 1559 knowing that he had left behind his half-sister Margaret of Parma, an illegitimate daughter of Charles V and a Flemish woman, as his trusted regent over the Spanish Netherlands. Margaret was assisted by Cardinal Granville (Antoine Perrenot) until 1564 and a hand-picked Council of State.[25]

While Margaret ruled the Spanish Netherlands in a relatively tolerant fashion, even relaxing the antiheresy laws to placate some of the reformist-minded nobles, Catholicism was clearly the dominant religion. The irony of the on-again, off-again religious struggles in the Netherlands was that Charles V had been the most liberal Catholic monarch in Europe. The threat posed by the Protestant Reformation had prompted the Burgundian-born emperor to initiate a series of reforms based on the teachings of Erasmus (1466–1536). The Dutch theologian promoted a humanistic form of Catholicism that had sprung up in the northern Netherlands during the last quarter of the fifteenth century and preceded the start of the Protestant Reformation (1516–1521). Erasmus's criticism of Catholic rituals, icons, and Latin was supported by liberal Catholics like Charles V, the Fuggers, and even the powerful archbishops of Toledo and Seville. However, problems arose after Erasmus and his fellow liberals parted ways with the more strident campaigns of Martin Luther and the Anabaptists.[26]

Erasmus's reluctant break with Luther had prompted Charles V to ban Luther's writings officially in March 1521 and establish an Inquisition in April 1523 against heretical Protestants of any kind. The emperor had no need for another Inquisition in devoutly Catholic Spain, but he feared the spread of German Lutheranism into the Netherlands. In July 1523, two heretical Dutch Augustinian friars were burned at the stake in Brussels. In the following year, eight suspected Lutherans were paraded around Amsterdam's main Catholic Church with burning candles affixed to their hands and then ordered to contribute ten thousand bricks to the city's public works office. However, it was the populist Anabaptists who accounted for most of the 1,300 or so Protestants who were executed for heresy in the Netherlands between 1523 and 1565. It got really ugly. The heads of nine executed Anabaptists were returned to Amsterdam in a herring barrel in November 1533. Forty protesters who dared to occupy Amsterdam's Town Hall in May 1535 were either killed, drawn and quartered, or hanged. Fifty-one Mennonites were either burned at the stake, beheaded, hanged, or drowned in or after 1536. Order was restored until thousands of French Calvinists began to stream into Flanders in search of a more tolerant environment. Calvinism was better organized than the fragmented forces of Lutheranism and was more appealing to the upper classes. An expansion of the Calvinist advance during the early 1560s was so threatening to Felipe II that he reaffirmed the antiheresy laws in October 1565 and reversed many of Margaret's reforms.[27]

Antwerp lay on the fault line of a brewing religious showdown. The spectacular profits derived from precious metals, spices, textiles, and public finance had been devastated by the bankruptcies of

1557–1560—but the city's economic capabilities were still in place. While the great metropolis boasted 89,996 permanent residents in 1568, the key was the 14,086 visitors categorized as foreign merchants. As capital returned to Antwerp, it stimulated competition in trade, merchant banking, military affairs, and technological innovation. These forces, in turn, catalyzed the development of larger-scale financial instruments, increasingly large transactions, and multinational intermediaries to handle them. European monarchs had raised debt obligations to unheard-of levels prior to 1557 only to discover that even a king could be held accountable by his creditors. Felipe II was still empowered to confiscate private treasure shipments, but the vast majority of the Crown's unserviceable juros had either been refinanced or restructured—they could no longer be written off for good.[28]

The Antwerp Bourse recovered from the state bankruptcies of 1557–1560, in part, by creating separate markets (bourses) for public finance and commercial credit transactions. The gnomes of Antwerp hadn't reached their lofty economic heights without skills and savvy. Trading in grain futures contracts and maritime insurance policies became so popular (and so lucrative) that many of the trading excesses had to be regulated by law in 1571. While foreign exchange trading (and arbitrage) was less prevalent in Antwerp than in Medina del Campo, Lyon, and Genoa, official exchange rates were eventually established to facilitate international clearing operations and to regulate money of account. The concept of treasury bills, pioneered by the otherwise notorious Gaspar Ducci, was followed by the introduction of modern discount theory. These and other Antwerp-based innovations were eventually consolidated and refined by the Amsterdam Stock Exchange in 1609. The recovery in financial services coincided with a recovery in the transatlantic trade, commercial upturns in Flanders and Brabant, and investments in local construction, real estate, and a host of start-up ventures (including silk). Antwerp also attracted the last remaining Spanish merchants from Bruges, regained the primary overland trade routes with Italy, and would have recovered most of the seaborne trade with the Baltic if not for fresh competition from extremely efficient Dutch grain freighters.[29]

Antwerp's recovery would be challenged by the emerging politico-religious conflict in the Spanish Netherlands. If the southern provinces of Flanders and Brabant were still solidly in the Catholic (Spanish) fold, Holland, Zeeland, and the five other northern provinces were moving toward Calvinism. Felipe II's reaffirmation of the antiheresy laws in October 1565 (over Margaret's protestations) turned out to be a big mistake. It triggered a petition by two hundred outraged nobles on April 5, 1566, calling for the abolition of both the antiheresy laws and the Inquisition in the Netherlands. The Calvinist revolt had now broken into the open. The rebellion unleashed a pent-up but frightening display of religious fervor in August when Calvinist mobs (so-called beggars) ransacked towns throughout the Netherlands, destroyed Catholic icons at will, and looted churches and monasteries. Antwerp herself was raided on August 20. When the rebels learned that Margaret was authorized to compromise, they smelled blood. A full-blown rebellion erupted in the north and would have succeeded if most of the state and local militias had not been controlled by Catholics. Some of the militias, including those in Amsterdam, refused to attack the Protestant insurgents and helped to restore order by March 1567.[30]

## THE DUKE OF ALBA

The violence of August 1566 ended any hopes for moderation. Felipe II responded by appointing Fernando Alvarez de Toledo, the third Duke of Alba, as captain-general of the Netherlands in December. Salaried at a whopping thirty-six thousand *escudos*, the sixty-year-old Alba arrived in Brussels on August 9, 1567, with ten thousand troops and authority to recruit fifty thousand more from the far-flung Habsburg Empire. Compromise was over. Alba's so-called Army of Flanders, backed by an Inquisition-styled tribunal called the Council of Troubles, was authorized to restore order by whatever means necessary. This meant that over one thousand Protestant rebels would be executed for treason or heresy during Alba's turbulent reign and another sixty thousand would flee to Germany or England. Even a prominent Catholic noble like William of Orange—the otherwise royalist-leaning Stadholder of Holland, Zeeland, and Utrecht—found all of his properties confiscated in 1568 when he was suspected of having Protestant sympathies. By then, Margaret of Parma had resigned her regency in protest, and Alba, promoted to governor-general, had been authorized to serve as a virtual military dictator in the Spanish Netherlands. Alba's political powers were enhanced by other developments. The controversial death of Felipe's only son Don Carlos, the passing of his wife Elisabeth of Valois, and the outbreak of the Morisco Revolt in Granada all occurred during the course of 1568. Felipe was forced to remain in Spain and leave the Netherlands to Alba.[31]

Alba focused mainly on the nobles. While a total of 12,302 ordinary citizens would be tried for heresy through 1573, the Council of Troubles targeted the nobles and wealthy merchants for publicity, taxes, and confiscations. The shocking execution of the Counts of Egmont and Hoorn in Brussels's main square on June 5, 1568, sent a powerful message to the Protestant nobility. It also coincided with Alba's need to raise funds for his so-called Army of Flanders. Margaret had incurred budget deficits of around five hundred thousand florins annually between 1561 and 1567, staunched by nearly eight hundred thousand florins in annual subsidies from the Castilian treasury, but Alba's army would need two million florins annually from Castile between 1568 and 1571 and funds from three new taxes in the Netherlands. In March 1569, he gained approval for a new 1 percent income tax, a 5 percent tax on real estate sales, and a 10 percent value-added tax (the Tenth Penny) on exports and other commercial transactions. The new taxes were attacked immediately by the upper classes. They could deal with the 1 percent income tax but argued successfully to suspend the two other levies in favor of an otherwise whopping two-year subsidy of four million florins. Alba's fundraising schemes had politicized an otherwise loyal sector of society. Even William of Orange, mocked as "William the Silent" because of his passivity, loyalty to Felipe II, and advocacy of religious toleration, had moved to the left. He began to attack Alba's policies after withdrawing to Germany in December 1568.[32]

The combined power of the government subsidies, the 1 percent income tax, and a reviving regional economy nearly balanced Alba's budgets in 1570 and 1571. This freed up cash for Felipe's military campaigns against the Moriscos in Granada and the Ottomans in the Mediterranean. But when Felipe instructed Alba to restore the controversial Tenth Penny tax in 1572, in response to an unexpectedly sharp decline in the Bolivian quinto, local resistance was even stronger than in 1569. So strong that thousands of "fence-sitters" joined the rebel cause and helped to launch a general Dutch Revolt in the spring of 1572. Since the Dutch Revolt, in turn, forced Alba to order a general mobilization and request a whopping increase in the annual Castilian subsidy to 3.5 million florins, one wonders if the Tenth Penny was really worth the effort. The other important implication is that a temporary shortfall from Cerro de Potosí helped to launch the Dutch Revolt of 1572.[33]

Blocked by the Army of Flanders on land, the Dutch rebels had turned to coastal raiding activities since 1570. One of these raids made history when a force of six hundred so-called Sea Beggars captured the southern Holland port of Brill on April 1, 1572. The unexpected triumph, attributable in part to Felipe's inattention to local naval defenses, allowed the rebels to gain control of most of Holland (excluding Amsterdam and a few other royalist towns), block access to the Scheldt River (Antwerp) from Zeeland, and persuade William of Orange to serve as their "Protector of the Netherlands." Orange did his part by raising a rebel army of as many as fifteen thousand men by late August. Of course, Alba responded with a series of brutal counterattacks in the north—he ordered the massacre of the entire town of Naarden on December 12, 1572, and besieged Haarlem for over six months. Alba's atrocities would come back to haunt him, but in the meantime, he confined the general revolt to the northern provinces of Holland, Zeeland, and Frisland.[34]

The Army of Flanders would deploy between 10,000 and 85,000 troops annually during the course of the Eighty Years' War (1568–1648). The States General had agreed to contribute two million florins annually in 1570–1571 to support a minimum standing army of 13,000 men, plus another 4,000 garrison troops, 4,000 reservists, and 500 cavalry. No allocations were made for a navy—France was tied up with her own Wars of Religion between 1562 and 1598, and Felipe was still concerned with the Ottoman threat in the Mediterranean after his great triumph at Lepanto in 1571. But if Felipe failed to supplement the Army of Flanders with a stronger naval presence in the North Sea, a failure that allowed the Sea Beggars to also capture over two thousand naval guns from the arsenal at Veere (Zeeland) in 1572 and gain a virtual control of the regional coastline through 1576, he was betting on Alba's ability to crush the Dutch rebels completely on land.[35]

Alba's military dictatorship was successful. He restored royal authority in the Netherlands, bolstered investor confidence in Antwerp, and reduced the rebels, a group of equally ruthless anti-Catholics, to nuisance-level guerrilla warriors and pirates. Antwerp and a new generation of Genoese merchant bankers were invigorated by the huge sums required to organize, equip, and maintain Alba's massive army. While Alba had raised local taxes to onerous levels, he revived the Antwerp money market and facilitated the immediate transfer of American treasure cargoes (via Genoese bankers) from Seville to Antwerp. This was very similar to the Lisbon-to-Antwerp spice model of the early 1500s. And with France on the sidelines, he countered the presence of English and Dutch privateers in the English Channel by sending troops, supplies, and troop wages along a protected overland route between Italy and Brussels. Spanish pay-ships could travel in relative safety between Barcelona and Genoa, then by wagon train to Brussels over a so-called Spanish Road through Lombardy, Piedmont, Savoy, Franche-Compte, Lorraine, Metz, and Luxemburg. The need was clear. Prolonged delays in troop payments and provisions were likely to prompt mutinies, lootings, and plunder. As it were, the Army of

Flanders was plagued by low pay and low morale—extended absences without leave, desertions, and mutinies were common even when wages were paid on time.[36]

Antwerp was still caught between Catholics and Protestants, between royalists and rebels, and between Spain and England. The death of Felipe's English wife Mary in 1558, coupled with France's internal struggles, had freed Protestant Queen Elizabeth to assist the nascent Protestant rebellion in the Netherlands. Gresham had reformed the Crown's shaky finances by revaluing (strengthening) the pound sterling in September 1560—nearly destroying exports of English cloth in the process—and establishing a London Exchange in 1566 based on the Antwerp model. When Felipe began to confiscate American silver cargoes in Seville that had been destined, in part, for English creditors, Elizabeth increased her direct support to the Dutch Sea Beggars. She also encouraged English privateers to raid Spanish merchant ships on both sides of the Atlantic. Felipe's tit-for-tat response, establishing a trade embargo against England through March 1573, was counterproductive. Trade embargoes were usually unenforceable, and this one hurt Antwerp more than it did London. Business was simply diverted north to Amsterdam and Hamburg. The escalating conflict between Spain and England, supplemented by the removal of war-torn French ports from European commerce, laid a foundation for Amsterdam's spectacular growth in the 1590s.[37]

Sea Beggars, English privateers, and shaky royal finances forced Felipe II to pay closer attention to his treasure fleets. In 1564, he had reorganized the transatlantic *flotas* into two separate sailings from Seville—one to Veracruz (via Havana) and one to Nombre de Dios in Panama (via Cartagena)—and established an Armada de la Guardia de las Indias to protect them. While a merchant-funded *averia* (convoy tax) had been introduced as early as 1518 to finance patrols in the Canaries and West Indies, Jean de Fleury's hijacking of two royal treasure ships in 1523 had been a wake-up call to the shipping industry. Every transatlantic trading vessel was required to sail with an averia-funded escort. By 1552, a six-ship escort service between Havana and Seville was being funded by a 2.5 percent averia assigned to the value of inbound and outbound cargoes. A typical flota comprised four vessels of 250–300 tons, two 80–100-ton caravels, and 360 armed sailors. After 1564, between two and eight heavily armed galleons sailed with the flotas and made sure to land at Caribbean ports that had been refortified (at local expense).[38]

The system worked. When English privateer John Hawkins attempted to exchange three hundred West African slaves for sugar and hides in Santo Domingo in 1562, a potentially profitable exchange resulted in the royal confiscation of most of Hawkins's slaves and half of his hides. Hawkins was outraged, but Elizabeth and her Privy Council were so impressed with his privateering model that they invested in Hawkins's next project—a slaves-for-goods exchange in Venezuela and Panama. Hawkins sailed into the Gulf of Mexico in September 1568 with a five-ship pirate fleet high-lighted by the *Jesus of Lubeck*, a massive 1,050-ton hulk that had been purchased by Henry VIII back in 1544 and recently upgraded with nine 25-pounders and six culverins. Unfortunately for Hawkins, Felipe's Armada de la Guardia was lying in wait. Hawkins's gunners managed to sink three vessels, including the Spanish flagship, but the *Jesus* and two other English ships were lost off Veracruz. Hawkins and the up-and-coming Francis Drake would have to come up with a new strategy.[39]

English privateers shifted their attention to smaller-scale prey. Not only did Thomas Cobham and Martin Frobisher enjoy some modest success in the Spanish shipping lanes in the North Sea, a squadron of Spanish pay-ships, in search of a safe English harbor to evade a French Huguenot raiding party, was confiscated by Elizabeth in December 1569. Felipe was especially furious because his troops had had to evict a Huguenot colony from Florida in 1565. When the Duke of Alba countered by seizing English properties in the Netherlands, Elizabeth raised the stakes. She closed the English Channel to Spanish shipping in the same fashion that English ships had been barred from the Spanish Netherlands in 1568. These actions may have hurt English merchants, but they devastated Spanish trade and the city of Antwerp. It was small consolation to

Felipe that his forces would subsequently foil an attempt by Drake to raid Nombre de Dios (Panama) in 1572.[40]

The trade embargoes were especially painful to Spain because the home country was importing virtually all of her outbound trade goods from abroad and sending virtually all of her American treasure shipments to royal creditors and military contractors. The embargoes promoted smuggling, privateering, and pirating in the Spanish Americas and constricted a Spanish economy that was already being propped up by taxes, tariffs, trade guilds, and monopolistic prices. Sevillian businessmen, having persuaded the Council of the Indies to establish a consulado to protect their merchant guilds, defeated a proposal to open up the transatlantic trade to nine northern (Cantabrian) ports in 1573. The limitation of shipping through Seville (and Cadiz) and a few American seaports—Veracruz, Havana, Nombre de Dios, Cartagena, Acapulco, and Lima—was counterproductive. It only served to inflate business costs, reduce competition, stifle business development, and promote contraband activities. Spanish America was forced to become more self-sufficient, but even this was problematic—the viceroyalties were captives of a medieval mind-set that promoted traditional institutions, mining magnates, and a small group of hacienda owners.[41]

Seville was a protected boomtown. The city's "wild west" atmosphere attracted thousands of adventurers and transients and raised the local population to 100,000 or more. This total included thousands of displaced Moriscos from Granada, following the failed Morisco Rebellion of 1568–1570, and an increasingly large number of African slaves. Hundreds of Portuguese Conversos arrived after the Inquisition was revived in Portuguese Asia. The expanding transatlantic trade made Seville a better bet than Lisbon. Of course, smuggling was rampant in Seville, and the circumvention of Casa inspectors and averia collectors became an art form. Casa inspectors on both sides of the Atlantic were skilled at cutting lucrative side-deals. Officials discovered that only one-third of the 450,000 silver pesos aboard a shipwrecked vessel off Cadiz in January 1555 had been legally registered. It wasn't until 1592 that a bonded silver master (*maestre de plata*) was appointed to supervise the process. The deal was that the silver master received 1 percent of all registered silver in return for covering all logistical costs—notarizing, packing, loading, unloading, and transporting cargoes—and losses due to theft.[42]

The growing disconnect was that the general economies of Spain and Spanish America were not benefiting significantly from the transatlantic treasure shipments. The colonies were substituting wildly expensive imports like wheat, wine, salt, oil, tools, and horses with locally produced goods, but importing virtually all of their value-added manufactured products—finished textiles, books, paper, hardware, metalware, timber, naval stores, cereals, tin, and linens—from non-Spanish sources in Europe. American treasure shipments that evaded the royal creditors and military contractors were being exchanged for foreign products that contributed nothing to the home economy. Warnings had been sounded as early as 1558, when a government report estimated that the value of Spanish imports exceeded that of Spanish exports by a factor of nearly ten to one. Felipe's decision to raise the export duty on wool in 1558 and again in 1564 nearly priced the product out of the marketplace. The averias raised prices even higher. As increasingly large volumes of precious metals were being retained illegally in the Americas, Spain's balance of trade deficits rose and ever-higher quantities of American bullion and debt were needed to fill them.[43]

With all of these problems, Spain still managed to contain her enemy's smuggling, privateering, and embargo-running enterprises and reduced the Dutch Revolt to a nuisance-level affair for the time being. No one knew in 1572 that the Dutch had started an Eighty Years' War with Spain that would ebb and flow—continuing through 1609, relaxing during a twelve-year truce period (1609–1621), and then resuming through the Peace of Westphalia in 1648. However, the key point was that the Dutch rebels who launched this war were far more successful than they should have been. The otherwise brutal Alba declined to destroy their dikes, dams, and fields when he had the chance. He could have devastated the Dutch economy for generations. This was possible because few towns in the Netherlands had adopted the trace italienne design and the crucial Dutch river system operated without walls, earthen ramparts, redoubts, and blockhouses. One of the few towns that did adopt the trace, Maastricht, withstood Spanish frontal assaults in 1579 and drew the Army of Flanders into a costly siege. Otherwise, there was little to stop Alba's massive army during the 1570s.[44]

Unfortunately for Alba, his general mobilization of 1572 happened to coincide with an unexpected decline in Bolivian silver output and a rising financial commitment to the Mediterranean fleet. Felipe received word from Viceroy Toledo in late 1572 that Potosí was on the road to recovery but, in the meantime, the royal quinto was headed for a trough of 105,000 pesos in 1573. The shortfall mattered because Bolivian silver secured many of the royal asientos that were financing military operations in Flanders and in the Mediterranean. Felipe had followed up his 1.1 million-ducat contribution to the Holy League in 1571 with a commitment to send an average of two million ducats annually to his Mediterranean navy between 1572 and 1574. The objective was to maintain the post-Lepanto naval balance of power after reports were received that Selim II was rebuilding the Ottoman fleet and the unreliable Venetians were planning to drop out of the alliance.[45]

Istanbul was very much aware of the situation in the Netherlands. Alba had received his first contingent of Spanish troops from the Italian front in 1567 only because the Ottoman threat had ebbed temporarily by the death of Suleyman in 1566. The restoration of order in the southern Netherlands and the receipt of huge subsidies from the States General had allowed the Netherlands to pay for her own defenses in 1570 and 1571. This achievement had freed up royal funds for the Mediterranean. The Sea Beggars, in turn, were very much aware of the situation in the Mediterranean. It had been Felipe's expanded campaign against the Ottoman navy that encouraged the Dutch to go on the offensive in the spring of 1572. Louis of Nassau even planned to invade the southern provinces with allied French Huguenots. The scheme was aborted after a Huguenot army was massacred in August 1572.[46]

The ante was raised after William of Orange crossed the Maas River in late August 1572. Alba countered with a general mobilization that raised the Army of Flanders to sixty-seven thousand men. Of course, the financial pressure on Felipe II was mounting. The *monthly* cost to place fifty-seven thousand Germans and other mercenaries in the field was a whopping 1.2 million florins—if the funds were there to pay them. They were not. Most of the Army of Flanders was unpaid on September 1, 1572, and would become increasingly unhappy as unpaid wages rose to 7.5 million florins by August 1573. No matter that the Spanish treasury and the States General had each dispatched 7.2 million florins to cover the first twelve months of the campaign. It got even worse for the Crown. Alba scrambled to add another eighteen thousand troops after Louis of Nassau had opened a third front in March 1574. The Dutch held their own militarily and would be rewarded when arrears owed to the Army of Flanders reached a staggering 17.5 million florins in July 1576. At that point, as many as sixty thousand unpaid soldiers had deserted, mutinied, or reduced Europe's greatest land army to chaos.[47]

The devolution of the Army of Flanders, an enterprise that drained twenty-one million florins from the Spanish treasury during the 1572–1576 period, carried enormous consequences. The outlays represented over one–fourth of royal expenditures (eighty million), over one-third of royal revenues (sixty million), and the entire deficit (twenty million) incurred by the Crown during this five-year period. The warning signs were crystal clear in 1574, a year in which royal expenditures (twenty-four million) were more than double royal revenues (eleven million), but Felipe refused to cut his four-million-florin commitment to the Mediterranean fleet. Instead, he opted to slash the budget for the Army of Flanders to 7.4 million, over half below plan, and finesse his way through a royal budget deficit of thirteen million. No matter that most of the Army of Flanders would remain unpaid.[48]

## THE GENOESE PLAY HARDBALL

Since the States General contributed no more than one-third of the army's budget in 1574, the bulk of the financial burden was left with Spain. It was, after all, Felipe's war. The 7.4 million florins that did make it to the front in 1574 were arranged primarily by the Genoese in the form of *asientos de las provisiones generales*—anticipatory short-term loans secured by American treasure or a specific stream of extraordinary royal revenue. The Genoese had become so skilled at transferring American bullion at Besançon and Genoa during the 1557–1560 crisis that they dominated the market. Their version typically included an unsecured short-term loan, a transfer payment, a currency exchange agreement, and an attached *juro de resguardo* (security bond). The security bond would pay the issuer 5 percent from the day that the borrower failed to repay the underlying asiento. In theory, the delivery of American silver to Medina del Campo would usually trigger the issuance of a 7–8 percent asiento that, in turn, would release an equivalent value of gold in Antwerp or Genoa. Gold was over eleven times more valuable than silver and could be delivered to the Army of Flanders with fewer wagons. Delayed (or unpaid) asientos were typically rolled over into a larger issue at a much higher rate of interest, reaching 16 percent in 1573 and sometimes in excess of 20 percent.[49]

The looming problem was the juros de resguardo. The attached security bonds were needed to add a measure of security to the concurrent flood of asientos that were being issued to finance the Netherlands campaign. Not only did the juros add a potential 5 percent charge to the interest rate on the underlying asientos, they added to Felipe's risk exposure. Royal silver receipts represented nearly 25 percent of royal expenditures, almost precisely the amount of funds (two million ducats) allocated annually to the army in Flanders, and secured most of the juros de resguardo that, in turn, secured most of the underlying asientos. Chaos would reign if American production came in below plan, if the treasure fleet was delayed (or halted) by enemy privateers, or if anything else happened to disrupt the bullion delivery system. Felipe should have been concerned that virtually all of his American silver was being handed over to the Genoese to service the asientos that financed the war machine in the Netherlands.[50]

Felipe's more basic problem was that he was fighting a three-front war against the Dutch, the Ottomans, and his own unpaid troops between 1572 and 1576. A military investment of 11.4 million florins in 1574 was insufficient to save Middelburg (Zeeland) from William of Orange in February, defend Tunis, and pay thousands of soldiers. Nearly five thousand unpaid mutineers held Antwerp hostage for six weeks in April–May until a ransom payment of over one million florins was received. Law and order broke down so severely in mid-1574 that the loyal provinces of Flanders and Brabant joined the rebellion as well. Alba's growing list of atrocities, the mutineers' occupation of Antwerp, and the economic pain of the rebel blockades were too much to bear. The loss of the southern provinces, in turn, forced Felipe II to cave in in October 1574. He replaced the Duke of Alba with Luis de Requesnes, abolished the Council of Troubles, and extended an olive branch to the moderates. It was not enough. Felipe refused to restore all of Margaret's religious reforms, and Requesnes was handed an army that was breaking at the seams. A new layer of chaos sprang up when the heart of the unpaid army (mainly Germans) decided to mutiny. Another group abandoned their posts in Holland in December 1575 until 130,000 florins in back pay was delivered. Severe damage was done in the interim. The Dutch rebels gained control of every town in Holland and Zeeland except Amsterdam and Haarlem.[51]

The Crown's always-suspect financial capabilities had been weakened by the recent deaths of Diego de Espinosa, the royal exchequer and inquisitor general, and his assistant Martin de Velasco. A proposal by the Cortes in late April 1573 calling for the alcabala and its associated debts to be shifted from the Crown to the Cortes for a thirty-year period was considered and rejected. Instead, the crucial position of exchequer was transferred to Juan de Ovando y Godoy, president of the Council of the Indies and inquisitor of the Seville Tribunal. Ovando may have been skilled at administering Spain's transatlantic trade, but he underestimated the complexities of the Spanish treasury system. The financial projections of his interim report of April 1574 varied considerably from reality. Ovando followed up with a half-baked plan to raise an additional two million ducats in royal cash flow, figuring that the Crown could service an additional eight million ducats in royal debt if the Cortes would approve an increase in the alcabala, the salt tax, and the servicio. Ovando didn't stop there. He also proposed to cut the interest rate on existing juros from 5 percent to 3.3 percent, hoping to save one million ducats in annual interest expense, and to reassign various pledged revenue streams from existing debt obligations to new ones.[52]

In his revised report of August 1574, a month in which the Army of Flanders was devolving into chaos, Ovando estimated that he could free up as much as 8.5 million ducats in annual cash flow. The Cortes bolstered Ovando's arguments with a warning to the Genoese that they would have to separate a whopping eight million ducats worth of juros de resguardo, which had been attached to their asientos since 1570, from the underlying asientos themselves. Of course, the Genoese were outraged by a proposed cancellation of their security bonds—especially since a dramatic recovery was taking hold at Potosí. Their outrage prompted Felipe II to back off. The king needed the Genoese badly now that Bolivian silver was finally on its way. The Genoese were set to deliver nearly four million ducats to the Army of Flanders in 1574, more than double the volumes delivered in 1572 and 1573, and were not to be alienated. Felipe agreed to raise only the alcabala on September 22 under the assumption that the move would add another 2.5 million ducats in revenues in 1575. The decision ignored the findings of a royal task force (*junta*) that had predicted correctly in July 1573 that overtaxed Castilians were incapable of supporting any increase in the 10 percent tax. The increase in the alcabala would have to be reversed to its normal rate in 1577.[53]

A declaration of bankruptcy was delayed by the recovery at Cerro de Potosí and a temporary boost from the hike in the alcabala. By August 1575, however, the arrears on unpaid asientos had soared to nearly three years' worth of state revenues and the annual debt service on the juros de resguardos was spiraling upward. The Crown was in a real fix. A last-ditch offer to securitize the arrears on the asientos with a new issue of higher-yielding juros was also a nonstarter—revenues from the inflated alcabala were coming in well below plan and there were no unpledged revenues left in the realm to service a single new juro. A declaration of bankruptcy (*decreto*) appeared to be inevitable.[54]

Felipe must have figured that he could negotiate a debt-restructuring agreement along the lines of the *medios generales* that had resolved the crises of 1557–1560. During those bankruptcy proceedings, the Crown had suspended (not canceled) payments on existing debt obligations and forced her creditors to exchange huge volumes of high-interest loans for lower-cost juros. A royal decreto was intended to send a message to the royal creditors. Those who refused to issue new loans or roll-over (write-down) existing ones were being advised to restructure their outstanding claims in a fashion that benefited the Crown—or run the risk of being paid nothing. Creditors lucky enough to survive the ordeal, like the Fuggers, Welsers, and Genoese, incurred massive write-downs and vowed to be more wary in the future. So while the Genoese came to dominate the purchase of royal asientos during the 1560s and 1570s, they were careful to attach security bonds—juros de resguardos—to the underlying loans.[55]

Felipe had no reason to think that the Genoese, despite their kicking and screaming, would rebel against "accepted practice." They would be required to convert their 16 percent-plus, short-term asientos (secured by extraordinary revenues) into newly issued 5 percent, long-term juros al quitar (secured by ordinary revenues) in the same fashion as in 1557–1560. Felipe viewed the 16 percent-plus asientos as a fluke, reflecting the fact that the amount of outstanding asientos had soared to an eye-popping 200 percent of total state revenues. He knew that the savvy Genoese, fearing the worst, had been swapping unpaid asientos for newly issued juros since

1573 and reselling these same juros to unsuspecting Spanish nobles and officials. The traditional juro investors were confident even after the amount of juros al quitar had soared to thirty-five million ducats. The Crown had almost always honored her juro commitments, and all juros were serviced by specific district taxes collected in specific districts. Local officials could usually be persuaded to raise taxes to guarantee a juro's safety.[56]

Unfortunately for all concerned, there was simply not enough collateral in the system to handle the rising tide of juros and asientos. The resulting Spanish state bankruptcy of September 1575 was a seminal event in world history—not because of the temporary pain inflicted against Felipe, the Genoese creditors, the Spanish investors, and the general Spanish economy. What really mattered was the fallout from delaying payments to the Army of Flanders during the three years (1575–1578) of difficult negotiations between Felipe II and the Genoese. The Genoese refused to cave in. In fact, they responded to Felipe's decreto of September 1, 1575, with a triple-barreled action to raise the likelihood that their ten million ducats in asientos and eight million ducats in juros de resguardos would not have to be written down at all. The Genoese imposed a debt ceiling on royal borrowings, halted the issuance of further loans, and most critically, placed an embargo on the delivery of precious metals (troop wages) to the Army of Flanders. Their resolve was stiffened by the royal response. Ovando called for an official investigation of all asientos issued between 1560 and 1575 in order to recoup "usurous" or "illegal" profits that had exceeded a "fair" return of 12 percent—comprising a "legal" interest rate of 10 percent and a 2 percent brokerage fee. No matter that most of the royal debt was the result of roll-overs and most of the charges and penalties had been agreed to contractually.[57]

Felipe raised the stakes by convening a "Usury Tribunal" on November 14, 1575. A small group of merchant bankers, highlighted by the Genoese Spinolas and a branch of the House of Fugger, refused to open their books to Ovando's inspectors and demanded special status during the work-out negotiations. Felipe refused. He didn't care that these creditors were stuck with as much as five million ducats (nearly 25 percent) of the royal debt and that nearly one-third of the Fuggers' two million-ducat claim had been due since 1559. Nor did he care that Lorenzo Spinola had helped to finance the construction of Felipe's beloved El Escorial between 1569 and 1575 and was owed 720,000 ducats from this project alone. He also forgot that Spinola had advanced Felipe 200,000 ducats in 1574, based on a personal pledge, to assist the Mediterranean fleet. The Crown dug in by issuing a second decree on July 15, 1576, that reaffirmed the 12 percent "fair return" ruling and ordered the immediate delivery of all financial accounts to royal officials. There would be no exemptions.[58]

If the Genoese were outraged, the timing of their actions—especially the embargo on the delivery of precious metals to the Army of Flanders—was excellent. The embargo would allow William of Orange to roll back some of the army's recent victories by the summer of 1576 and retain fragile control of the northern provinces. The Dutch Revolt could have been crushed for good in 1575. The removal of the Duke of Alba in October 1574 had helped to mend fences in the southern provinces and the recovery at Potosí had led to a doubling in bullion deliveries with much more to come. The deliveries kept tens of thousands of Spanish troops at their posts and allowed the Army of Flanders to regain the military offensive. Another four million ducats in 1575 and 1576 might have snuffed out the Dutch rebellion for posterity. But the war was being financed by the Genoese, and they stood to lose over half of their seven million ducats of unsecured claims against the Spanish Crown. The Genoese viewed the revival at Potosí and the Crown's recent string of military victories as potential leverage—by halting the Army of Flanders in its tracks in 1576, they would be less likely to incur a massive write-down on their claims and would even have a decent opportunity to be repaid in full.[59]

The army's territorial advances in 1575, achieved on a relatively modest budget of 3.5 million florins, had been a last-ditch effort to regain Holland and Zeeland before a state bankruptcy was declared.

Flanders and Brabant were back in the royal fold, and a full-scale invasion of the north looked very promising. However, the bullion embargo was as successful as the Genoese could have hoped. It halted the asientos that financed the Army of Flanders and slashed troop wages to less than two million florins in 1576. The chaos of 1573–1574 was repeated. One group of mutineers abandoned their Dutch posts in December 1575, followed by an even larger wave of unpaid Spaniards after the surrender of Zierikzee in July 1576. As most of Holland and Zeeland reverted back to rebel control, thousands of unpaid Spanish troops marched into Brabant in search of plunder.[60]

The great disconnect was that Cerro de Potosí had recovered mightily between 1572 and 1576. Production turned up in 1573 and was on track to reach 406,121 marks (3.5 million pesos) in 1577. If the value of this output was just a modest fraction of total state revenues, silver was the most important source of security for the flood of asientos and juros de resguardos that had financed the Army of Flanders and the Mediterranean fleet. So the depressed royal quinto of 1573 had undermined the entire financial system. Existing asientos lost their underlying security and the issuance of new loans grounded to a halt. The timing of Potosí's revival, occurring prior to the bankruptcy declaration of September 1, 1575, was also momentous. It allowed Felipe and the Genoese to play hardball against one another—collateral was on its way—until a mutually acceptable agreement could be hammered out in December 1577. It was this two-year-plus delay that allowed the breakaway northern provinces to consolidate their gains and to shortly launch a history-making Dutch Republic.[61]

In the meantime, the Genoese were in no hurry to conclude a medio generale with Felipe II. Most of the unpaid Army of Flanders had either mutinied or deserted by August 1576 and were engaged in the extortion of southern towns for funds and supplies—pay up or be sacked. While a series of royal edicts branded the mutineers as outlaws, the edicts were unenforceable. Royal tax officials were classifying plunder lost to mutineers as military tax contributions. The death of Luis de Requesnes in March 1576 added to the chaos.

A number of Catholic nobles in the south organized private militias of their own to restore order, resorting in one case to arresting the royalist members of Brabant's Council of State. Others opened up discussions with the breakaway northern provinces (Holland and Zeeland). The new militias did little to halt the waves of mutineers who were ransacking the countryside. Frustration erupted on November 4, 1576, when Antwerp, already devastated by the various embargoes, was sacked by thousands of unpaid, unhappy mutineers. During an eleven-day period, the "Spanish Fury" inflicted five million guilders worth of local property damage, two million guilders worth of plundered cash, metals, and booty, and over seven thousand dead. An estimated 2,600 tons of Antwerp plunder was transported to Italy in May 1577.[62]

The Sack of Antwerp was too much. When the Catholic-dominated States General met in Ghent on November 8, they demanded that the Spanish Army leave the Netherlands immediately. They also called for religious toleration and the return of all confiscated properties. The show of unity and moderation displayed at the so-called Pacification of Ghent might have salvaged a compromise solution with the Crown, but the situation on the ground was out of control and the Crown was still negotiating with her creditors. The Calvinist activists who had helped to start the Dutch Revolt had their own ideas about pacification. They managed to briefly capture Antwerp, Brussels, and Ghent and declined to share political power with the local Catholic nobility. In fact, the Protestant rebels smelled blood. They refused to demobilize, they refused to restore Catholicism in the northern provinces, and they signaled their intention to make an outright declaration of independence.[63]

The Pacification of Ghent failed to rein in the Protestant rebels, but it prompted the Catholic nobility and the Crown to arrange a face-saving deal in early 1577. The Perpetual Edict of February 12, 1577, called for the southern provinces to restore Catholicism as the state religion in exchange for two actions: the removal of all Spanish troops from the Netherlands within twenty days; and the suspension of the antiheresy laws and the Inquisition. While the second action represented a major cave-in on the part of the Crown, it freed Felipe to return to the bargaining table with the Genoese in March. By then, Juan Fernandez de Espinosa had replaced the reviled (but deceased) Ovando, and the spectacular carnage wrought by the Spanish Fury had given the Genoese the upper hand. The largest holders of juros des resguardo—Lorenzo and Agostino Spinola, Nicolão Grimaldo, Agustín Gentil, Lucian Centurione, Estevan Lercaro—were even willing to resume lending if the Crown agreed to standardize some of the repayment schedules in a common currency. Felipe eventually agreed. He and incoming governor-general Don Juan of Austria figured that they would avenge their assorted humiliations at a later date.[64]

The general chaos had forced Felipe to back down. The formal medio generale that was signed on December 5, 1577, required a reluctant Felipe to acknowledge that 15.2 million ducats in royal debt had in fact been outstanding at December 31, 1575, and that 8.1 million of this total was represented by valid juros de resguardos. In return, the Genoese and the other creditors agreed to accept a less onerous write-down of 2.2 million—under a trumped-up charge of "illegal profit taking"—after two-thirds of their claims had been converted into 3.5 percent juros al quitar and one-third was repaid in cash and other assets. The key point for Felipe (and Don Juan of Austria) was that the Genoese agreed to advance five million ducats in Spanish and Italian gold coins ("de oro en oro") over a six-year period. The funds were intended to supplement the services of embargo-runners like Jerónimo de Curiel. An ex-royal factor in Antwerp, Curiel managed to scrape up 1.7 million florins in Paris in August 1577 as a bridge to the arrival of 1.8 million ducats in Genoese gold in April 1579. These and other funds allowed Don Juan of Austria to invade French Flanders and the southern provinces in due course.[65]

The combined benefits of the medio generale with the Genoese, the dramatic expansion at Potosí, and a truce signed with Istanbul in February 1578—the Ottomans were back at war with Persia through 1590—allowed Felipe to regain the offensive in the Netherlands and exploit an unexpected opportunity next door. But considerable damage had been done. The Genoese embargo had been effective because the Genoese controlled the delivery of money, men, and provisions along the Spanish Road, and Protestant privateers were blocking the English Channel. If carrying bullion along the Spanish Road was expensive and risky, sending bullion to the Netherlands by sea was even more dangerous. Half of an 800,000-ducat shipment to the Netherlands was somehow lost en route during the summer of 1577, leaving the Crown to secure an emergency loan of 150,000 at 33 percent.[66]

The Genoese lending cartel had demonstrated its economic clout. Felipe had sought to circumvent the Genoese by arranging deals with Curiel and other Iberian bankers at Medina del Campo, Seville, and Lisbon. Other than Curiel's last-minute heroics in Paris, the plan was a complete failure. No one had the financial power of the Genoese, and few non-Genoese lenders wished to jeopardize their future dealings with the cartel. The Portuguese Crown had cash flow problems of its own, and so did Habsburg Austria—Felipe had some sway through his marriage to Anna of Austria in 1570, but he was not Holy Roman Emperor. One of the few merchant banks that could defy the Genoese, the weakened but still formidable House of Fugger, managed to deliver only 333,333 ducats in bullion annually during the 1575–1578 period. This was a fraction of the two million ducats needed in the Netherlands. Small consolation to the Fuggers that the last of the great medieval fairs, Medina del Campo and Genoese-dominated Besançon, were nearly through. Anton Fugger had played a major role in converting the Antwerp Fairs into the world's first daily bourse and would have appreciated the gradual roll-out of modern bourses to London, Venice, and a host of other European towns.[67]

The absence of the Fuggers from the work-out negotiations of 1575–1578 signaled that the Genoese were now the true power in European finance. The former uncrowned kings of European commerce had been in decline since the death of Anton Fugger in 1560. An avalanche of unpaid royal claims and the incompetence of

Anton's heirs destroyed much of the family's wealth and reputation. Affairs had deteriorated so badly in 1563 that one nephew, the forty-seven-year-old Hans Jacob, was exiled from Augsburg with personal debts of over one million florins. After three other nephews passed away between 1569 and 1579, the family business was split into two branches—one controlled by Anton's sons Marx, Hans, and Jacob and one by his more capable great-nephews Philip Eduard and Octavian Secundus. The latter set of Fugger brothers, the grandsons of Raymund, formed a pepper-and-spice trading firm in January 1574 to exploit the family's extensive factory system.[68]

The House of Fugger remained a financial power in Antwerp and Augsburg, but the two towns had passed their peaks. Augsburg was devastated by the state bankruptcies and the declining fortunes of the Central European metals industry—over seventy Augsburg firms failed between 1556 and 1584 thanks to the steady depletion of the great silver mines and the gravitation of the European copper business to Sweden. Antwerp recovered from the state bankruptcies of 1557–1560 only to be rocked by the chaos years of 1572–1578. The heirs to Erasmus Schetz's estate were nearly destroyed during the process—no matter that one son (Gaspar) had been appointed royal factor in 1559 and the other (Melchior) had purchased the former residence of Gaspar Ducci. The Schetz brothers watched much of Antwerp's money market business flow to emerging banking (and insurance) centers like Cologne, Hamburg, and Frankfurt. A further insult was inflicted when Colonel Carl Fugger participated in the Sack of Antwerp on November 4, 1576.[69]

The state bankruptcy of 1575–1578 also damaged the Spanish Empire. Overtaxed Spaniards on both sides of the Atlantic did not need an increase in the alcabala during the recession of 1570–1578, the temporary crisis at Potosí, and a deadly epidemic in Mexico that killed nearly half of the already-reduced Indian population in 1576. The economic recovery that followed the medio generale of December 1577 was a mixed blessing. The good news was that transatlantic tonnage would double between 1562 and 1608. The bad news was that non-Spanish manufactured goods represented most of the outbound volume. Sevillians were also concerned about the failure of native banks like the House of Espinosa and the growing influence of the already powerful Genoese. When the general banking industry recovered in the mid-1580s, the position of the Genoese would be stronger than ever. So strong that Sevillian public opinion would turn against the Genoese, scapegoating them for the apparent decline of Spain herself. One writer, Balthasar Gracias, wondered "if Spain would not have had the drag of Flanders, the blood-letting of Italy, the gullies of France, the leeches of Genoa, would not all of her cities today be paved with gold and encased in silver?"[70]

The "drag of Flanders" was extended by Don Juan. Although the Perpetual Edict of February 1577 had called for the removal of all Spanish troops from the "obedient" southern provinces within twenty days, the Army of Flanders was staying put. The thirty-three-year-old governor-general determined that a coastal evacuation route was too dangerous, the Spanish Road was too expensive, William of Orange needed to be stopped, and a financial deal was likely to be worked out shortly with the Genoese. Don Juan's analysis was accurate, but his betrayal of the southern provinces was repaid in kind—he died from typhus on October 1, 1578. Everyone in Flanders and Brabant was left to ponder the intentions of his replacement, Alexander Farnese. Farnese, the Duke of Parma, opted for a "carrot and stick" approach. Having weighed the relative merits of policies adopted by Margaret, Alba, Requesnes, and Don Juan between 1567 and 1578, Parma's first step was to reach an agreement with the Catholic nobles of Walloon in May 1579. The Treaty of Arras returned Hainaut, Artois, and French Flanders to the royal fold, removed all Spanish troops from Walloon, and offered the nobles a package of Castilian-style privileges that had been unheard of in the Spanish Netherlands.[71]

The Treaty of Arras covered Parma's southern flank and allowed the refurbished Army of Flanders to go on the offensive. The coast was clear. Record-breaking silver volumes were being generated from Bolivia, France was engulfed in her own Wars of Religion, a peace treaty had been signed with Istanbul, and Felipe's fourth wife,

Anna of Austria, had given birth to a long-awaited male heir. The arrival of future Felipe III on April 14, 1578, averted a possible succession crisis—Maria's unstable son Don Carlos had died suspiciously in 1568 during his imprisonment in Madrid's Alcazar and the daughters of the late Elisabeth of Valois, Isabella and Catherine, were not recognized as suitable heirs.[72]

The birth of a royal heir was overwhelmed by an unexpected development in the summer of 1578—the Portuguese Crown had just been vacated thanks to a foolish attempt by Sebastião I to recover a larger share of the caravan gold trade in Morocco. The king's untimely death in late July meant that the infant Felipe was second in line to occupy the unified throne of Spain and Portugal. The twenty-four-year-old Sebastião had hoped to place a different type of stamp on Portuguese history. He had resorted to international finance and extortion to raise funds for the Morocco campaign. This meant leaning on an Augsburg merchant banker, Konrad Roth, for a loan of 400,000 cruzados in exchange for an exclusive three-year pepper contract with the Crown's business partners in Malabar. It also meant extorting an additional 240,000 cruzados from Portuguese Conversos in exchange for recognizing a papal suspension of property confiscations under the Inquisition. These funds were still not enough to cover the cost of a seventeen-thousand-man expedition to Morocco. A collection of Portuguese, Spanish, German, and Italian troops consumed an investment of over one million cruzados.[73]

Sebastião's campaign was a very big mistake. Most of the Portuguese army, including the monarch himself, was destroyed near El-Ksar el-Kebir (south of Tangier) by a hard-to-believe force of forty-one thousand Moroccan cavalry and eight thousand infantry. The geopolitical consequences were profound. On one hand, the concurrent battlefield deaths of the two rival sultans transferred the Sultanate of Morocco to the neutral Ahmad al-Mansur—the restoration of a neutral Morocco helped to redirect Ottoman military attention to Persia, formalize a treaty between Istanbul and Spain in August 1580, and even freed al-Mansur to invade the Niger Bend region (unsuccessfully) in 1590. On the other hand, Portugal's staggering losses were magnified by the transfer of royal power to Sebastião's sixty-six-year-old great-uncle Cardinal Henrique. The youngest son of Manoel I was all that stood between Felipe II and the Portuguese Crown. Felipe II had a solid claim. He was a grandson of Manoel I, he had married João III's daughter Maria in 1543, and it was his sister, Juana, who had given birth to the unfortunate Sebastião. When Henrique died in January 1580, Felipe's claim exceeded that of António, an illegitimate grandson of Manoel, and provided an electrifying opportunity to unify the two Iberian Crowns.[74]

With the Portuguese army in ruins and the royal treasury in disarray, there was little to prevent Felipe II from asserting his legitimate claim to the Crown of Portugal. A merger between Imperial Spain and the undercapitalized Portuguese Empire held tantalizing economic potential in 1580—Potosí was setting new production records, Spanish Manila had emerged as a credible rival to (and partner of) Portuguese Macao and Nagasaki, and the Brazilian sugar industry was expanding. Portugal's boom-and-bust spice franchise was another valuable asset. The revival of the Venice-Levant trade (under Ottoman supervision) had essentially divided the global spice market into three profitable thirds. The latest spice boom, complemented by recent Portuguese investments at Kanara (pepper), Ambon (cloves), and Bantam (pepper), had attracted a new generation of European spice traders—Portuguese Conversos and descendants of the Fuggers and Welsers.[75]

In addition to the economic potential, Portuguese monarchs had been just as committed to Catholicism as Felipe II. The efforts of Cardinal Henrique—Portugal's inquisitor general (1539–1580), regent (1562–1568), and king (1578–1580)—had coincided with the remarkable exploits of the Society of Jesus. The modest, three-man Jesuit mission that arrived in Portugal in 1540 had gone on to establish a major presence in Goa, a noted university at Evora, and an aggressive Inquisition against suspected heretics at home and abroad. Portugal's Protestant population was virtually nil, Judaism and Islam were banned, and most Moriscos had already relocated to Morocco. That left a Portuguese Converso (New Christian) population of around sixty thousand to be either befriended, persecuted, or extorted—depending on the situation. Nearly twenty thousand suspected heretics were condemned by the Portuguese Inquisition between 1543 and 1684; 1,379 of whom were burned at the stake, and induced to leave home. The condemned were a sizable component of the thirty thousand or so Portuguese Conversos who immigrated to the Far East, Brazil, the Dutch Republic, and even Istanbul by 1604. Many of these emigrants started (or participated in) successful trading enterprises. Thousands of other Conversos joined the religious orders or took advantage of an emigration-for-fee policy between 1604 and 1610.[76]

In short, domestic Portugal looked a lot like domestic Spain in 1580, with one notable exception—Felipe's royal treasury was in unusually good shape, thanks to the revival at Potosí, the conclusion of the medio generale in December 1578, the truce with Istanbul, France's continuing internal struggles, and the ease with which Parma had regained French Flanders in 1579. There were ample funds available for a full-scale assault on Lisbon. After António, the Prior of Crato, had dared to claim the Crown of Portugal for himself in early 1580, Felipe suspended Parma's offensive and sent a heavily armed Spanish fleet to Lisbon in June. The fleet supported an overland campaign under the Duke of Alba, a man who was anxious to restore his tarnished but otherwise solid reputation. He did. The Spaniards routed António's seven-thousand-plus army at Alcântara on August 25, leaving António to flee to the Azores, and conquered the overmatched Kingdom of Portugal within months. Felipe was crowned king (in Tomar) in April 1581. The king won immediate plaudits for leaving Portugal's internal and external affairs in the hands of Portuguese and paying a ransom fee of 300,000 cruzados to gain the release of Portuguese captives held in Morocco since 1578. Of course, he needed these men to resume the military offensive in the Spanish Netherlands.[77]

Felipe's four-year fixation with the Portuguese Crown, coming on the heels of a timely bust-and-boom cycle at Cerro de Potosí, had allowed the Dutch rebels to further consolidate their gains in the north. The consolidation benefited from an otherwise uneasy political union between Holland and Zeeland since June 1575, the addition of neutral Amsterdam in the fall of 1578, then Utrecht, Guelderland, Overijssel, Groningen, and Friesland at the Union of Utrecht in January 1579. Finally, on July 26, 1581, the entire group issued an Act of Abjuration—an outright declaration of independence from Spain and the concurrent creation of a Dutch Republic that would last until 1795. Of course, seven backwater Dutch provinces were the least of Felipe's worries in 1581. They and their Protestant friends in England could be handled at a later date, after Portugal and the "obedient" southern provinces had been straightened out. But the Crown's failure to honor the terms of the Perpetual Edict of 1577 had hardened attitudes in the north for good. The new Dutch Republic would make no compromises with Felipe II or Felipe III. Not only that, the two Felipes would discover very painfully that the underestimated Dutch were capable of exploiting the unified Iberian Empire even more effectively than they were.[78]

# CHAPTER EIGHT

# CONVERGENCE IN THE FAR EAST

T he insignificant Dutch Republic established in 1581 would somehow fuse together the global achievements of the unified Iberian Empire—but not quite yet. In the meantime, the unification of the two Iberian Crowns in 1580 sent shockwaves around the world. Akin to a huge corporate merger, Spain brought the mineral wealth of the Americas and a massive military machine, while Portugal added a global commercial empire, an excellent (if overstretched) navy, and an exciting position in the Far East. There were plenty of challenges. While the Spanish Crown was squandering a spectacular series of treasures from Bolivia, Colombia, and New Spain on military affairs, Portugal was adjusting to the ebbs and flows of the global spice trade, dealing with the Ottoman Empire, defending her primary factories at Hormuz, Goa, and Malacca, and preparing to exploit her modest positions in Brazil, Angola, Mozambique, and the Far East.

The sixty-year period between Manoel's death (1521) and Felipe's accession had been rocky. Istanbul's conquest of Mamluk Egypt and Syria in 1516–1517 ended a Golden Age of profits from the metals-for-spices trade and the associated currency arbitrage opportunities—nearly every Portuguese ship had returned from Malabar with nearly one million cruzados in spices during the golden 1510–1518 period. A staggering 48,097 quintals (2,315 tons) of pepper was received in 1518 alone. However, pepper and spice imports sank to just 18,184 quintals in 1530 before stabilizing at 30,000–40,000 quintals during the second half of the century. While Portuguese Malacca held her own with Moluccan spices, the home country was left with conventional trading relationships that fluctuated with market forces. Lisbon's misfortune was directly related to Istanbul's restoration of the traditional Muslim spice routes through Aden, Cairo, and Aleppo and the inability of the Portuguese to block them from Hormuz. Suleyman's refurbished navy captured Aden in 1538 with a fleet of seventy-two Egyptian ships, thwarted Estevão da Gama's attempt in 1541 to destroy the Ottoman fleet at Suez, and extended Ottoman influence to as far east as Basra (Iraq) by 1546. Increasingly large quantities of Malabar pepper and Indonesian spices returned to the fabled bazaars of Damascus, Beirut, Aleppo, and Alexandria.[1]

The revival of the traditional Muslim trading routes carved up the global spice market into three rough pieces—a Venetian-Ottoman third, a Portuguese third, and a third for everybody else—and led to a surge in pepper and spice plantings in India and Indonesia. The fragmented market helped to stabilize prices and allowed most spice traders to prosper from rising demand in Europe and China. However, the Ottoman advance and Lisbon's cash flow problems prompted João III to rethink the entire enterprise in the 1540s. First, he ended a twenty-eight-year-old cooperative (tributary) arrangement with the sultans of Hormuz in 1543 by seizing control of the island's seventy-five thousand cruzados in annual customs revenues and converting a fortified naval base into a fortified trading emporium. Then he played hardball with the Affaitatis. The Crown had been happy to receive a large annual advance from Giovanni Carlo Affaitati's spice syndicate in Antwerp, leaving the risk and the reward with the contractors, but João nearly ruined his partner by failing to repay a 2.2-million-cruzado debt in a timely fashion. The king responded to his own incompetence by relocating Portugal's spice-marketing activities to Lisbon in 1549, shifting more of her pepper purchases from Malabar to northerly Kanara, and eventually terminating the Affaitati contract in 1554.[2]

João was being squeezed by irregular spice profits and the costs to maintain a powerful naval presence in the Indian Ocean. A four-masted, six-hundred-ton Portuguese carrack of 240 feet, armed with five-foot-long (or longer) artillery pieces capable of firing sixty-pound iron balls through the sides of an enemy hull, conveyed enormous power, prestige, and fear in the Indian Ocean. This firepower greatly surpassed the modest four- to eighteen-pound stone shots that had been fired against Calicut in 1503 and the light gunnery carried by ships like the *Santa Catarina do Monte Sinai* in 1520. But it did not come cheaply. By 1543, Lisbon was spending roughly 138,000 cruzados per year to protect the royal spice route between Goa and Lisbon and to replace vessels lost to shipwreck, piracy, and warfare. This was relatively small change to people like Suleyman, Charles V, and Anton Fugger, but it was very significant to a nation of less than 1.5 million. Despite occasional extraordinary subsidies from the Portuguese Cortes, the assumption of revenues from the two military orders (Santiago and Avis) in 1550, and the regular issuance of long-term government bond offerings (*padrões de juros*), João suffered through an accumulating string of budget shortfalls between 1548 and 1557. State revenues were at only 607,000 cruzados when a Portuguese state bankruptcy was declared in 1557.[3]

Lisbon's problems were Goa's problems. The viceregal capital of the Estado da India was forced to adjust to increasingly regular budget deficits, the transition from a Lisbon-centric business model to an inter-Asian one, and various geopolitical developments. However, the conversion of a second-tier horse depot into the cosmopolitan viceregal capital of Portuguese Asia had been remarkable. Conveniently located between the textile ports of Gujarat and the pepper orchards of Kanara and Malabar, Goa managed all Portuguese trade between Lisbon and the fifty-odd Portuguese feitorias situated between Mozambique and Malacca. The port was always undermanned, but Goa rarely had to defend herself when there were heavily armed Portuguese carracks in the vicinity. Tropical diseases and desertion were more dangerous to the Portuguese military than enemy attacks. The city was initially sited on the forty-eight-square-mile island of Tissuary, but the gradual addition of adjacent territories extended Portuguese Goa to 275 square miles. Produce from the surrounding farmlands was supplemented by imported foodstuffs from Bijapur and the northern ports of Cambay. The Jesuits played a major role in local affairs. They started the Cathedral of Santa Caterina in 1562 and established seven parish churches, sixty-two local missions, and the two-thousand-student São Paulo College. Although Christian Goans would outnumber the local Hindu population by 1600, very few Muslims were willing to convert to Christianity. The Jesuits' reestablishment of the Inquisition in 1561 was even more counterproductive.[4]

Goa resembled medieval Lisbon architecturally, but her prosperity was dependent on the jumble of waterfront businesses and warehouses that lined the mile-long Rua Directa. The Ribeira Grande, modeled on the Venice Arsenal and Lisbon's Armazém, served as Portuguese India's principal shipyard, repair center, arsenal, foundry, and mint. A port typically defended by only a few hundred soldiers was swelled by many hundreds when the fleet was in (and by thousands during the summer rainy season). Minimal defenses and limited agriculture forced Goa's ten thousand Portuguese and mixed-race (*mestico*) population to cooperate more fully with neighboring Bijapur after Goa's primary ally, the two-hundred-year-old Hindu kingdom of Vijayanagar, was destroyed by a Muslim League in 1565. Relations were also improved with Gujarat. Scores of Portuguese merchants had shifted their focus from Malabar pepper to Gujarati textiles long before the Moguls swooped into Gujarat from northern India.[5]

The southerly advance of the Moguls was beneficial to Portuguese interests. Having inherited some of the fragmented domains of Timur's legendary Mongol Empire of the early fifteenth century, Mogul emperor Akbar the Great projected his own imperial ambitions after assuming the Mogul throne in 1556. He managed to conquer the commercially powerful Sultanate of Gujarat in 1572 without upsetting Persia, Gujarat's traditional trading partner, and skillfully avoided a possible clash with Portugal, Istanbul, or Calicut. By then, the Ottomans were in no position to move against anyone—their 1.7-million-ducat surplus of 1567–1568 had been spent, their navy had just been crushed by a Christian League at

Lepanto (1571), and Muscovy (Russia) was threatening their territories along the Black and Caspian Seas.[6]

The Portuguese reached a mutually beneficial "understanding" with Akbar the Great in 1573. Having just withstood attacks by the Calicut-led Muslim League in 1569 and an Ottoman fleet in 1571, Goa needed allies. So did Akbar. The Portuguese could continue to trade with Mogul Gujarat in exchange for a free *cartaz* and Portuguese naval protection. Regional trading surged after most of the coastal shipping business was licensed out to private Portuguese. As many as 350 small trading vessels sailed every four to six months in convoy fashion between Goa and Portugal's Cambay ports—Diu, Daman, Bassein, and Chaul. Several hundred Portuguese Converso families based in Goa and Cochin invested roughly eight hundred thousand cruzados annually in the Gujarati textile trade and worked effectively with local Muslim enterprises. Conversos and *mesticos* drove a new merchant class that was more entrepreneurial than the bloodlines-centric model found in Lisbon and official Goa. Akbar eventually designated Surat as his primary Gulf of Cambay port to control the bribes and kickbacks that were embedded in the Portuguese system. Even after Goa (and Calicut) had been surpassed by Surat, the annual trade between Goa and Portuguese Cambay was valued at over four million rupees (over 1.7 million cruzados) in 1600. This is what attracted the English East India Company to Cambay in 1610.[7]

When Jan Huyghen van Linschoten visited Goa in 1583, he commented on the power of the viceroys, noting the placement of their portraits in the great hall of the royal palace, and described Goa's morning-only marketplace as a combination Antwerp-like bourse and a trade fair. He observed the scores of authorized hawkers, money changers, and arbitrageurs who set up shop there:

> [They dealt in] all kinds of costly jewels, pearls, rings and precious stones . . . many sorts of captives and slaves, both men and women, young and old . . . Arabian horses, all kind of spices and dried drugs, sweet gums . . . and many curious things, out of

Cambay, Sinde, Bengal and China. . . . [Some earned large profits by purchasing] rials of eight, when the ships come from Portugal . . . and keep[ing] them until the month of April, which is the time when ships sail to China. . . . [Others purchased Persian larins via Hormuz, because merchants] must have these larins with them to Cochin to buy pepper and other wares. . . . [As for the viceroys, they] have great revenues, they may spend, give and keep the King's treasure, which is very much, and do with it what pleaseth them, for it is their choice, having full and absolute power from the King.[8]

Goa's relationship with Malacca, the crown jewel of Portuguese Asia, was in transition. The Portuguese had begun to wear out their welcome. Portuguese traders who had initially paid premium prices for Asian goods had become just as cost-conscious as their Muslim competitors. When peaceful competition failed, the Portuguese captains resorted to forced discounts, inflated customs duties, and extortion. The corruption scared away hundreds of Asian traders and invited military action. Malacca was attacked twenty-five times during the sixteenth century, fourteen of those by the feisty Achinese of northern Sumatra. Sultan Alauddin Riayat Syah al-Kabar (1527–1571) schemed to capture the tin mines of Perah and the fragmented pepper and rice markets of Java after gaining control of Sumatra's northern pepper and gold fields. The sultan attacked Malacca unsuccessfully in 1537, 1547, 1551, and 1568, but he did manage to sack Portuguese-friendly Johore (opposite Singapore) in 1565 and avoided the Portuguese fleet by sending pepper ships through the underdefended Sunda Straits. Backed by naval assistance from Suez and the always-cooperative Zamorin of Calicut, the sultan and his successors converted the Straits of Malacca into a virtual war zone. Malacca remained impregnable, but she was forced to improve her relations with neighboring Sumatrans and Javans.[9]

The regional spice trade that had made Malacca's fame and fortune was fragmented. Religion was no longer an important business factor because Islam had come to dominate Indonesia. Islam promised to replace Hinduism's rigid caste structure with something more egalitarian. The Portuguese traded officially with the Hindus of western Java and unofficially with the Muslim ports on the northern side of the island—Demak, Japara, Bantam, Mataram, and Grise. Some ten thousand quintals of pepper were received annually from Bantam alone during the 1560s. Other Portuguese were trading textiles with the Sumatran pepper traders of Jambi and with the Chinese traders who frequented it. Jambi was selling as much as 1,250 tons of pepper to Malacca (and China) on a regular basis. Portuguese traders reexported Chinese goods to Timor (and Solor) to obtain roughly two hundred tons of sandalwood for their Chinese customers. Try as they might, the Portuguese failed to exploit the cinnamon riches of Sri Lanka. It was not until after 1610, nearly a century after a Portuguese factory had been placed at Colombo, that the captain of Colombo had solidified Portuguese control, raised annual cinnamon output to four thousand *bahars* (eight hundred tons), and broken up the hereditary Salagama peeling monopoly. In the meantime, Linschoten observed in 1583 that Sri Lanka's (Sinhalese) elephants "are esteemed the best and sensiblest of all the world [since even] elephants of other countries do reverence and honor to [them]."[10]

Linschoten estimated that three hundred ships visited Malacca annually during the 1580s but added that East Indian textile traders from Coromandel, Masulipatnam, and Pulicat had supplanted most of Malacca's spice merchants in town. Annual customs revenues had fallen to less than forty-five thousand cruzados when Englishman Ralph Fitch visited the port in 1588. Foodstuff-deprived Malacca allowed two hundred Javanese rice ships to land duty-free and lost other duty opportunities to contraband trade—customs duties were applied to only half of the four thousand *bahars* (eight hundred tons) of nonfoodstuff goods. While Fitch was hard-pressed to separate royal business from private enterprise, Malacca was still well defended. The Portuguese continued to repulse the Achinese and would thwart a series of Dutch blockades in 1601. The unification of the Iberian crowns in 1580 meant that the Malacca–Macao–Nagasaki trade route was now protected by the Spanish fleet in

Manila. And no one was capable of blockading the Straits of Taiwan (opposite Fujian).[11]

Malacca's ties to the Spice Islands had become frayed. In Ternate, Jesuit missionaries created more enemies than they did converts. Sultan Baab Ullah finally evicted the Portuguese from the island between 1575 and 1578 and even invaded the neighboring island of Tidore. These actions prompted the Portuguese to arrange a peace treaty with the Sultan of Tidore, angering the Banda traders who had monopolized Tidore's clove crop to date, and to expand their fledgling clove industry on the island of Ambon. Ambon's clove crop would exceed one thousand quintals per year in due course. The Bandas responded by prohibiting the Portuguese Crown from their mace and nutmeg business and promoting Islam on Ambon—the Ambonese port of Hitu traded so heavily with the Muslim merchants of Java that Hitu emerged as a Muslim center of note and led the eviction of local Christian missionaries.[12]

Private Portuguese merchants filled the void left by the Portuguese Crown. Sixteen-year-old King Sebastião had even been persuaded to privatize the sixty-seven-year-old royal pepper monopoly in 1575. Since scores of Portuguese textile entrepreneurs had offset the collapse of the royal monopoly in India, a proposal to split the royal pepper monopoly on a fifty-fifty basis with a syndicate managed by the obscure Konrad Roth of Augsburg made sense. Entrepreneurship triumphed. The so-called Europa Contract of 1575 was so successful that Sebastião authorized Roth and other licensed contractors (*contratadores*) to import a fixed amount of pepper and spices at a fixed price. The so-called Asia Contract signed on March 1, 1577, was subsequently combined with the Europa Contract of 1575 to supply the Crown with as much as thirty thousand quintals of pepper (valued at 200,000 florins) on an annual basis. The catch was that Roth was obligated to lend Sebastião a whopping 400,000 cruzados for an upcoming military campaign in Morocco.[13]

The talented Roth presold 17.5 million shares in the Asia Contract of 1578 to a syndicate of Portuguese and Italian investors, prior to Sebastião's demise at El-Ksar el-Kebir, to match the buying power of the Venetian-Ottoman-Egyptian spice alliance. The Venetians were also importing thirty thousand quintals of pepper via Cairo and Alexandria on an annual basis. Unfortunately for Roth, only four of his five ships managed to return safely to Lisbon in August 1580, and their disappointing cargo, twenty thousand quintals of Malabar pepper, was insufficient to cover the total costs. The ten-thousand-quintal shortfall, on top of the unpaid 400,000-cruzado loan to the late Sebastião, bankrupted Roth immediately.[14]

The outbound leg of Roth's voyage was covered by *Fugger News-Letter* (#30), filed from Cochin on January 10, 1580. The private news service had been launched by Eduard Fugger in 1568 to integrate nonconfidential intelligence from the family's global network of agents with reports (and rumors) filed by scores of worldly travelers. The Cochin reporter stated that Roth's outbound fleet had left Lisbon on April 4, 1579, with cash, wine, oil, cheese, fish, paper, and other goods in anticipation of a profitable exchange of pepper at Goa on October 10, 1579. However, the reporter noted the struggles of India's independent traders:

> The country [India] is no longer as it was formerly. . . . Our viceroy imposes so many new taxes that all commerce diminishes. . . . Buying and selling here is more profitable than sending many wares to Portugal. German merchandise has no market here and is useless for this country. Writing tables split in the great heat, clockwork or anything else made of iron deteriorates at sea . . . there is nothing to send to Portugal, for pepper, ginger, maces, cocoa-nut fat have all been brought for the contractors.[15]

## CHINA AND JAPAN

The gathering collection of threats in the Indian Ocean prompted the Portuguese Crown to relax the onerous tribute system and to step up commercial activities in the Far East. The shift from a Lisbon-centric model to an inter-Asian one allowed scores of former soldiers, government officials, and clergymen to enrich themselves with the type of smaller-scale, Asian-style partnerships that had been practiced for centuries by native rulers and native merchants. The new model also had a new center. Goa retained the title of administrative capital, but the centers of Iberian commerce were transferred to three former fishing villages: Macao (1557), Nagasaki (1570), and Manila (1571). Portugal's gradual penetration of the Far East was a logical extension of the complicated series of port-to-port transactions that had operated between West Africa and East Asia since 1500. Malacca had been created as a Chinese satellite in the early fifteenth century and had returned the favor for Portugal since 1511. The growing pressures against Malacca had already shifted Portuguese (and Spanish) attention to the biggest commercial prizes in Asia—China and Japan.

Of course, a direct trading relationship with China had been a European (and papal) objective since the mid-thirteenth century. By 1266, two Venetian merchants, Niccolò and Maffeo Polo, had followed an earlier round of missionaries to the new Mongol capital at Cambaluc (Beijing). The Polos hoped to pursue trading opportunities with emperor Kublai Khan, and the emperor hoped to pursue trading opportunities with the wider world. The Polos were so encouraged by their discussions that they returned to China in 1271. Niccolò's son Marco was left behind to chat with the emperors and to chronicle the wonders of the Far East. Marco Polo's astonishing (and mainly accurate) reports included Europe's first details about the mysterious islands of Cipangu (Japan), Sumatra, and the spice markets of the Indian Ocean. After Marco's return to Venice in 1295, another round of Christian missionaries was dispatched to the Orient in search of converts and geographic information. A "Michelin-like" guide to South Asia, *Libro di Divisamenti di Paesi*, was shortly in the hands of Italian merchants.[16]

Unfortunately for the West, the exploitation of this work was halted by the Black Death, the Ottoman advance, and the accession of the insular Ming dynasty in 1368. No matter that the Mings sought briefly to foster global trading relationships between 1402 and 1436. During this brief opening, Admiral Zheng He (Cheng

Ho) sailed powerful junk fleets into the western Indian Ocean in pursuit of trade and tribute. The shift to isolationism in 1436 closed China's trading doors officially but inspired scores of Guangzhou-based merchants to find ways to circumvent the system. Unofficial trade between China and Southeast Asia could almost be described as regular by the time that the Portuguese captured Malacca in July 1511. Since the Portuguese had exploited West Africa, East Africa, India, and Indonesia in stepping-stone fashion, it was only natural that China (and Japan) would be penetrated in the same way. Portuguese Macao, a nine-square-mile peninsula that remained in Portuguese hands through 1999, became an extension of Guangzhou, earlier Portuguese activities in Malacca (and Japan), and the willingness of emperors Longqing (1567–1572) and Wanli (1573–1619) to reopen China's trading doors officially.[17]

The establishment of Macao in 1557 was the product of over forty years of effort. Portuguese adventurers had visited the Chinese coastline since 1513, trading directly with Neilingding Island through 1521, and even hoped to construct an unauthorized fortress on Tamao (Hong Kong). Beijing may have been unaffected by Portugal's seizure of Malacca in 1511, but the displaced Sultan of Malacca complained regularly to Chinese authorities. Viceroy Lopo Soares de Albergaria attempted to clear the air in 1517 by dispatching Tome Pires as an ambassador to the Chinese court at Beijing. The new ambassador had served as a pharmacist in Malacca between 1512 and 1515 before returning to Cochin. Pires and a four-ship Portuguese fleet under Fernão Peres de Andrade arrived on Tamao in August 1517 and was permitted to sail up the Pearl River to Guangzhou. Pires noted the presence of exotic Japanese junks along the river, but his stay turned out to be longer than expected— he would be held up in Guangzhou for seven years.[18]

Andrade's brother Simão managed to destroy whatever goodwill remained between the two powers. Not only had he begun construction on an unauthorized fortress on Tamao during his expedition of 1518–1519 but he outraged local authorities by enslaving Chinese children. The actions carried grave consequences for Pires after he had finally been invited in January 1520 to meet with Emperor Zhengde unofficially in Nanjing (Nanking). But when Zhengde died unexpectedly in May 1521, with Pires en route and ignorant of the traditional period of official mourning, Pires had a real problem. The new fifteen-year-old Emperor Jiajing (1521–1566) was not interested in meeting a Portuguese envoy. The unfortunate Pires was arrested on his return trip to Guangzhou and died in a Guangzhou prison in May 1524.[19]

Skirmishes began to break out between Portuguese privateers and Chinese junk fleets. Sixty Portuguese sailors were captured in 1522 and died in possibly the same Guangzhou prison that held Pires. In 1523, Emperor Jiajing threatened to ban all trading between China and Portugal unless Malacca was returned to the sultan. The Portuguese responded by establishing a new trading post at Lappa, an island near Macao, and traded from there illegally during the 1520s. Chinese officials returned the favor by reopening the port of Guangzhou to any foreign flag except the Portuguese—this was the first official opening of trade since 1436. Undaunted, the Portuguese continued to trade illegally with the provinces of Zhejiang and Fujian despite a new political storm tied to the Crown's failure to punish atrocities committed by Captain Jorge de Meneses in the Moluccas in 1527–1530. When the Portuguese established a second unauthorized trading post at Ningbo (opposite Shanghai), Chinese authorities outlawed trading there in 1544 and then sacked the port in 1548. Even with these setbacks, Portuguese Malacca continued to trade regularly with Chinese merchants (in Malacca) and served as a base for the scores of enterprising Portuguese who traded irregularly (and illegally) with China and with China's unofficial business partners in Japan.[20]

Japan, an archipelago comprising the main island of Honshu, three major islands (Kyushu, Shikoku, and Hokkaido), and scores of minor ones, turned out to be Portugal's primary gateway to China—this despite the fact that no European had set foot on Japanese soil until 1543. Japan and China had been irregular trading partners since ancient times but had resumed their latest relationship (unofficially) with the upgrading of Japan's primitive mining industry during the 1530s. While Chinese merchants prized Japanese copper and silver, Japan's feuding warlords needed gold, a metal that mineral-poor China obtained in trade, to pay their warlord armies. Rumors of unofficial copper-and-silver-for-gold exchanges eventually reached Malacca via the so-called Gore traders who arrived irregularly from the Ryukyu Islands. The southern Japanese archipelago (including Okinawa) was a mystery to most Malaccans. Of course, so was Japan. The expansion of Japan's otherwise dormant mineral industry was a direct result of the bloody civil wars that had engulfed Kyushu and Honshu since the 1460s. Scores of regional warlords (*daimyos*) fought to control Japan's scattered deposits of precious metals. The warlords inadvertently created a modern Japanese mining industry that, in turn, facilitated the gradual consolidation of power under Oda Nobunaga (1534–1582), Toyotomi Hideyoshi (1536–1598), and Tokugawa Ieyasu (1542–1616).[21]

Japan was a latecomer to standard mining, metallurgical, and minting practices that had existed in China and Korea for centuries. Iron making was not introduced to Japan until around 200 CE, followed centuries later by rudimentary copper and gold mining and the importation of copper coins from China and Korea. The introduction of gilded bronze images and copper vessels was delayed until the arrival of Buddhism in the early seventh century. Casted bronze images of Buddhist divinities, fabricated from the copper deposits of Musashi province and the adoption of a Chinese mercury coating process, were commissioned by the imperial court at Nara in 709. Modest amounts of silver, known as *gin* or *shiro-kane* (white metal), were mined in Tsushima and Iyo provinces after 674 and used occasionally to decorate swords, scabbards, necklaces, rings, armlets, and bowls. Silver was never applied to ceremonial Buddhist objects.[22]

While modest quantities of gold had been mined in Mutsu province at the northern tip of Honshu since 700, it wasn't until richer deposits of Mutsu gold were discovered in the twelfth century that a significant metals trade developed between Japan and China.

Samples of Mutsu gold were presented to the Southern Song dynasty's Imperial Court in 1175, followed in later years by as much as 15,000 *ryo* (7,800 Troy ounces). Trading with China expanded with the start of the Kamakura period in 1185 but ended with the Mongols' failed invasions of Japan in 1274 and 1281. The mines of Mutsu province, noted secondhand by Marco Polo in the 1290s, were constrained by primitive mining and metallurgical techniques. All of the copper coins in circulation were of Chinese origin. The shortage of precious metals even prompted the introduction of Chinese-styled bills of exchange in 1282 to facilitate official business between the Imperial Court in Kyoto and the military headquarters in Kamakura. No official coins were minted in Japan until the seven-inch-wide, gold-silver *oban* was introduced in the 1570s.[23]

Japan's fifteenth and sixteenth centuries were plagued by an endless series of civil wars among hundreds of feudal warlords. Since most warlords needed precious metals to pay their troops, control of regional mineral deposits became a primary military objective. The problem was that Mutsu's gold deposits had been nearly depleted, there was a general paucity of silver, and no one knew what to do with the silver-bearing copper ores uncovered in the 1430s at Dogamaru in Iwami province in southwestern Honshu. Like the copper ores of upper Hungary, Dogamaru copper contained a relatively large amount of silver that could not be separated efficiently with existing metallurgical practices. Japanese warlords were unaware of separation processes that had been introduced in China and Korea, leaving fifteenth-century Japanese "mines" to be excavated as pits and shallow trenches. Needy warlords simply exported these silver-bearing copper ores to China and left the separation and minting to the Chinese. An estimated 154,500 *kin* (1,450 quintals) of untreated copper ore was sent to China in ten ships in 1451.[24]

Relations between Japan and China deteriorated with the depletion of Japan's copper deposits in the second half of the fifteenth century. China's always-tenuous supplies of precious metals were also in decline. The population of the Ming Empire had exceeded

100 million, and Chinese merchants were starved of silver to transact business internally and externally. Domestic silver production had peaked at around thirty-four tons per year during the epic 1409–1434 period. The Ming treasury in Beijing received 100,000 *taels* (four tons) of silver in only four years between 1436 and 1520 and averaged less than 50,000 *taels* (two tons) during the entire eighty-four-year period. A modest silver strike made at Yuwang Gorge east of Beijing in 1556 was insignificant. As Portuguese Malacca emerged as an unofficial source of imported European silver, Japan was short of silver herself and was a prohibited trading partner of China.[25]

Chinese (and Korean) metallurgists had at least been cupelling silver for centuries by either separating silver from argentiferous lead (galena) ores or by adding lead to treat lead-free ores. If lead-based cupellation was practiced at all in Japan, it was primitive and probably consumed too much charcoal to be profitable. So eyebrows must have been raised in 1526 when an entrepreneurial Hakata-based merchant named Kamiya Jutei hired two Korean mining engineers to introduce the so-called *haifuki* (ash-blowing process) to the otherwise untreatable silver-bearing ores of Iwami province. Silver was somehow separated by "blowing ash" into a molten mixture of lead-silver compounds, possibly along the lines of the temperature-raising smelting methods employed by native Bolivians and Mexicans. Following a successful demonstration at the Omori mine between 1526 and 1533, the warlord Ouchi was thrilled to see Iwami's annual silver production soar to possibly eight tons between 1533 and 1540. This was insignificant compared to a Joachimstahl, Potosí, or Zacatecas, but large enough to be monitored closely from Ouchi's perch at Yamabukiyama Castle. Jutei transferred the new process to the Ikuno mine in Tajima—located, like Iwami, within one hundred miles of Kyoto—and encouraged the miners to deploy iron hammers, chisels, picks, geological surveys (*sunpo-kiru*) and civil engineering techniques that had been used to construct warlord castles. Ventilation systems, hand-operated pumps, and wooden supports were adopted in later years.[26]

The *haifuki* process was eventually leapfrogged by the two-step (*Yamashitabuki*) smelting process, a technique imported either by Chinese merchants, Koreans, or Portuguese, but Jutei deserves credit for catalyzing a Medina-like smelting revolution in Japan. The spread of *haifuki* and other processes to other sites raised Japanese silver and copper production significantly during the 1540s. It also reopened the trading doors to China. Japanese production figures are murky, but three vessels bound for China in 1542 were captured with three tons of Japanese silver aboard (worth around eighty thousand cruzados). Annual Japanese silver exports to Zhenjiang, Fujian, and Guangdong provinces may have reached eleven tons (worth 265,000 cruzados) during the 1550s. The surge in copper output was reflected in the 2,800 quintals (179 tons) shipped to China by three Japanese cargo ships in 1539. The load was worth around 67,200 cruzados. The eventual adoption of Thurzo's separation technology would raise annual Japanese copper production to a hard-to-believe 1,172 tons in the early seventeenth century.[27]

If the scale of Japanese silver exports to China during the 1540–1580 period is unclear, China's narrow 4:1 bimetallic ratio meant that three tons of captured silver in 1542 was worth nearly nine tons of the European variety. Metal-starved Chinese silk merchants in Ningbo, Haicheng, and Guangzhou were elated by a mining revival catalyzed by Japan's warring daimyos as well as by the modest amounts of Central European silver being delivered by unofficial Portuguese traders. On-again, off-again trading bans were relaxed whenever prized cargoes of silver were at stake. In addition, it was the expansion at Omori and Ikuno that drew the first Portuguese traders (the first Europeans) to Tanegashima Island in 1543. Their visit to this small, iron-making island off the southeastern tip of Kyushu was a typical ad hoc affair. Francisco Zeimoto and António da Mota were two of one hundred passengers aboard a Chinese trading junk that was shipwrecked off the eastern coast of Tanegashima on September 23, 1543. The *nanban* ("southern barbarians") were received warmly by the island's ruler, the fifteen-year-old Tokitaka, who appreciated their Chinese trading goods and their exotic European arquebuses (muskets). The Portuguese were allowed to return to Ningbo in the repaired junk after a stay of around five months. When other Portuguese traders returned to Tanegashima in 1544, they delivered enough arquebuses, gunpowder, and manufacturing expertise to help launch one of the first gun-making enterprises in Japan.[28]

Portuguese traders visited the larger island of Kyushu irregularly until a Chinese junk landed at Kagoshima on August 15, 1549, with Francis Xavier, two other Jesuit priests, and three Japanese Christians. Their arrival (from Malacca) was no accident. The remarkable Xavier (1506–1552) had relocated to Goa in 1542 after cofounding the Society of Jesus with Ignatius Loyola during the 1530s. Catalyzed by the reform-minded Councils of Trent (1545–1563), the Jesuits offered a humane form of Catholicism to thousands of Hindus, Buddhists, and heathens in South Asia. They established churches, hospitals, schools, charitable institutions, and business enterprises. The results were mixed. Goa was a success story, supported by a formal bishopric in 1534, and Malacca was home to as many as five thousand Christian converts by 1550. However, most Muslims, Hindus, and Buddhists were happy with their own religion and were turned off by the regular attacks against suspected heresy. But Japan (and China) presented exciting opportunities for missionary work in 1549, and the Philippines would become as devoutly Catholic as Spain herself.[29]

The Jesuits also helped to fill a void created by an increasingly stodgy and corrupt viceregal system. While the number of Portuguese nationals serving in Asia rarely exceeded fifteen thousand—a total comprising *casados* (married settlers), *clerigos* (clergy), *soldados* (soldiers), and unauthorized *lancados* (entrepreneurs)—the Jesuits and the other major religious orders provided these nationals with an attractive alternative. More than a few frustrated groups—Portuguese captains, Conversos, military deserters, retirees, and non-Muslim Asians—were persuaded to join the Jesuits when it became clear that the order was a more dynamic (and more lucrative) institution than the Portuguese Crown. Yet the Jesuits helped to strengthen royal authority abroad by providing energy, spiritual assistance, and charitable institutions of lasting value. The reestablishment of the Inquisition in Goa in 1561 was counterproductive, but the Jesuits became unusually influential in the Far East and provided excellent "on the ground" intelligence to monarchs and merchant bankers alike.[30]

In the meantime, the cofounder of the fledgling Society of Jesus had landed on an island kingdom that had received its first European visitor only in 1543. Kagoshima was selected as Xavier's starting point only because one of his fellow passengers happened to have family there. While the first Christian mission in Japan was limited to around one hundred Kagoshimese converts, Xavier's outbound letter of November 5, 1549, provided the first on-site description of Japan. He left Kagoshima in the summer of 1550 for a two-month stay in Hirado, a northern Kyushu port that was already home to a number of Portuguese traders, and then struggled his way to the imperial capital of Kyoto. Xavier's eleven-day stay in Kyoto in January 1551 was tarnished by a regional civil war and Xavier's failure to deliver appropriate presents to Kyoto's gift-happy rulers.[31]

Following his return to Hirado, Xavier made sure to offer suitable presents to the daimyos of Yamaguchi, Ouchi, and Yoshitata in late April 1551. The grateful daimyo of Yamaguchi received a grandfather clock, music box, glass mirror, three-barreled musket, books, brocades, spectacles, crystal vases, Portuguese wine, and other European exotica. The daimyo was so impressed with Xavier (and his gifts) that he authorized the establishment of Jesuit missionary activities in Yamaguchi. Xavier proceeded to convert over five hundred Japanese to Christianity over a two-month period and would have been even more successful if not for intense Buddhist opposition and the death of the friendly daimyo on September 30. An inconsequential visit to Bungo province prompted a disappointed Xavier to return to Goa on November 20, 1551, and prepare for the first Jesuit mission to China.[32]

By 1551, most of the Portuguese traders and missionaries operating in Japan had learned that Japanese silver output had expanded significantly at the mines of Omori (Iwami) and Ikuno (Tajima). Despite their eviction from Ningbo in 1549, the Portuguese were sending increasingly large amounts of Japanese silver to China illegally in exchange for gold, raw silk, and muskets. The latest ban on official trading between China and Japan presented Portuguese traders (and coastal Wako pirates) with a lucrative business opportunity—they could serve as middlemen between the two powers. Portuguese merchants visited Hirado every year between 1550 and 1561 and began to dovetail their efforts with Jesuit mission-building activities on Kyushu and possibly China.[33]

The scattered (illegal) Portuguese trading communities of southeast China were an unstoppable force. Some of them had conducted business with Chinese merchants since the 1520s. The subsequent arrival of Jesuit missionaries proved to be decisive. No matter that local Buddhists had been so hostile to Francis Xavier in 1549–1551 that he was forced to leave Japan somewhat abruptly. The undaunted Xavier left Goa in April 1552 to try his hand in China. The forty-six-year-old missionary arrived on the island of Shanghuan in mid-September but found no one willing to carry him up the Pearl River to Guangzhou. It was too dangerous. Xavier filed his last letter from Shanghuan on November 21 before being stricken with a deadly fever. He passed away on December 3, 1552, knowing it would be up to his successors to penetrate Japan more fully and to enter China at all. As Xavier's remains were being transferred to Goa in 1554, the Portuguese were already negotiating with the mandarins of Guangzhou to permit a Portuguese trading station at Lampacau Island. It was the five Jesuit priests who arrived at Lampacau in 1555 who helped to establish Portuguese Macao in 1557.[34]

Beijing had shifted gears when it became clear that Portuguese traders and Wako pirates were making too much money at the expense of the imperial treasury. China's acute metal shortage and the birth of a genuine Japanese silver mining industry had altered the regional landscape. Portuguese traders had relocated again to the islands of Lampacau and Shangchuan, sited southwest of Macao in Guangdong Province, and traded from there between 1549 and

1557. Quasi-legal trade resumed in Zhejiang Province following the death of hard-line governor Chu Yuan in 1549 and the transfer of at least twenty thousand cruzados in bribes in 1553. When trading activity emerged in Macao's Hao-jing Bay in 1555, the Portuguese were working on a new (Japanese) angle to enhance their position with Chinese authorities. The business potential in Japan was so attractive to João III that he had issued a royal patent in 1550 to manage an annual sailing between Goa, Malacca, and Kyushu. A massive seventeen-ship fleet under Leonel de Sousa arrived on Kyushu in the summer of 1555 to support the scores of Portuguese traders and Jesuits who had operated there since 1549 and to send a powerful message to Beijing: if China was disinterested in Japanese silver, the Portuguese had plenty of buyers in the Indian Ocean.[35]

An enterprising Ming official in Guangdong Province, Wang Po, was suddenly open to negotiations. Wang Po was very much aware of the silver boom in Japan and recognized that the otherwise detested Portuguese could serve as legal middlemen to gain access to this highly prized metal. Wang Po also saw an opportunity to enlist armed Portuguese vessels against the Wako pirates who were terrorizing China's coastal waters. The Wakos made thirty-four separate raids against the Chinese coastline in 1555 alone. So Wang Po agreed to Sousa's ambitious proposal of January 15, 1556: to formalize trade between China and Portugal and to allow the Portuguese to establish a feitoria on the Macao (Hao-jing) Peninsula. When the landmark deal became official in 1557, the Portuguese were required to pay a 24 percent customs duty, license fees, gifts, and rent and open their books to Chinese authorities. In return, Portugal gained access to a lightly populated fishing village that had no agricultural base and featured a few shrines and temples dating from the thirteenth century. Wang Po was immediately impressed by the sharp reduction in Wako activity in the South China Sea. Armed Portuguese carracks reduced the number of pirate raids to twenty or less in the decade ahead.[36]

China was light in precious metals but heavy in high-quality silks. According to legend, the ancient silk-making process had been established by Lady Xiling, the wife of the so-called Yellow Emperor, during the 2600s BCE. The secretive process involved the cultivation of mulberry trees and bushes, feeding the mulberry leaves to domesticated silkworms, retrieving the silk from the worms' cocoons, and then spinning and weaving the silk into bolts of fabric. Mulberry leaves from the Yangtze River valley, especially the white leaves from eastern Shangdong province, were eventually determined to be the highest quality foodstuff for the revered silkworms. This led to the establishment of the first imperial workshops in Shangdong during the Han dynasty (206 BCE–220 CE) and the birth of the legendary Silk Roads. The most famous Silk Road of the third to ninth centuries—dotted by an array of Buddhist temples that provided spiritual sustenance, rest-stops, and warehouses—ran between the Tang dynasty's cosmopolitan capital of Chang'an (present-day Xi'an), via the fabled Turkestani silver depots at Samarkand, Bukhara, and Merv, and the western termini at Antioch and Tyre.[37]

This same silk-for-silver trading network was maintained by the later Song, Yuan (Mongol), and Ming dynasties. However, it was the Mings who concentrated China's silk industry in the eastern coastal provinces. Production yields soared from plantings of small mulberry bushes (rather than trees) and increasingly ambitious quotas from Beijing—by 1605, Emperor Wanli was demanding 180,000 bolts of silk production annually. The imperial workshops at Suzhou and Hangzhou had become world-renowned during the Song and Yuan periods for their silk-weaving skills, twills, brocades, satins, and specialty weaves. The Mings added another twenty-two workshops in eight eastern provinces, including Beijing, Nanjing, Fujian, and Guangzhou, and encouraged the establishment of urban workshops at Shaoxing, Ningbo, Jinhua, and Taizhou. The three hundred looms and three thousand silk workers employed at Nanjing, for example, turned out high-volume silk fabrics for the Chinese masses and the export trade.[38]

The unexpected expansion of the Japanese silver industry created a mutually beneficial business opportunity for three parties—Chinese merchants prized Japanese silver, Japanese merchants prized Chinese silk (and gold), and Portuguese merchants prized middleman profits. The establishment of Macao in 1557 legalized the three-party enterprise on a permanent basis. Macao's potential was highlighted by João de Barros's publication of *Asia* in 1563. An ex-factor at the Casa da India, Barros collected the observations of earlier Portuguese visitors and became the first European writer to note the Great Wall of China, gunpowder, printing, and the official prohibition of direct trade between China and Japan. Barros underestimated the large volumes of Japanese silver and copper that were being shipped to Macao, but he noted Japanese exports of swords, pikes, lacquerware, screens, and human slaves (mainly captured Koreans). The bronze cannon works at Goa and a later foundry at Macao were eventually supplied by Japanese copper and Malaysian tin.[39]

The last major pieces of Macao's economic foundation were set in place when Beijing reasserted imperial control over Fujian (1564) and Guangdong (1566) to stamp out the lawlessness, corruption, and pirating that had plagued these provinces for decades. Emperor Longqing (1567–1572) even licensed fifty Chinese ships to trade with Southeast Asian ports (excluding Japan) in 1567. China's actions prompted Japanese warlord Oda Nobunaga to authorize (order) Portugal to consolidate her scattered Kyushu trading operations at a single factory at Nagasaki in 1570. Visited by Xavier in 1549, Nagasaki had been an insignificant fishing village until Christianity took hold in the 1560s and a Christianized daimyo, Omura Sumitada, transferred the port to Jesuit control in 1571. Despite a steady expansion of China's licensed merchant fleet to 137 ships by 1597, the trade between Portuguese Nagasaki and Portuguese Macao would be dominated by Portuguese carracks.[40]

Silver production, Japanese unification, Chinese openness, and Jesuit influence went hand-in-hand. Jesuit missionaries played a major role in establishing a virtual Portuguese trading monopoly between Nagasaki and Macao. Xavier had converted no more than one thousand Japanese to Christianity during his travels in 1549–1551, but his successors raised the Japanese Christian population of Nagasaki to nearly fifteen thousand by 1568 and spread the message to Honshu after Nobunaga had captured Kyoto in that same year. Nobunaga promoted Christianity to weaken the formidable power of Japan's Buddhist communities. A number of daimyos converted to Christianity to gain favor with Nobunaga and to participate in the Jesuits' extensive business activities. Jesuit missionaries like Luis Frois and Alexander Valignano helped to raise Japan's Christian population to around 150,000 in 1582, including over twenty thousand converts in the Kyoto-Osaka region alone, making it the largest Christian community in Asia.[41]

Nobunaga's success was due to the fragmentation of rival daimyos, an ability to capture many (but not all) of the important silver mines, and the adoption of European military tactics—he was the first daimyo to deploy musket-bearing infantry and cavalry-stopping pikemen. Nobunaga had launched the unification process with huge territorial conquests in central Honshu and the capture of the Ikuno mines in 1568. Not only did his conquests help to end the so-called *Sengokujidai* (Age of Upheaval) that had engulfed Japan since 1450, but he opened up the domestic economy, liberated Japan's downtrodden peasantry, encouraged foreign trade, and promoted Japan's underdeveloped shipbuilding and construction industries. While Nobunaga was careful to upgrade the mines and smelters of Ikuno and Kai province with the latest technology, possibly including the so-called *nambambuki* smelting process, he failed to conquer the mines of Iwami and the northern mines of Sado Island, Ashio, and Ani. It was the rival Mori clan, having seized Iwami in 1563, who installed the *nambambuki* processs at the Omori mine in 1581.[42]

Nobunaga also attempted to reform Japan's chaotic monetary system. While an estimated 92 percent of all Japanese business transactions were effected with copper coins in 1573, virtually all of these coins were imported from China. Bags of gold and silver dust had begun to compete with copper coins after 1550, followed by the introduction of silver bars that could be sliced into smaller pieces as *kirigin* (cut silver). When Nobunaga captured Kai province in 1568—a region that had minted gold and silver coins since the 1550s and was stockpiling a gold hoard that would reach 300,000 ryo (5.4 tons) when it was seized by Tokugawa Ieyasu in 1582—he declared the acceptance of gold and silver coins for large transactions. Unfortunately for Nobunaga, Kai's gold output was in decline, the gold deposits of Tajima were only modest, and the deposits of Mutsu, Hitachi, and Sado Island were owned by rival daimyos.[43]

Nobunaga's monetary struggles were no barrier to his military successes. He could be brutal. After capturing Kyoto in 1568, he plundered Buddhist temples and massacred 1,600 Buddhists at the monastery at Mt. Hiei. He deployed over three thousand musketeers against the underarmed Takeda clan in 1575, prior to their destruction in 1581, and captured or killed as many as thirty-five thousand members of the Ikko clan during the conquest of Echizen and Kaga provinces. Nobunaga's conquest of central Honshu had spread to thirty-two of Japan's sixty-six provinces when one of his lieutenants, Akechi Mitsuhide, rebelled against his authority in June 1582. A humiliated Nobunaga chose to commit suicide in a besieged Kyoto temple.[44]

## THE PHILIPPINES

The partial unification of Japan by 1582 was financed mainly by the Japanese silver industry and Nobunaga's skillful use of Portuguese middlemen to manage the exchange of Japanese silver for Chinese gold and silk. The emergence of a quasi-trading monopoly between Nagasaki and Macao provided a jolt to the otherwise fading fortunes of the Portuguese Empire. If the great achievement of the early 1500s—the assemblage of an Indian Ocean axis of Mozambique, Hormuz, Goa, and Malacca—was now under pressure, newly crowned Felipe I of Portugal was handed an intriguing opportunity to supplement his recent initiatives in the Philippines. Spanish Manila provided an alternative gateway to China, in this case with Spanish American silver, and the timely discovery of large silver deposits in Japan was likely to raise the scale of Portugal's Macao-Nagasaki trade.

Portugal's ability to penetrate the huge Chinese market through Macao and Nagasaki had been monitored by Felipe for decades. Felipe was keenly interested in the Far East. He had commissioned a long-delayed follow-up voyage to the Philippines in the bankruptcy year of 1559 in an attempt to cash in on the same silver-for-silk trading pattern that had launched Portuguese Macao in 1557 and had attracted scores of Portuguese traders to Kyushu. If silver-starved China was prepared to open up commercial doors that had been essentially shut since the 1430s, Felipe of Spain was prepared to revisit a Pacific archipelago that had received his name in 1542. He recognized that the recent transfer of Medina's amalgamation process to Zacatecas (and then to Potosí) presented a huge opportunity to market surplus American silver to China once Spain's royal finances had become under control.

Ignored by Charles V for over two decades (and by Portugal completely), the Philippines had been dominated by the Chinese for over five centuries. The first Chinese accounts of the archipelago appeared in 972, with mention of trading activities at Butuan (northern Mindanao) and Mai-i (Mindoro), followed by the gradual importation of Filipino products—wax, resin, hemp, cotton, tortoise shells, pearls, coral, and human slaves. Slave trading, especially on Visaya, was a standard feature of Filipino commerce. By 1405, Emperor Yongle was claiming control of Luzon, a sixty-junk fleet under Zheng He had explored the entire archipelago, and Chinese immigrants were competing (or partnering) with a host of recently arrived Muslim traders (Moros). The Moros penetrated as far north as Luzon by 1500 and traded regularly with Malacca, Borneo, and Timor. They also converted thousands of native Filipinos to Islam. The future site of Manila (Tondo), a Muslim-ruled village that was very familiar to Chinese and Moro traders, presided over the exchange of Chinese porcelains, metals, glass, and pottery for Filipino goods.[45]

The Portuguese and Spaniards who followed found the Philippines to be devoid of minerals or spices other than a low-quality cinnamon crop on Mindanao. The new attraction in 1559, however, was

## MAP 7.  THE FAR EAST, 1521–1571

Courtesy of George Strebel, © author.

that the islands might serve as a Pacific way station to manage the exchange of amalgamated Mexican silver for Chinese silks and porcelains. There was even a large Chinese presence on the islands to expedite matters. The global Jesuit network had alerted Felipe II to these facts and added that the "Silver Islands" of Japan were likely to expand their fledgling trading activities with Macao. Felipe wanted in. Spain's silver resources in Mexico and Bolivia were believed to be many times larger than those uncovered in mysterious Japan, and the silver-starved Chinese market was viewed as a huge business opportunity. There was also an opportunity to capture a piece of the Portuguese spice trade now that the Portuguese monopoly was under attack by the Ottomans, Achinese, and some of the Moluccans.[46]

While the Philippines were located clearly within the Portuguese zone demarcated by the Treaty of Zaragoza, Portugal had no interest in them, and the demarcation line, fixed at seventeen degrees west of the Moluccas, was never enforced. Saavedra's ill-fated voyage of 1527 was eventually followed up by Ruy López de Villalobos in 1542. Villalobos, a nephew of Viceroy António de Mendoza, commanded a six-ship fleet to establish a trading colony on Mindanao. The plan failed miserably, but it was Villalobos who named some or all of the archipelago for Charles's fifteen-year-old son Felipe. Whether it was the name or a sincere interest to tap into the China trade, Felipe ordered Viceroy Luís de Velasco in 1559 to organize an expedition to colonize the islands in much the same way that Columbus had been instructed to colonize Hispaniola. Where Hispaniola held the promise of gold, the Philippines held the promise of trade with a silver-starved Chinese market and access to South Asian spices. Of course, Felipe was flat broke in 1559, and the expedition funds were provided by Mexico City. Velasco was also encouraged to find a new and improved return route back to Acapulco—the projected silver port for this transpacific enterprise. The Philippines were subsequently colonized by New Spain, not by the mother country.[47]

Velasco's expedition to the Philippines was delayed until 1564. On November 21 of that year, a four-ship fleet under Miguel López de Legazpi, another unusually competent Basque nobleman, set off from the Mexican village of Navidad (north of Manzanillo) with the assistance of an Augustinian friar named Andrés de Urdaneta. Urdaneta was an excellent geographer, navigator, and strategist and had visited the Moluccas many years earlier. It was he who helped to champion a proposed trade route through which Seville, Veracruz, Mexico City, and Acapulco would be linked (via Mexican silver) to the Chinese trade (via the Philippines). If that were not enough, Urdaneta managed to discover and chart a more direct return route to Acapulco that would last for 250 years.[48]

After nearly three months at sea, Legazpi's fleet—with Urdaneta, treasury officials, five Augustinian priests, and 350 sailors—finally sighted the Philippines in February 1565. For whatever reason, Legazpi decided to place Spain's first Asian colony on Cebu in late April despite the fact that Cebu was just as unfriendly in 1565 as it had been during Magellan's earlier visit. Cebu had been attacked in 1563 by one of Portugal's occasional raiding parties and had no interest in converting to Christianity. Legazpi decided to take the island by force, despite the fact that the island, like most of the Philippines, was essentially mineral free, spice free, and inhabited by natives who were nearly as primitive as those encountered by Columbus on Hispaniola. The floundering Spanish colony on Cebu served mainly as a staging area for the subsequent conquest of the Philippines. The fragmented chiefdoms provided little military resistance to Legazpi's well-managed activities, allowing the conquest to be nearly completed by 1572. The Moro traders of Mindanao would remain a thorny problem, however, and it would take time to understand the regional marketplace—when a Moro trading ship was captured in 1565, the Spaniards discovered that the vessel was owned by a Portuguese trader in Borneo and contained cargo that was owned by the Sultan of Brunei.[49]

The logistical challenge presented by the vast Pacific Ocean was eased by Urdaneta's discovery of a suitable return route to Mexico during the summer of 1565—this was the first easterly crossing of the Pacific Ocean in recorded history. By sailing north to thirty degrees latitude, Urdaneta discovered that Mexico-bound ships could take the prevailing winds to the coast of California before heading south to Mexico. He arrived in Acapulco on October 8, 1565, with samples of Mindanao cinnamon and the discovery of an elongated return route that would still require four to seven months. While the same wind patterns blocked a direct route between Manila and Peru, the prevailing westerlies reduced the outbound trip to Manila to around three months. The later Manila Galleon would usually leave Acapulco in late February and follow a south-westerly path to a latitude between the tenth and thirteenth parallels. From there, sixty to seventy days of westerly sailing were required to reach a way station on the Ladrone Islands and then eighteen to twenty days more to reach Manila.[50]

Back in the Philippines, Legazpi sought to replace inhospitable Cebu with a more suitable headquarters for New Spain's first overseas possession. The search was over when Martin de Goiti and 2 ships, 100 Spaniards, and around 250 allied natives sailed into Manila Bay on May 3, 1570. Not only was Manila an attractive deep-water port, but there was an existing Muslim trading village (Tondo) and acres of cultivated farmlands. The Spaniards were in no mood to negotiate when Tondo's ruler, Suleyman, appeared to be hostile—Goiti burned Tondo to the ground in preparation of Legazpi's arrival. When Legazpi sailed into Manila Bay in May 1571 as the newly appointed governor of the Philippines (at an annual salary of two thousand ducats), he was happy with Goiti's work. Local Muslim chiefs, with or without Suleyman's blessing, were ready to submit.[51]

Manila was officially established on June 24, 1571, on behalf of New Spain to govern an archipelago nation of around seven hundred thousand people. Legazpi's reign was a short one. The sixty-seven-year-old governor died of a stroke in August 1572, but his decision to relocate the capital from Cebu to Manila provided Spain with one of the finest natural harbors in Asia and a gateway to China. Having had centuries of trading experience with Moros and a more recent experience with Portuguese, Manila received Chinese merchant ships almost immediately in 1572. Two Spanish galleons sailed back to Acapulco in 1573 with 712 pieces of Chinese silk and

22,300 pieces of porcelain. Of course, Moros, slavers, and pirates were less than happy about the arrival of the powerful Spaniards. The assistance of thousands of native Filipinos was required to repulse an assault by the Chinese pirate Limahong and over two thousand confederates in November 1574. While the triumph facilitated a formal commercial treaty between Manila and Beijing in 1575, the Moros were another matter. The Spaniards were forced to attack Moro ships, facilities, and trading routes on a regular basis between 1578 and 1582. This left the Moros to fight guerilla wars from Mindanao, Sulu, and northern Borneo. A later party of Jesuit missionaries found this out the hard way—they were killed by a Moro ambush in inland Mindanao in February 1596.[52]

Modeled on Santo Domingo and Mexico City, Manila served as Spain's answer to Portuguese Macao and Nagasaki. Most of the Chinese goods that had been imported into the Philippines for centuries were utilitarian porcelains, bowls, ceramics, and conventional silks. Yet products that were very common to the average Filipino were prized greatly in New Spain (and in Europe). Sensing a lucrative profit opportunity with the "ignorant" Spaniards, Chinese merchants began to demand payment in American silver rather than in the low-quality cinnamon and raw materials that had been bartered by native Filipinos for years. Sensing a lucrative profit opportunity with the "ignorant" Chinese, Mexican merchants began to trade increasingly large supplies of American silver for underpriced Chinese goods that could be marked up many times over in New Spain.[53]

Attempts were made to restrict the trading and distribution of imported Chinese goods to a short list of licensed Manila-based enterprises. In late December, scores of Chinese junks would arrive in Manila from Fujian Province to unload porcelains, silks, pearls, paper, lacquer, damasks, velvets, and other items and then reappear in March to receive their payment (in silver) from licensed Spanish merchants. Representatives from New Spain and Peru were prohibited from purchasing Chinese products. The royal objective was to protect the monopolistic galleon trade with Mexican merchants, traders, and government officials and to ensure that the outbound

(Chinese) component of this trade was not disrupted by market forces. If the silver wealth of Potosí and New Spain and the importation of prized Chinese goods were too important to be left entirely to private enterprise, private enterprise of all varieties ultimately found a way to circumvent the royal monopolies. Spanish officials received commissions and kickbacks to ensure that the process ran smoothly—these practices were no different than those employed in Malacca and elsewhere in Asia.[54]

While the entire enterprise was supervised theoretically by the Council of the Indies, the Philippines were managed as a colony of New Spain. The islands were carved into thirteen administrative provinces with hundreds of thousands of natives converted forcibly to Christianity. Then they were assigned either to 280 encomiendas, cattle ranches, public-works projects, timber-clearing activities, or the shipyard on Manila Bay—the legendary Manila galleons were built from local hardwoods like teak and molave. Resistance was minimal. The viceregal coffers in Mexico City benefited from the licensing revenues and duties collected at Manila and Acapulco. Royal decrees of 1582 and 1593 prohibited direct trade between the Philippines and Peru and attempted to limit the annual Acapulco trade to two 300-ton galleons or 500,000 pesos (fourteen tons) worth of American silver and 250,000 pesos worth of Chinese goods. The restrictions were unenforceable. Manila's trade with Acapulco was complemented by the receipt of thirty or so Chinese junks annually from Fujian, the arrival of hundreds of Japanese merchants, and an explosion in contraband trade. The royally mandated figures would be rendered meaningless.[55]

The scale of these transpacific volumes was indicated by Thomas Cavendish's capture of an Acapulco-bound galleon, the *Santa Ana*, off the California coast in 1587. This single galleon carried Chinese silks and merchandise valued at over two million pesos. Since Potosí's annual output soared to two-hundred-ton levels during the 1590s, the amount of silver shipped annually from Acapulco to Manila may have averaged two million pesos (fifty-six tons). Manila-based Chinese junks were sending twenty-two registered tons to their customers in Fujian and Guangdong and an equivalent amount of contraband silver. A viceregal report of 1602 came up with a much larger estimate—it pegged the annual volume of silver sent from Acapulco to Manila at a hard-to-believe five million pesos (150 tons). This staggering total apparently included over two million pesos (over sixty tons) of American silver that had recirculated illegally via Buenos Aires, Lisbon, and Seville. Portuguese merchants in Brazil were annually sending as much as thirty-three tons of Bolivian silver either to the slave markets of Angola or around the cape to Malacca, Macao, and Manila. Contraband trade dwarfed the official variety. The official value of Manila's trade rose to 1.8 million pesos in 1612, based on an interpolation of royal customs duties (*almojarifazgo*) on registered imports, but occasional viceregal inspections suggested that the actual value was closer to five million.[56]

Manila was only marginally profitable during the early decades. The cost to administer and defend the Philippines averaged around 200,000 pesos per year, roughly equivalent to the islands' annual contribution to the Crown, and there were limited mining and spice revenues to soften the blow. Official revenues from licenses, custom duties, and tribute were overwhelmed by an explosion in unofficial (untaxed) trade. The encomenderos complained that population losses from disease, warfare, and overwork exceeded 33 percent in some regions, often under the eye of religious orders who had accumulated huge agricultural estates and were exploiting natives as badly as the reviled encomenderos. Mexico City was forced to contribute an annual subsidy to maintain order. After a proposed swap of the Spanish Philippines for Portuguese Brazil was rejected, the Jesuits chimed in with an outrageous proposal to use Manila, Macao, and Nagasaki as staging areas for a full-scale invasion of China—there were millions of Chinese to be converted to Christianity. A detailed invasion plan of 1585 called for as many as twelve thousand European troops (Spaniards and possibly Italians) to link up with six thousand allied Japanese. Fortunately for all concerned, the ridiculous plan was shelved.[57]

China was invaded commercially. The Ming dynasty, having

abandoned the impractical paper currency system during the fifteenth century, was thrilled to receive American silver (via Manila), Japanese silver (via Macao), and European and American silver (via the Cape of Good Hope route). The Acapulco-Manila-China leg was set to explode with the ramp-up of American silver production to fourteen million pesos annually in the 1590s. China received an average of around fifty tons of silver per year between 1572 and 1600—roughly twenty-four tons from the Americas, eighteen from Japan (via Macao), and another eight from the Cape—but as much as one hundred tons of American silver (worth four million pesos) annually between 1590 and 1631. Middlemen of all nationalities, but especially Manila's Chinese community, schemed to arbitrage the silver-starved Chinese market against the Chinese goods–starved markets in Spanish America and Europe. China had eased trading restrictions (except with Japan) in 1567 and continued to draw silver from a bimetallic ratio (5:1–7:1) that was half the global average. Whether or not the Ming treasury held a reported 248 tons of silver reserves in 1582, the flood of imported silver clearly opened up new outlets for huge volumes of Chinese silk. Manila was linked with a new class of wealthy silk merchants on the Chinese mainland.[58]

As Manila began to rival the great Portuguese entrepôts at Malacca, Macao, and Nagasaki, Felipe was careful to maintain the Portuguese status quo in the Far East. Portuguese Asia would remain a Portuguese enterprise. Other than a temporary moratorium on all customs duties between Spanish and Portuguese interests (until 1593), he left the Goa-Malacca-Macao-Nagasaki axis alone. He refused to allow direct trading between Manila and Macao (until 1619) and even offered Portugal the services of the Spanish fleet in Manila. Felipe's capable twenty-three-year-old nephew, Archduke Albert of Austria, the first of a series of regents who ruled Portugal between 1583 and 1640, was instructed to maintain separate currencies and state budgets and to leave the day-to-day management of the Portuguese Empire to Portuguese officials. In return, Portugal agreed to stay away from Manila and contribute thirty-one ships to the Invincible Armada in 1588.[59]

Felipe's problem was that most of the Asian profits were staying in Asia, most of the military costs were being borne by Lisbon, and no Portuguese had uncovered a Potosí in either Brazil or Africa—expeditions into the wilds of Brazil, Mozambique, and Angola had come up empty. The economic dynamism of Macao, Nagasaki, and Manila was lining the pockets of almost everyone involved except the Iberian Crown. Smuggling, pirating, extorting, and self-dealing were rampant. If that were not enough, the official viceregal system was being wracked slowly by incompetence, nepotism, and corruption. The appointment of a lazy nephew or brother-in-law rarely enhanced the status quo. When this same pattern was extended to over fifty Portuguese factories worldwide and then repeated for a few generations, it became virtually impossible to restore the type of creative energy that had built the empire in the first place. As talent was crushed, the energies of Portugal's most talented men sought out opportunities outside of the official system—inter-Asian trade, textiles, precious stones, and nonmonopoly spices. Some Portuguese captains diverted military resources to further their personal ends. This left behind a crumbling Indian Ocean system that was propped up only by a credible fleet of armed carracks and easy profits from pepper, customs duties, and the sale of royal licenses.[60]

The far-flung Portuguese Empire was spinning out of control. Lisbon had to import almost everything to feed her 100,000-plus residents and to manage the overseas trade. Meanwhile, a collection of licensed Portuguese entrepreneurs was paying the Crown less than 200,000 cruzados annually for the right to generate annual revenues of as much as two million. Wealthy Portuguese merchants and nobles preferred to reinvest their Asian profits either in Asia, juros, or Iberian real estate rather than in native manufacturing enterprises. The home country was stuck with wine, fish, salt, oil, and other low-margin commodities. In addition, most of the outbound trade goods associated with the fledgling slave trade between West Africa and the sugar plantations of Brazil—copperware, brassware, wheat, textiles, manufactured products, weapons, and luxury items—had to be imported from Europe.[61]

Unlike the Spanish experience in the Americas, native Portuguese represented just a tiny fraction of Portugal's overseas population. Portuguese colonists were limited to a handful of settlements in Brazil and even less in Angola and Mozambique. In Angola, repeated efforts to find and conquer the rumored silver mines of Cambambe between 1570 and 1604 were blocked by hostile Ngolas, cannibalistic Imbangalas, and the painful fact that there were no silver mines of Cambambe. Three Portuguese fortresses defended a slave-trading colony surrounded by hostile natives, a poor climate, and agricultural lands that were poorer than advertised. In Mozambique, Portuguese treasure hunts conducted along the upper Zambezi River during the 1570s had also ended in failure. No one could find the fabled gold mines of Monomotapa, if they existed at all, leaving a few hundred Portuguese to acquire agricultural properties, royal land titles (*prazos*), and slaves along the lower Zambezi. Even in Goa, fewer than 10 percent of the city's 100,000 inhabitants were native Portuguese. There were just six hundred official Portuguese households in Malacca and less than one thousand in Macao. The flip side was that six hundred unlicensed Portuguese *lancados* had launched a 700,000-cruzado trading business at Hughly (upriver from Calcutta) because the enterprise was not subject to royal monopolies.[62]

Corruption raised the costs of defending the primary shipping lanes and naval bases between Mozambique and Malacca. Budget shortfalls occurred regularly in the second half of the sixteenth century. As in Spain, the higher the overseas revenues, the higher the domestic debts. The shortfalls were met less by subsidies from the Portuguese Cortes, which convened only in 1562–1563 and 1579–1580, than by an expansion of *padrões de juros*. Nine separate juro offerings were concluded between 1555 and 1588 to convert expensive, unfunded obligations into lower-coupon, long-term debt. While this Spanish-styled refinancing process enabled Lisbon to survive the economic turmoil of 1576–1582, the unfortunate royal creditors were out the difference. Overseas merchants also had to deal with currency issues. Silver cruzados minted at Goa and

Malacca were debased so regularly that most Portuguese merchants preferred to stick with Spanish reales and Persian *larins*.[63]

Portugal's shaky presence in the western Indian Ocean was buttressed by a resumption in the Ottoman-Persian conflict between 1578 and 1590. The viceroy in Goa presided over three sprawling governorships—East Africa to Sri Lanka, Sri Lanka to Pegu (Burma), Pegu to Japan—and was only too happy to provide naval assistance to Persian shah Abbas I (1568–1628). The expanded Persian fleet stalemated the Ottoman navy and, in turn, forced Istanbul to devalue the venerable silver *akce* by 44 percent in 1586. The devaluation, from 65 akces per gold sultani (or ducat) to 120, was intended to relieve the Ottoman treasury, offset the near collapse of the timar system, mollify thousands of unpaid Ottoman troops, and compete with the growing flood of American silver that was entering the eastern Mediterranean. Shah Abbas (the Great) also expanded Persia's sea-borne exports of native silk, a product introduced to the Caspian Sea region by Chinese traders in the seventh century, and schemed of ways to circumvent the Ottoman system (and Portuguese Hormuz). While the stalemate helped to preserve the status quo in the Indian Ocean, the looming problem for Portugal was that the annual five-or-so-ship Carreira de India was becoming vulnerable to English privateering, corruption, and undermaintenance. Losses would reach one–third during the tumultuous 1590s.[64]

The failure of the Asia Contract of 1578–1580 prompted Felipe to rewrite the terms of the next ones in his favor. Although spice purchases averaged only twenty thousand quintals per year during the 1581–1586 period and one of the ships, the *São Luis*, was lost to bad weather in 1583, the five-year contract was reasonably profitable for the Crown. The key was that Felipe received a fixed price of thirty-two cruzados per quintal for pepper purchased in India for only eight. The Asia Contract of 1586–1591 was co-managed by Philip and Octavian Fugger, great nephews of Anton who had reestablished the family firm in 1574. Despite a fall-off in annual average pepper purchases to only fourteen thousand quintals, due primarily to the loss of the *Madre de Dios* in 1589, a stepped-up price of forty cruzados per quintal helped to deliver annual royal profits of around 177,000 cruzados. The problem was that the *Madre de Dios* and over six thousand quintals had been captured by English privateers off the Cape of Good Hope in 1589. Even the well-connected Fuggers were unable to persuade Queen Elizabeth to release a captured vessel during a state of war. The affair prompted Felipe to separate the two-year Europa Contract from the Asia Contract and provide armed escorts for the outbound spice fleets of 1591—commissioned, no less, from shipyards in the Netherlands. Purchased by Tomas and Andrea Ximenes, the Fuggers, Markus Welser, and others, the Europa Contract returned a disappointing 4.7 percent and raised the probably that the Crown would have to assume direct control of the Indian pepper trade. It did so between 1598 and 1607.[65]

The Asia Contract of 1591 fared even worse. The 1,500-ton *Santa Maria*, four other pepper ships, and an eleven-ship naval escort left Lisbon in early April 1591 to complete the sixth and final year of the contract. If the outbound segment was uneventful, the unescorted return trip from Calicut became somewhat legendary. The fleet left Calicut in January 1592 with a spectacular haul of 47,100 quintals of pepper and spices, plus the viceroy of India, ninety Portuguese sailors, and over one hundred passengers. Only one of the five ships reached Lisbon in the fall of 1592—English pirates managed to capture, burn, or wreck the other four and stranded the viceroy and the surviving passengers in the Azores. The privateers preserved the *Santa Maria*, sailed her north to Portsmouth harbor, and unloaded her cargo with great fanfare. The value of the ship herself, her armaments, and precious contents—two hundred barrels of pearls, jewels, gold bullion, pepper, cinnamon, nutmeg, mace, linen, and silks—was estimated at a staggering two million English pounds. Given the direct involvement of the Welsers and Fuggers, the heist of the *Santa Maria* signaled a possible changing of the guard in the high seas.[66]

## PORTUGUESE ASIA

Setbacks notwithstanding, the unification of the Iberian Crowns was benefiting Portugal more than it was Spain. The evidence is that the redefined, easterly focused Portuguese Empire even enjoyed a second Golden Age between 1590 and 1620. Trading in precious metals, Indian textiles, Chinese silks, and traditional spices would raise Estado da India's revenues to two million cruzados in 1607. The royal mints at Goa and Malacca were receiving roughly 2.5 tons of gold annually from Africa, Sumatra, Java, and Borneo on top of their importation of silver (and copper). The gold volumes were modest, but they equated to around twenty-eight tons of silver based on Europe's 11:1 bimetallic ratio. The real story, however, was the surge in private trade and the focus on Macao and Nagasaki. If the Crown wanted to stick with customs duties and monopolies on pepper, slaves, and precious metals, a new genera-tion of Portuguese entrepreneurs was happy to focus on profits. Underappreciated products like textiles, indigo, cinnamon, and pre-cious stones carried huge profit margins, and there was a lucrative middleman role to be had between China and Japan. The best part was that the Crown continued to pick up the tab for the naval defense of these private trading activities.[67]

The involvement of Portuguese Conversos, like the Ximenes brothers in the Europa Contract of 1591, spoke volumes about the new order of things. While extended overseas families had partici-pated in Portugal's global business activities (Genoese-style) for over a century, a small group of Converso merchant families had accu-mulated enough capital to compete with the venerable Italian and German merchant houses. The descendants of the tens of thousands of Iberian Jews who had converted to Christianity after Manoel's convert-or-leave decree of 1497 still had to deal with restrictions, confiscations, and even an Asian Inquisition—eighty-four convicted heretics were burned at the stake in Goa between 1561 and 1590. But they bolstered Portugal's global presence. Portuguese merchant families, New Christian and Old, seized the entrepreneurial initiative in the second half of the century. Some merchant families shifted to private (nonroyal) trade with emerging markets like Angola, Mozambique, and Brazil, while others focused on the more glamorous Carreira trade in the Far East. The simple formula was that merchandise leaving Goa should be worth three times its value upon arrival in Lisbon. The wealthiest merchant families even began to provide direct financial assistance to the Iberian Crown.[68]

Their timing was exquisite. Global trade was exploding on all fronts thanks to the injection of huge amounts of American (and Japanese) silver. By 1600, entrepreneurial Portuguese families were investing as much as two million cruzados annually to support an annual Carreira trade that had reached five million cruzados. While the total enterprise surpassed European trade with the Levant (three million) and approached Iberia's annual volume in the transatlantic (possibly ten million), the Crown stuck to her pepper monopoly, customs duties, and the carriage of one million or so cruzados of silver bullion (*cabedal*). The entrepreneurs focused on products with the highest profit margins. Private cargo that occupied only 38 percent of the Carreira's annual tonnage capacity—mainly textiles, but also indigo, cinnamon, and diamonds—accounted for a staggering 93 percent of the Carreira's annual market value between 1580 and 1640. The 770,000 pieces of cloth and silk received by Lisbon annually between 1600 and 1609, a volume that accounted for over half of Portugal's total trading value, was over twenty times the scale of Dutch textile imports during the 1620s. Profits from cloves, cinnamon, and other lightly taxed spices were also impressive.[69]

Private Portuguese trade was overwhelming the royal system. Private textile imports of four million cruzados in 1588 was eight times the value of royal pepper shipments. When annual textile imports fell to less than three million cruzados between 1600 and 1609, Converso merchants shifted to high-end silks for Europe and low-end cloth for West Africa, Brazil, and Spanish America. The horrific exchange system of the era saw a slave acquired for twenty cruzados worth of cloth in Angola and resold for one hundred cruzados in cloth in Brazil. Chests of cloth, cotton bales, indigo, cinnamon, and precious stones fetched prices in Lisbon that were between three and five times their original purchase cost. Lightly taxed indigo became a 600,000-cruzados-per-year business. The value of diamonds, rubies, and sapphires from Golconda (India) and Borneo, carried in standardized lacquered wooden boxes (*bizalhos*), were untaxed because they were so easy to smuggle. Viceregal authorities estimated in 1613 that private Portuguese merchants were generating two million cruzados in revenues from precious stones alone. When the *Nossa Senhora da Luz* ran aground off the Azores in 1615, it was carrying one million cruzados worth of precious stones.[70]

Portuguese merchant families were also active in Spanish Manila, a market otherwise split between Mexico City and the fifteen-thousand-member Chinese community. The extended Converso family of Diogo Fernandes Vitoria applied capital of 450,000 cruzados to manage a web of business dealings among Mexico, Manila, Macao, and Malacca between 1580 and 1598. A sizable quantity of American silver was exchanged in Malacca for textiles, precious stones, spices, and slaves. Vitoria was too successful for his own good. After he and his partners had possibly assembled a 30 percent position in the 1.5 million-cruzado Manila trade, viceregal authorities used the pretext of the Mexican Inquisition to confiscate 150,000 cruzados of his Manila-based inventory in 1598. One of Vitoria's Macao-based cousins, Leanor de Fonseca, had already been tried under the Inquisition and burned at the stake in Goa in 1594.[71]

Portuguese Macao was more tightly controlled than Manila. A multinational creation of Portuguese initiative, Jesuit energy, Chinese need, and Japanese silver, Macao was the most dynamic entrepôt in the Far East. The enterprise was driven by three principal trade routes—Lisbon-Goa-Malacca-Macao, Guangzhou-Macao-Nagasaki, and unofficial Macao-Manila-Acapulco—and the arrival and departure of the annual Great Ship. If less than one thousand of Macao's households were Portuguese nationals, the Jesuit community had helped to cultivate an array of mutually beneficial relationships among Japanese warlords, Chinese mandarins, and the Portuguese traders of Malacca. Macao was also impregnable. Fortifications had been constructed after a massive Wako pirate raid had been repulsed in July 1568. Nervous Chinese authorities walled off the north side of town in 1573. These and other projects were funded by a tributary system modeled on a fourteenth-century Muslim trading colony in Guangzhou. When news of the unification of the Iberian crowns reached Macao in May 1582, the response was underwhelming. The union's primary benefit was the availability of naval support from Spanish Manila. By 1588, Macao was upgraded to "city" status and the Bishopric of Macao was given authority over all of East Asia. This included the recently closed Dioese of Funai in Japan.[72]

Profits from the silver-for-silk trade helped to build a more powerful Christian presence in Macao than in Malacca. The energetic Jesuits established a Santa Casa de Misericordia (1569), a formal Diocese of Macao (1576), a Bishopric of Macao (in 1583), a municipal council (Senado da Câmara) in 1586, and scores of charitable institutions. The bequests-driven Misericordia, launched in 1498 as the Holy House of Mercy to care for Lisbon's poor, had already been transferred successfully to Cochin, Goa, Mozambique, Diu, Hormuz, Malacca, and Tidore. The Misercordia complemented the efforts of Father Matteo Ricci to establish a Jesuit College, introduce a printing press with moveable type in 1588, and organize a Jesuit mission to Beijing in 1601 (where he died in 1610). The Senado da Câmara, a council based on the Evora model, represented Macao's six hundred Portuguese casados, financed her military defenses, managed agricultural deliveries from the Chinese interior, and controlled occasional political interference from Spanish Manila.[73]

The Jesuits were in the middle of everything. A representative report filed by Father Manoel Dias in April 18, 1610, assured the director of the Jesuit Order in Rome, Claudio Aquaviva, that the corporate agreement (*armacão*) governing the export of Chinese silk to Japan allocated shares (*baques*) to the citizenry of Macao in an equitable fashion. Many Jesuits developed proficiency in the Japanese and Chinese languages and gained access to well-placed government officials, warlords, and merchants. The Japanese valued Jesuit services so highly that official bans of Catholicism (in 1587 and 1614) were ignored when it involved the high-priority trade with Macao. Portuguese traders were always willing to smuggle Jesuits priests into Japan to expand Portuguese influence.[74]

Like everyone else in Macao, the Jesuits were heavily dependent on the annual Great Ship—the *Não do Trato* or "Black Ship" (because of the color of its hull). The carrack usually left Goa in April to exchange precious metals and Indian textiles for spices in Malacca. From there, it sailed on to Macao to receive Chinese goods obtained from the trading fairs at Guangzhou. Portuguese merchants sent small lighters up the Pearl River to purchase raw and finished silks, gold, porcelains, and other goods during the two principal fair periods (January and June) and to present gifts to Guangzhouese officials. While the 2,000 *piculs* (133 tons) of Chinese silk received annually at Macao was just a fraction of the 12,000 piculs made available to export markets, the Portuguese were especially focused on the 1,000–1,500 piculs of white raw silk that was prized so highly by Japanese elites. The Great Ship typically laid over in Macao through the June fair, which was reserved for Japanese goods, and left for Nagasaki in July. It returned to Macao laden with Japanese silver in late October. While the Portuguese earned huge profits from the silver-for-silk exchanges, profits from mundane porcelains and imported sandalwood (via Timor and Solor) were also attractive.[75]

The Great Ship evolved into a massive one-thousand-ton-plus affair that was twice the size of the Manila Galleons. Although the captain major and chief pilot were always Portuguese, the ship was usually manned by a polyglot crew of Gujaratis, Eurasians, Chinese, and African slaves. No matter that official trade was still banned between China and Japan. The Portuguese had been authorized to operate legally and had protected this authorization by sharing the business with well-placed Chinese Mandarins and Japanese warlords. The number of sailings fluctuated. There were one or two departures annually between 1575 and 1590, but rising cargo values (and rising naval threats) prompted the viceroy of Goa to limit the Great Ship to a single defensible voyage (or none at all) between 1591 and 1617. The Crown eventually abandoned the sluggish one-thousand-ton carrack in favor of smaller (but faster) galliots, pinnaces, and Chinese junks. The switch helped to secure the Macao-Nagasaki sea lane from Dutch raiding parties, but it shifted Dutch privateers to the Malacca Straits and India.[76]

Of course, it was the Japanese side of the transaction that had catalyzed the entire enterprise. The Japanese silver boom had fostered Japan's unification process, reopened trade with China (via the Portuguese), and raised the Jesuit influence in Japan to great heights. The unification process started by Oda Nobunaga (1534–1582) and continued by Toyotomi Hideyoshi (1536–1598) was financed by expanding silver output and profits earned from the exchange of Japanese silver for Chinese gold and silks. Hideyoshi nearly completed the unification of Japan under a single ruler. The son of an Owari peasant, he had somehow risen from the position of "sandal-holder" for Nobunaga to one of the daimyo's most skilled commanders. Hideyoshi came to power after discovering that a fellow commander, Akechi Mitsuhide, had rebelled against Nobunaga in June 1582 and was besieging Nobunaga in Kyoto. Nobunaga's suicide prompted Hideyoshi into action. He abandoned a difficult campaign against the Moris, a powerful clan that had controlled the mines of Iwami since 1563, and crushed Mitsuhide completely. Hideyoshi assumed the title of regent (*kannaku*) in 1585 and reduced all but one of Japan's major daimyos to vassaldom by 1590—the one exception was the eight northeastern provinces held by Tokugawa Ieyasu from his castle at Edo (Tokyo).[77]

Hideyoshi knew as well as Nobunaga (and the Portuguese) that silver was prized more highly in China than in Japan and that gold (not silver) was the preferred method of paying warlord armies. The arbitrage opportunity had become even more attractive by 1590. The expansion of Japan's silver industry had halved the value of one *tael* (cruzado) of gold in Japan to around 1.5 *kan* in silver between 1555 and 1580 and doubled Japan's bimetallic ratio to around 10:1. Silver was abundant and gold was not. Portuguese traders returned to Nagasaki in 1589–1590 with around two tons of Chinese gold (worth 400,000 ducats) after having exchanged roughly twenty tons of Japanese silver in Macao. The gold was commissioned to support Hideyoshi's pending invasion of Korea. While the scale was modest compared with future transactions— Ieyasu imported a hard-to-believe five tons of gold (worth one million ducats) through the Portuguese in 1605—it made a huge financial contribution in 1590.[78]

The precise value of the Macao-Nagasaki trade is unclear. If Portuguese vessels shipped between sixteen and twenty tons of Japanese silver to Macao annually between 1546 and 1597 and Chinese junks contributed another thirteen tons, China received somewhere between twenty-nine and thirty-three tons of Japanese silver annually during the period. These projections are somewhat higher than contemporary estimates. Ralph Fitch estimated that the Portuguese were annually exchanging two thousand piculs (133 tons) of raw Chinese silk for 600,000 cruzados of Japanese silver (24 tons). Sebastião Goncalves pegged annual Japanese silver exports at nearly 500,000 cruzados between 1550 and 1600—or midway between Alexander Valignano's 333,000-cruzado estimate for 1593 and Diogo do Couto's one-million-cruzado estimate for the 1575–1600 period. Whatever the actual total was, Macao was probably receiving around 500,000 cruzados worth of silver (twenty tons) from Japan, 300,000 (twelve tons) from assorted Chinese traders, and another 200,000 (eight tons) from the Cape during Hideyoshi's reign (1582–1598). The raw Chinese silk that returned to Nagasaki went to Kyoto; the imported gold and muskets went to the generals. One-third of Hideyoshi's 150,000-man Korean invasion force was armed with muskets in 1592.[79]

While nearly everyone made money between 1580 and 1600— Japanese, Chinese, and Portuguese—Hideyoshi invested a metallic war chest of 360,000 ryo (nearly seven tons) on warlord campaigns, the construction of Osaka Castle, a refurbished six-hundred-ship navy, and two disastrous invasions of Korea. The cost to construct the first phase of Osaka Castle in 1583–1585, based on the value of capital, materials, and a fifty-thousand-man workforce, was probably in excess of two million cruzados. Placed atop the ruins of the recently destroyed temple town of Ishiyama, Osaka Castle boasted stone walls that were eighteen- to twenty-one-feet thick (without mortar), iron-sheeted roofs, and a set of eight-story towers. After additional rounds of plunder, tribute, and taxes, Osaka Castle served as a linchpin for Hideyoshi's final territorial conquests in Japan in 1590. An estimated ten thousand gold coins and thirty thousand silver coins were stored at Osaka in 1592 while Hideyoshi carried eighty thousand ryo of gold (1.5 tons) across the Straits of Tsushima to support the Korean invasion of 1592. Japanese lenders were advised to watch their profit margins. Ten gold merchants were executed for lending gold at usurious rates in 1591.[80]

Hideyoshi's ill-fated invasions of Korea in 1592 and 1597, analogous to the failed campaigns of Charles V and Suleyman, squandered huge volumes of precious metals. Hideyoshi had prepared to invade Korea, China, and even the Philippines as early as 1586 to cap off his pending triumph in Japan. With control over many (but not all) of Japan's mining districts and the Japanese side of the Nagasaki-Macao trade, he schemed to capture the principal trade routes to the Chinese mainland and remove the Spanish (and their competing silver) from the Philippines once and for all. No matter that Korea's Yi dynasty had served as a Chinese vassal since 1392 (and would remain so until 1910) and was backed by a powerful navy—so powerful that Hideyoshi attempted to acquire two large Portuguese galleons in exchange for an agreement that would allow the Jesuits to establish Christianity in a postwar China. The politically savvy Jesuits passed on the deal.[81]

When the Portuguese opted for neutrality, Hideyoshi was forced to invest substantial amounts of plunder on the Japanese navy. He expanded the fleet to over six hundred ships, armed them with the latest European weaponry, and ignored the fact that Japanese maritime skills had become rusty since the arrival of the Portuguese. The subsequent invasion of Korea was a big mistake. The upgraded fleet managed to deliver the 150,000-man Japanese army to the abandoned port of Pusan on May 23, 1592, plunder Korea's western coastline, and support the army's northward advance to Seoul and Pyongyang (July 23). That was the peak. Tens of thousands of Chinese troops moved to the Korean border and waited for Admiral Yi Sun-sin's eighty-five-ship fleet to go into action. Outweighing Japanese naval gunnery by forty to one, the heavily armed Korean navy was highlighted by twenty-four oar-powered, iron-clad "turtle" ships lined with ten guns per side and a smoke-belching dragon figure at the bow. They routed the Japanese fleet in seven major engagements. Japan lost nearly seventy ships in Hansan Bay on August 1 and another one hundred after the Japanese naval base at Pusan was attacked on October 21. By February 1593, Pyongyang was abandoned to the Chinese and Hideyoshi's stranded army was lucky to return home safely under the guidelines of a three-party armistice.[82]

A huge infusion of Japanese silver (and gold) would be needed to offset these devastating losses—and Hideyoshi would get it. His annual report of 1598 indicated the receipt of one ton of gold combined from Echigo, Sado Island, Mutsu, and ten other Japanese provinces. Another one ton or so of tributary gold was received from Noto, Etchui, and Kaga. Hideyoshi would just miss the huge strikes made on Sado Island in 1600, but he helped to convert the island's erstwhile penal colony into the foremost mining center in Japan. The island's gold, silver, and copper deposits had been mined professionally only since 1542. In that year, Sado's first modern mine was opened at Tsurushi and worked with a huge army of convict labor. Gold deposits were uncovered at nearby Nishimikawa in 1564. While it is unclear whether Sado received the advanced *nambambuki* smelting process during the 1580s, the island contributed to the twenty or so tons of silver that were exported annually to China. Hideyoshi's major contributions were to seize Sado's fading Tsurushi mines in 1589 and to recruit professional mining engineers to the island in 1594.[83]

Hideyoshi replenished his Osaka war chest with enough precious metals to invade Korea once more in 1596 following the collapse of ongoing negotiations with China. History repeated itself. The refurbished Japanese navy enjoyed a series of early victories in August 1597 only because Yi Sun-sin had been demoted after his naval triumphs in 1592—the talented admiral had apparently posed a political threat to Korea's shaky Choson regime. But after being restored to command, Yi Sun-sin destroyed a 144-ship Japanese fleet in October and pushed the reeling Japanese navy back to Pusan (with Chinese assistance). The stranded 140,000-man Japanese army could boast only that it left a devastated Korean countryside in its wake. A face-saving retreat, ordered after Hideyoshi's death in August 1598, was interrupted by Yi Sun-sin. The admiral destroyed another two hundred departing Japanese ships during a nighttime raid on December 16 and would have sunk more if not for his receipt of a fatal musket ball. Japan's humiliating defeats prompted the succeeding Tokugawa shogunate to adopt a Mogul-style naval strategy in the future—pay European navies for protection and save financial resources for trade and land-based defenses.[84]

The Korean campaigns cost Hideyoshi a huge amount of money and a more prominent position in Japanese history. The final unification of Japan and the spectacular mining boom that followed was left to Tokugawa Ieyasu (1542–1616). In the meantime, Hideyoshi's free-spending ways had generated large profits for Portuguese merchants in Nagasaki and Macao and had raised the influence of the Society of Jesus. No matter that Hideyoshi had attempted to ban missionary activities entirely in 1587, citing an illegal slave-trading enterprise involving Portuguese and Japanese children as a pretext, and had reassumed control over Nagasaki in 1588. The ban was ignored because the Jesuits were crucial to his fundraising efforts. Rome was so impressed with the commercially minded Jesuits that a papal bull had been issued in 1585 to prevent the less-inclined Franciscans from interfering in Jesuit affairs. However, Hideyoshi shifted back to persecution after his disastrous invasions of Korea. Someone had to be blamed. He also feared a possible Spanish invasion following the arrival of Manila-based Franciscans in 1593. Large numbers of missionaries and Japanese converts were persecuted between 1593 and 1598, including the execution of scores of Franciscans in the Kyoto-Osaka region in 1596 and the crucifixion of twenty-four Japanese Christians in Nagasaki in February 1597. Hideyoshi's death in 1598, a year in which Japan's Christian population was approaching 300,000, was welcomed by the Iberians.[85]

# CHAPTER NINE

# UNIMAGINABLE WEALTH
# AND SQUANDER

Spanish Manila had emerged as a credible competitor to Portuguese Macao and Portuguese Nagasaki because Peruvian viceroy Francisco de Toledo's investment program had paid off spectacularly. Cerro de Potosí supported Manila's economic position in the Far East, bailed out the Spanish treasury after the bankruptcy of 1575–1578, financed Felipe II's conquest of Portugal in 1580–1582, and allowed Parma's Army of Flanders to resume the offensive in the southern half of the Spanish Netherlands. But financial machinations involving Potosí's slump and recovery, fueled in part by Felipe's recruitment of the hardball-playing Genoese, had allowed the seven northern provinces in the Netherlands to declare their independence from Spain in 1581. No one in Spain would have guessed that a continuing string of production records at Potosí would be insufficient to halt the underestimated Dutch and their English allies.

Felipe's global empire was critically dependent on a huge cash machine that had somehow been launched in the wilds of Bolivia. Toledo had worked wonders at Potosí since 1570. The conversion to water-powered amalgamation, supplemented by a series of technical refinements to the process itself, launched an incredible run of success during the 1580s and 1590s. A complete conversion to Velasco's amalgamation process raised Bolivian silver production to a record 858,017 marks (7.3 million pesos) in 1585 and then to another new record of 887,448 marks (7.5 million pesos) in 1592. Potosí's two hundred tons of annual silver output was the equivalent of ten

Joachimstahls, ten Japans, four Ottoman Balkans, and twice the output of Habsburg Europe at its peak.[1]

Toledo had encouraged everyone to think creatively. Given the cost of royal mercury, no effort was spared to recover excess material during and after amalgamation. After the amalgamated mixture had been washed to remove waste and recover the amalgam, excess mercury was placed in linen bags and twisted to squeeze out remaining material. The amalgam was then placed in a triangular wooden mold, hammered to force out additional mercury particles, and then heated to recover the last remaining droplets. In 1576, Juan Capellin introduced an improved mercury condenser that recovered more of the mercury vapor that was released from the amalgam. Bernadino de Santa Cruz added a stone *buitrone* ("fire chest") in 1580 to speed up the amalgamation process and consume lower quantities of mercury. The mercury works at Huancavelica were also improved. In 1573, Rodrigo Torres de Navarra experimented with low-cost Andean grass (*ichu*) as a smelter fuel to replace charcoal and wood. Henrique Garcés, the star-crossed founder of the Peruvian mercury industry, followed in 1581 with a reverbatory furnace that had been used in Central Europe for decades. The two lateral kilns heated the central furnace via reverberation, providing more uniform yields and reducing fuel requirements.[2]

Toledo controlled labor costs by replacing encomiendas with Inca-styled mitas. The viceroy had stumbled onto the idea while traveling between Cuzco and Potosí in late 1572. The Audiencia de Charcas (Bolivia), autonomous from viceregal Peru since 1558, had been subdivided into eighty-two encomiendas to reward Spaniards who had fought against the Pizarros. These encomiendas supported the mining works at Potosí and the six primary Spanish towns— Sucre, La Paz, Cochabamba, Santa Cruz, Tarija, and Potosí (city)— until disease and overwork began to take their toll on native Aymaras. Experiments with slavery and paid labor were generally unsuccessful. The problem was solved by reestablishing the Inca's traditional mita system of tribute in 1572. One of every seven Indian males between the ages of eighteen and fifty were recruited from a 490-mile-wide radius of Potosí for mining duty. A workforce of 13,500 barely paid men, organized into three rotating groups of 4,500, mined ore, carried supplies, or constructed facilities. Another 3,500 mitayos (from Peru) were assigned to the mercury works at Huancavelica. To ease the recruitment process, Indians from depopulated rural villages were relocated forcibly to a few large towns. The larger communities could be administered (and taxed) more efficiently. In one case, 129,000 Indians from nine hundred rural villages were resettled into forty-four enlarged communities.[3]

The mita system shifted a somewhat voluntary (barely paid) process to a required (barely paid) process. Toledo subjected every adult male to a one-year stint in the mines every six years, working on three-week-on, three-week-off shifts. Transportation, food, and the welfare of the deserted family members were the responsibilities of the mitayo's village. Mitayos received less than one half peso per day or an equivalent share of ore—providing they could survive the overwork, the harsh winter climate, diseases, cave-ins, and asphyxiation. Few mitayos survived more than four years of labor. The shift to deep excavation work was increasingly dangerous to the poor souls who dug the shafts, galleries, adits, and tunnels. Climbing leather and wood-runged ladders with nearly ninety pounds of ore strapped to one's back was grueling. So was the waist-deep trampling and mixing work required for amalgamation. Mercury poisoning was often lethal. And if the work didn't kill you, the extreme temperature shifts could induce a fatal bout of pneumonia.[4]

Working conditions at the mercury mines of Huancavelica were even worse than those at Potosí. At an altitude of 12,500 feet, wood supplies were limited, and there were no wheat or maize fields within fifty miles. The mercury furnaces began to be fueled by *ichu* grass in 1573, but shortages occurred regularly in the 1580s. Huancavelica's mitayos discovered that the process of roasting, vaporizing, and collecting mercury (after condensation) was extremely hazardous to one's health. The fumes were especially lethal. Finished mercury was also dangerous. The storage bags often leaked as the llama and mule trains wound their way to the Pacific outlet at

Chincha—Toledo had replaced the direct (but mountainous) 995-mile delivery route between Huancavelica and Potosí with a circuitous (but faster) land-and-sea scheme that relied on a coastal run between Chincha and Arica. Despite the many challenges, mercury production at Huancavelica tripled from 1,830 quintals in 1574 to a level that met Potosí's annual needs of 5,000 to 6,000 quintals. Production averaged 7,403 quintals between 1586 and 1595.[5]

Mercury was not the only item to carry an inflated price tag. The cost to build and operate Potosí's network of stamping mills, flotation basins, amalgamation (patio) floors, sieves, roasting ovens, furnaces, warehouses, offices, living quarters, and chapel—all of which were located within an enclosed compound—was daunting. Iron was imported from Spain at a cost of seventy pesos per quintal. Wood supplies were hauled from remote areas of Peru. The cost of a single twenty-one-by-two-foot wooden log could be as high as five hundred pesos. A twenty-three-foot-long wooden axle came in at 1,500 pesos. To raise productivity, mineros and ingenio owners contracted out much of the refining operations to private companies in return for shares of production. The private companies recruited smelting workers from the displaced huayra workforce, leaving the last of the primitive huayras to work down the last remaining deposits of *plata blanca* and *plomo ronco* (argentiferous galena).[6]

Trial-and-error experimentation continued. The most significant technical contribution of the period was from Carlos Corzo, an ex–mercury factor and the nephew of a Sevillian merchant. Among many attempts to reduce the "combination period" of the amalgamation process from weeks to days, Corzo discovered in December 1586 that the addition of iron scoria (iron filings) caused the silver to separate more rapidly and even reduced mercury usage. The upper layers of Potosí's silver sulfide zone had apparently contained enough weathered copper and iron minerals to avoid the need for additional iron. But these iron-rich layers were long gone by the early 1580s. Corzo tweaked Velasco's formula by mixing a fifty-quintal load of ground silver-bearing ores with brine, water, three quintals of mercury, and five to eight pounds of iron filings. The mix was trampled by Indian workers for five days and heated during the nights. The results were shocking. Corzo's revised amalgamation process could be completed on the sixth day with a huge 50 percent reduction in mercury consumption. No matter that the Crown had rejected any attempt to reduce mercury usage—there was a royal monopoly to be considered and ten thousand quintals of royal mercury equated to around 400,000 pesos in annual sales for Felipe II's war chest. The Crown prohibited Corzo's revolutionary "iron process" by royal decree in January 1588.[7]

However, the productivity benefits of Corzo's iron process were clear to nearly everyone in Bolivia, especially since the last of the highly profitable huayra slag heaps had been recycled by 1587. The royal ban was ignored and the mercury savings prompted an explosion of business activity at Potosí. The number of mine operators tripled to six hundred as mineros learned to process low-quality mulatto and negrillo ores with Corzo's iron filings recipe (or a tin-amalgam). The grounded ores were then roasted in reverbatory furnaces fueled by dry sheep dung (*taquias*). Depending on the quality of the ores and size of the ore heaps (*cuerpos*), which varied between twenty-five arrobas (6.25 quintals) for richer ores and one *cajon* (50 quintals) for poorer ones, the amalgamation process could be completed in six to twelve days. This "supply-side" expansion, in turn, had the unintended effect of increasing the volume of royal mercury delivered to Potosí—from around 4,150 quintals in pre-Corzo 1587 to a peak of around 6,073 quintals in post-Corzo 1589—and nearly doubling iron imports from Vizcaya (Spain) to 5,500 quintals. An embarrassed Felipe II removed his short-sighted prohibition. Where Medina had failed financially, Corzo left a spectacular estate valued at 1.6 million ducats at his death in 1597.[8]

The magic recipes developed by Corzo and his successors generated much greater volumes of silver even as Potosí's ores became deeper, harder, and silver poor. The value of a quintal of excavated ore in 1594 was just a fraction of that generated in the late 1570s. Between 1594 and 1603, however, the technological improvements produced an average of 166 marks of Potosí silver for every quintal of consumed mercury—this was 66 percent higher than the accepted ratio of losing 100 quintals of mercury to produce 100 marks of amalgamated silver. The 166 marks were also 46 marks per quintal more than the yields realized at Zacatecas. The only problem was that Corzo's stimulation of mercury consumption had inadvertently pressured mercury supplies. The amalgamation works in New Spain were also consuming around 5,000 quintals of mercury per year, 3,000 of which from unreliable Almadén, leaving a 2,000-quintal shortfall to be filled by Huancavelica and Idria (Slovenia). A Royal Decree of 1591 even called for 1,500 quintals of Peruvian mercury to be shipped north to Mexico. Incoming viceroy García Hurtado de Mendoza seized the opportunity to gradually raise contract prices at Huancavelica from forty pesos per quintal (based on 4,000 quintals) to fifty-two pesos (based on 5,000–6,000). Felipe was delighted.[9]

These and other challenges were overwhelmed by Bolivia's spectacular wealth. The seventy-six million pesos of Potosí silver that had been sent to Spain between 1545 and 1575 were a mere fraction of the massive volumes that followed—a staggering 222 tons (worth 7.5 million pesos) were produced in the peak year of 1592. A total of 820 million pesos worth of silver would be cranked out through 1783. In the meantime, Villa Imperial de Potosí had emerged as the greatest boomtown in the Western Hemisphere. Reincorporated to "Villa" status in 1561 with the help of a forty-thousand-peso "donation" to the Lima treasury, Potosí claimed 120,000 residents in 1585 (more than double the level in 1555) and reached 160,000 by 1603. The latter population figure included around 66,000 Indians, 40,000 Spaniards, 35,000 creoles (American-born Spaniards), and as many as 6,000 African slaves.[10]

Virtually everything in Potosí had to be imported. No matter that the city was located in one of the most remote sites of South America. In addition to mining supplies, basic necessities, and manufactured goods, a chain of middlemen operating between Lima and Buenos Aires imported luxury items from all over the world and made sure to mark up their prices en route. Silks, carpets, perfumes, crystals, ivories, and precious stones were delivered from Africa and

South Asia. Spices and porcelains were imported from Ternate, Malacca, and Goa. The expression "as rich as Potosí" circulated around the globe to convey the incredible sums that were being tossed around town. Local residents had bragged about their one-million-peso celebration in honor of Felipe's coronation in 1556. By 1579, however, a dowry estimated at 2.3 million pesos was arranged by a General Pereyra to marry off his daughter. The price of a good bottle of wine (200 ducats), trousers (300), coat (1,000), and a fine horse (4,500) were off the charts. A single theater ticket was priced at fifty pesos. Potosí's "wild western" lifestyle also included an estimated 600 to 700 common criminals, 120 prostitutes, 15 gambling houses, and 14 dance halls.[11]

The local mining industry required huge quantities of everything. An official report of 1603 estimated that Potosí annually consumed 60,000 sacks of coca, 85,655 quintals of charcoal, 115,136 pesos worth of wood, 5,750 quintals of mercury, and 450 African slaves (at around 225 pesos each). Thousands of small-scale factories (*obrajes*) were established in local villages to manufacture goods for the mining industry (e.g., mercury bags) and cheap textiles for the native population. Over five hundred obrajes were operating in the Quito district alone between 1590 and 1620. As for basic agricultural products, the adoption of a Spanish-style haçienda system with animal-drawn plows helped to develop wheat and maize fields along the Cochabamba River and other major valleys. The town of Cochabamba was founded in 1571 to manage the new agricultural estates (and to reorganize the local Amayras). Tarija was established in 1574 in the Andean region south of Potosí to raise agricultural products (grain and wine), to serve as a staging area for future exploration activity, and to defend against incursions by hostile Indians.[12]

The emergence of Spanish businesses and colonial towns coincided with the devastation wrought by diseases and overwork. Bolivia's total population was halved to roughly 500,000 by 1600. As in New Spain, enterprising Spaniards were forced to develop enormous haçiendas to address the labor shortage. Haçiendas were less labor-intensive than scattered farms and villages, but they still absorbed as much as one-third of Bolivia's dwindling labor supply. Like nearly everything else in Bolivia, they were intended to provide food supplies and coca leaf for the mines of Potosí. Coca was a big business. Some sixty thousand sacks of the chewable stimulant were delivered to Potosí in 1603 because most Indian workers refused to work without it. The savvy Hernán Pizarro had attempted to monopolize the leaf-producing areas around Cuzco prior to the product's spread to the Yunga valley near La Paz. The fields were worked by Aymaras and black slaves.[13]

There was little incentive to export Bolivian-grown wheat, maize, and coca to other markets. The transportation costs were daunting, and New Spain produced a much wider variety of agricultural products and trade goods. Silver was viceregal Peru's only significant export product. Not only did this constrain the native populations of Peru and Bolivia from developing local industries unassociated with mining but it resulted in the economically unhealthy exchange of Potosí silver for products manufactured in Mexico, Spain, Europe, and even China. Potosí silver even helped to finance the development of rival shipbuilding facilities in Huatulco, Mexico, and Guayaquil, Ecuador. When Potosí suffered through one of her regular cyclical declines, the regional economy had little to fall back on. New Spain, on the other hand, was far more diversified economically.[14]

Many of the two hundred merchants operating in Potosí by 1601 discovered that royally controlled silver and mercury were less lucrative than conventional merchandise. Attention was also paid to the fate of Gonzalo Fernández de Herrera, a contractor who was sentenced to ten years in prison for conspiring to pocket 1,500 quintals of royal mercury in 1593. Profits from regionally grown wheat, maize, fruits, wine, sugar cane, vegetables, melons, pigs, and llamas were unregulated. Potosí's remote location was no barrier to the daily arrival of a variety of goods from Cuzco (five hundred miles to the northwest) and Tucumán (six hundred miles to the southeast). A report in 1586 stated that 2,200 mita Indians from the village of Chuquito arrived with a herd of as many as 40,000 llamas—the delivery helped to maintain Potosí's llama population at around 100,000. The 5,000-llama train that worked the 350-mile trail to Arica was managed by 312 Indians (at one peso per day) and assumed that half the train would die en route. Llamas were not limited to freighting requirements—nearly half were slaughtered annually for their meat, skin, and other by-products.[15]

Since Potosí was nearly as close to Buenos Aires (1,500 miles away) as it was to Lima (1,150), enterprising merchants had hauled trade goods along an unofficial (illegal) route via Tucumán, Rio Paraná, and Rio de la Plata well before the two Iberian crowns were unified in 1580. The economic potential of this "back door" route had been missed by Cabeza de Vaca and his predecessors. Portuguese Brazilians became experts in managing a lucrative black-market trade in sugar, slaves, and cloth between Brazil and the Spanish Americas. Silver was off-limits theoretically, but increasingly large volumes of Potosí silver were diverted to purchase slaves for the sugar plantations of Brazil, the gold fields of Colombia, and even the mines of Bolivia. A large number of Portuguese merchants, many of them Conversos, became active in the Lima market after 1600. The extended transatlantic family of one such merchant, Manuel Baptista Peres, had access to 600,000 cruzados in investment capital.[16]

The "back door" route itself had been pioneered by Father Francisco de Vitoria, a member of a powerful Portuguese merchant family who had established the religious, ranching, and agricultural town of Tucumán (Argentina) in 1580. Buenos Aires, established for good in 1583, was a much longer sail than the Panamanian ports, but the "back door" route to Potosí avoided the onerous customs duties at Panama, Callao (Lima), Chincha, and Arica. It also avoided the cumbersome mule trains that were required to carry merchandise across the Isthmus of Panama. The silver was exchanged for slaves (for Brazil) and a wide variety of merchandise for Potosí. The price mark-ups could be tremendous—a mule that might cost five pesos in Buenos Aires would find a buyer in Potosí for forty pesos—but traffic surged during Brazil's sixty-year union with Spain (1580–1640). Less-adventuresome merchants preferred to trade via Panama (City), Nombre de Dios, and Portobelo. The three Panamanian ports handled nearly half of all cargo shipped between Seville and Spanish America for nearly a century. Southbound goods were shipped to one of the three Peruvian ports and hauled by llama (or mule) trains to Potosí.[17]

If Brazil (via Buenos Aires) and Spanish Manila (via Acapulco) were consuming an increasingly large volume of Bolivian silver, there was plenty more where that came from. Lost in the shadow of Cerro de Potosí were three other major silver mines in Bolivia— Porco, Vilcabamba, and Oruro—and the prolific gold fields of Colombia. The combined silver output from the three other Bolivian mines nearly matched Potosí's production during the heady 1596–1610 period. Venerable Porco, jump-started by Hernán Pizarro in the late 1530s, was eclipsed by the opening of the Castrovirreina mine at Vilcabamba in 1590. The site of the last remaining Inca kingdom (1571) produced between two and three million pesos of silver between 1596 and 1610 and emerged as the second-largest mine in Bolivia. In 1605, another bonanza was uncovered at the San Miguel mine near Oruro. The rowdy boomtown of Oruro expanded to around thirty thousand inhabitants by 1607 and averaged a very impressive 185,000 marks (1.6 million pesos) of silver between 1607 and 1612—this was double Joachimstahl's peak level in Central Europe. Oruro received a Real Caja in 1607 and served as a mining supply center for all of Bolivia because it was conveniently located between La Paz and Potosí. Oruro's warehouses stored most of the Huancavelica mercury that was delivered from Arica. Silver from Peru proper was limited to a modest strike made at Cerro de Pasco, a 13,673-foot-high site located 135 miles from Lima.[18]

Colombia was an even brighter story for Felipe. Gold was at least eleven times more valuable than silver, and Colombia's gold fields were conveniently located near the Caribbean port of Cartagena. The Cauca and Magdalena rivers generated substantial wealth between 1560 and 1600, highlighted by the 150,000 gold pesos registered annually in the Cali-Anserma-Cartago region of Popayán. While most of Colombia's placer gold continued to be recovered by old-fashioned crushing and panning, 150-foot (thirty-estado) mining shafts were occasionally sunk to access deeper-seated veins. An estimated two thousand Indians and black slaves were working the veins at La Concepción, south of Popayán, between 1575 and 1600 and generating thirty thousand gold pesos per year. It wasn't all peaceful. Spanish miners had to contend with regular Indian attacks during the second half of the sixteenth century. Mines located in the dangerous Choco region, at Novita and Toro, had to be abandoned after an Indian revolt in 1586. The Choco remained untouched until an estimated 540,000 pesos of gold were extracted there annually in the late 1700s.[19]

Prospecting activities along the lower Cauca River had been stepped up by Governor Gaspar de Rodas in 1570. Buriticá was still home to two thousand workers in 1583, but her annual production fell steadily to around twenty-five thousand pesos during the 1586–1598 period and shifted Rodas's attention to the gold deposits at neighboring Caceres—twenty miles downriver from Buriticá. Caceres provided a staging area for the underexploited gold works at Zaragoza de las Palmas. Discovered east of Caceres along the Rio Nechi and its tributaries in 1579, Zaragoza emerged as the single greatest American gold discovery to date. It was designated as a formal mining district (with a Real Caja by 1582) to exploit the terrace and sand-bar deposits of Rio Nechi more intensively. Gold output reached 300,000 gold pesos in 1595 before expanding to as much as 500,000 pesos in 1615. By then, Zaragoza was being worked by three hundred Spaniards and four thousand black slaves who had been delivered to the site from Cartagena.[20]

The Magdalena River gold fields along the eastern slope of Colombia's Cordillera Central were not as prolific as the Cauca River region, but Spaniards prospected virtually every one of her tributaries for placer gold. Despite recurring native Indian attacks, starting with the eviction of Spanish miners from upriver Neiva by local Pijaos in 1543, the sixty-mile middle stretch between Mariquita (1551) and Ibagué became a significant source of vein gold and even silver during the 1580s. Silver had also been discovered between Neiva and La Concepción during the 1550s. At Remedios, located southeast of Zaragoza along an upper tributary of the Magdalena, the rediscovery of vein and placer gold launched a massive gold rush between 1594 and 1598. Some two thousand black slaves were engaged in the annual production of 150,000 gold pesos worth of yellow metal. Annual output at Remedios was halved during the 1599–1601 period, however, and averaged less than twenty thousand gold pesos between 1602 and 1622.[21]

The extraordinary efforts to find, develop, and enhance the mines and smelters of Colombia, Bolivia, Peru, and New Spain were amply rewarded. Spanish America generated unimaginable volumes of precious metals during the last two decades of the sixteenth century. The otherwise staggering 7.5 million pesos of silver produced at Cerro de Potosí in 1592 was supplemented by 2 million pesos from the other Bolivia mines, 2.5 million from New Spain, and at least three tons of gold (worth 1.2 million silver pesos) from Colombia. As the annual output of these mines approached 14 million pesos, a total that would have left Charles V, Jacob Fugger, and Suleyman absolutely speechless, Felipe was preparing to a squander large portion of this spectacular treasure on his latest initiatives—the reconquest of the entire Spanish Netherlands and an invasion of England.[22]

## PROTESTANT ALLIES

Felipe was growing tired of Queen Elizabeth. The Protestant successor to Felipe's ex-wife Mary Tudor had been interfering in Iberian affairs for nearly two decades and had assembled an increasingly powerful navy. Elizabeth's support for the Dutch rebels and trading enterprises like the English Company of the Levant were painful, but Felipe was especially outraged by a series of overseas expeditions—some fanciful, some small scale—that Elizabeth had commissioned to disrupt the flow of funds to Felipe's war machine in Europe. One royal expedition of note was completed in spectacular fashion by Francis Drake between 1577 and 1580. A cousin and protégé of the talented John Hawkins, Drake had privateered around the Spanish Caribbean during the early 1570s and had even hijacked a load of Potosí silver that was being carried across the Isthmus of Panama by mule train. Since Potosí was just recovering from a lengthy slump, Felipe was outraged by the heist and the acclaim that greeted Drake upon his return to England in the summer of 1573.[23]

Drake's next assignment was even more ambitious—Elizabeth instructed him in 1577 to revisit the treacherous Magellan Straits and respond to the recent execution of fellow Englishman John Oxenham in Peru. Oxenham, a relatively obscure figure because he was captured by the Spanish in 1576, condemned by the Inquisition, and then hanged in Lima, had attempted to plunder the underdefended Peruvian coast. He had the right idea. Drake's five-ship fleet reached the Magellan Straits in just sixteen days only to discover that the extremely dangerous route had been ignored for good reason. Drake lost two vessels in the straits and was fortunate to sail his remaining three ships up the western coast of Chile. After surveying the coasts of Peru, Mexico, and even California, he crossed the Pacific Ocean and captured an outbound Spanish galleon laden with Potosí silver. The entrepreneurial Drake traded some of the silver for a load of Moluccan cloves, rounded the Cape of Good Hope, and completed his circumnavigation to Plymouth in a single ship in September 1580. The achievement, plus the 600,000 English pounds worth of gold, silver, and spices aboard the *Golden Hind*, earned Drake a knighthood. The windfall allowed Elizabeth to pay off all of England's foreign debt and invest a surplus of 42,000 pounds in the Levant Company. Since the profits from the Levant Company, in turn, helped to establish the later East India Company, one could argue that the catalytic effect of Drake's plunder was far more important than the nominal value of the silver bullion.[24]

The psychological impact of Drake's exploits outweighed the financial gains and losses. With expanding volumes of precious metals to protect, Felipe instructed the merchants of Spanish America and Seville to invest in regional fortifications and offshore patrol fleets. Admiral Pedro Menéndez de Avilés organized two modestly sized fleets to patrol the Caribbean out of Santo Domingo and Cartagena. Viceregal authorities invested much larger sums to refortify Havana, Veracruz, and San Juan (Puerto Rico) and protect the gold port of Cartagena with an Armada de Tierra in 1582. While most of these investments had been catalyzed by English raiding parties, Dutch privateers were regarded as a much greater naval threat to Spanish interests. England was still regarded as a second-tier power, and the Dutch fleet was twice the size of England's.[25]

Felipe's conquest of Portugal in 1580–1582 had at least freed up resources to deal with England and her Protestant friends in the Netherlands. The Netherlands were a mess. A series of armed campaigns, financial crises, and political chaos had devastated business conditions, split the nation in two, and reduced Antwerp to its lowest position in recent history. Having survived the bankruptcies of 1557–1560, Antwerp had suffered through regular blockades of the Scheldt River, the six-week mutiny-occupation of 1574, the state bankruptcy of 1575–1578, and the horrific sacking of 1576. The "Spanish Fury" of November 4, 1576, had been the last straw for many Antwerpers. Business activities shifted to the north, and hundreds of local merchants simply left town—Hanseatic merchants for good, Portuguese merchants to Cologne, English merchants to Bruges, and Protestant-leaning merchants to backwater Amsterdam.[26]

With little to lose from Spain and a population that was more Calvinist leaning than Amsterdam, Antwerp sided reluctantly with the Dutch Revolt in August 1577—this was nearly a year before the more cautious burghers of Amsterdam and Utrecht joined the great cause. It didn't matter. By January 1579, Europe's greatest entrepôt had been drained of much of her investment capital and many of her most energetic businessmen. There were still twenty-three Italian merchant houses in town, but only ten remained from Portugal, and just four from Spain. Antwerpers watched helplessly as the Genoese handled the asientos that financed Parma's military operations and the currency exchanges of silver reales for easier-to-transport golden escudos. Antwerp's murky future, sitting on the fault line between the northern and southern provinces, relied primarily on the Mediterranean trade and a series of royal pepper contracts issued to the Fuggers and Welsers in the early 1580s.[27]

Antwerp's demise affected Felipe's ability to deliver troop wages to the Army of Flanders. During the 1568–1576 period, blockades by Dutch rebels and English privateers had forced the Genoese to redirect the carriage of American silver to neutral French ports like Hendaye, Nantes, or Bordeaux until these ports were engulfed by France's latest religious war. The Genoese returned to the Spanish Road after the medio generale was signed in 1578. No matter that roughly four months were required to ship American silver to Seville, exchange it into gold at Genoa (or Venice), and then haul the coins up the Spanish Road to the Netherlands. In happier days, a bill of exchange transaction between the Spanish treasury and Antwerp could usually be concluded in around four weeks. Bills of exchange deals were unusually risky during times of war. Italian merchants preferred high-quality reales for their business dealings in the Levant, and the Army of Flanders refused to be paid in anything other than gold coins. However, a rising bimetallic ratio of 12:1 meant that a single horseman could carry 9,600 escudos worth of gold but only eight hundred escudos in silver. While these exchanges converted Genoa and Venice into major distribution centers for American silver, some merchants dared to ship American silver illegally to Antwerp. These were needed to settle outstanding asientos and other bills of exchange.[28]

Antwerp was shocked further by the States General's issuance of its Act of Abjuration on July 26, 1581. The act superseded the Union of Utrecht of January 23, 1579, by converting what had been mainly a religious dispute into an outright declaration of independence. By then, however, Portugal was in Spanish hands, Felipe was receiving increasingly large quantities of American silver, and Parma was preparing to retake the mainly Catholic southern provinces. Asientos backed by American silver funded the expansion of

Parma's Army of Flanders to over sixty thousand men by October 1582 and supported Felipe's half-baked plan to invade England in due course. The two campaigns were linked. Following a collapse in diplomatic relations between England and Spain in January 1584, Felipe's hastily prepared invasion plan assumed that the recently successful schemes in Portugal (and the Azores) could be repeated and that Parma was capable of securing the Flemish coastline.[29]

The Army of Flanders had been bolstered by the return of thousands of Spanish troops from the Portuguese front in September 1582. Their absence had constrained royal authority in the southern provinces since 1580. Coupled with an expanding war chest from the mines of Spanish America, Alexander Farnese, the Duke of Parma, was empowered to restore the Army of Flanders to its former glory. Parma recaptured most of Flanders and Brabant between 1581 and 1584 and squeezed recalcitrant Antwerp by blockading the Scheldt River in 1583. The talented Parma deployed a combination of sieges, pitched battles, and economic deprivations to roll up the last remaining pockets of southern resistance. Ypres, Bruges, Ghent, Brussels, and Antwerp were all targeted for reconquest in 1584 and 1585. Retributions directed against southern Protestants were guided by the Reconquest and Inquisition—convert to Catholicism or leave the country. In the meantime, the so-called Dutch Republic was ignored. The backwater northern provinces were perceived to be far less important strategically, and there was little urgency to retake them. England loomed as Parma's more significant adversary. No matter that England boasted a very credible navy and had not been invaded since the Norman Conquest of 1066. If Elizabeth wanted to support the Dutch rebels, Felipe would scheme to restore Catholicism in England under Mary of Scotland.[30]

Tensions escalated after William of Orange was assassinated (by gunshot) in Delft on July 10, 1584, and it was discovered that the assassin, a Burgundian named Balthasar Gérard, had been hired by Italian Jesuits for thirty thousand pounds and ennobled (in advance) by Felipe II. The Dutch stiffened their resolve after Gérard was captured and quartered in Delft on July 24, 1584. But the outrage failed to halt Parma's march through the southern provinces. Bruges and Ghent were recaptured by year-end 1584, Brussels fell in March 1585, and Antwerp was besieged between April and August. The *Asientos de Flandres* issued in Antwerp to support the siege suggested that the town's few remaining investment companies were profiting from their own destruction. The future was looking so bleak for the southern rebels in May 1585 that sovereignty was offered to (and rejected by) religiously tolerant Henri III of France and Protestant Elizabeth of England.[31]

The situation in the Netherlands was ugly. Between the capture of Maastricht in July 1579 and the seizure of Antwerp in August 1585, the Army of Flanders regained control of over thirty major towns in the southern provinces and banned Protestantism entirely. Protestant-controlled town councils in the north returned the favor by prohibiting the practice of Catholicism—following a ban in Antwerp on July 15, 1581. Parma supplemented his demonstrated military skills by bribing local nobles with property, pensions, and royal pardons. In later years, he won over turncoat English mercenaries with old-fashioned cash. The defection of the southern nobles, plus indemnities received from the recaptured towns, helped to finance the Army of Flanders locally. The royal treasury had to contribute "only" twenty-two million florins for the entire campaign of 1579–1585.[32]

By the time that Antwerp had surrendered officially to Parma on August 17, 1585, thousands of Antwerpers had already fled with their capital, craftsmanship, entrepreneurial skills, and business contacts. Most of these refugees resettled temporarily in Hamburg, Bremen, Frankfurt, Cologne, and Lubeck before reconsolidating their positions in Amsterdam. The closure of the Scheldt River and Spain's application of trading embargoes against Dutch and English shipping between 1585 and 1589 only added to Antwerp's woes. An estimated thirty-eight thousand Antwerpers, half of the town's remaining population, abandoned the town by 1589 and reduced one of the world's greatest entrepôts to second-tier economic status.[33]

In the meantime, a rising tide of American silver was securing the asientos that financed Parma's army, Felipe's navy, and the estimated cost of the English invasion plan. Incoming royal treasure from the Americas would average 1.9 million ducats per year between 1580 and 1590—well above the average 1.5 million ducats sent annually to the Netherlands between 1567 and 1585—excluding even larger sums from royal confiscations. An expansion of the treasure-for-juros scheme delivered nearly five million ducats to the Spanish treasury in 1587 and helped to support an annual military budget of five million ducats between 1586 and 1590. The more money Felipe had, the more he spent. The Army of Flanders consumed a whopping 13.5 million florins in 1587 alone, over three times the amounts received in 1585 and 1586, in preparation of the invasion of England. The army would receive an average of 9.6 million florins annually between 1588 and 1597.[34]

## THE INVINCIBLE ARMADA

Felipe prepared to invade England only after Parma had restored royal authority in the southern Netherlands. Backed by record production volumes from Potosí, he determined that a single massive armada—like his father's legendary Danube flotilla of 1532—would either deter English (and Dutch) attacks at sea or provide a platform for a full-scale invasion. The bold invasion plan commissioned on December 29, 1585, was modeled on the two-pronged Portuguese campaign of 1580–1582 and the successful invasion of the Azores in 1583. The Invincible Armada would rendezvous with the Army of Flanders in the spring of 1587 and somehow land Parma's massive army on an English beach head in Kent.[35]

The funds to organize the Invincible Armada were there in 1585. According to Luis Capoche, annual silver output from Potosí alone had averaged roughly four million pesos during the 1580–1584 period. These impressive totals were nearly matched by the mines of New Spain and Colombia. In addition, the beleaguered Spanish economy was benefiting from a transatlantic trading boom brought on by the influx of precious metals. One Veracruz-bound convoy in 1585 totaled fifty-one ships and seventeen thousand *toneladas* in trade goods. The 213 Spanish merchant ships that sailed across the Atlantic in 1586 were nearly double the level in depressed 1576. No matter that Felipe's ambitious plans threatened to undermine the transatlantic trade. An armada fleet comprising 100,000 tons would remove Spain's highest quality vessels from the business, divert huge volumes of private treasure shipments that fueled this trade, and leave Spanish merchants with a capacity shortage and an expanding *averia* (8 percent in 1591). The invasion of England would also allow foreign flags to capture an estimated 21 percent share of the transatlantic shipping market between 1588 and 1592—over three times the share held by foreign vessels in 1579–1587.[36]

Geopolitical events shifted the Invincible Armada from a transatlantic defense scheme to a platform to invade England. The unusually capable Duke of Parma reconquered Flanders and Brabant by 1586 and was heading up the Maas and Rhine rivers in preparation of a full-scale assault against the United (Dutch) Provinces. A military triumph over the fledgling Dutch Republic, a missed opportunity in 1575–1578, would have been a near certainty in 1586 if not for the intervention of England and a temporary threat from France. Although the Peace of Fleix had ended France's seventh War of Religion in November 1580, the subsequent death of the Duke of Anjou in June 1584 removed Henri III's primary heir and promoted the claim of Protestant-leaning Henri (Bourbon) of Navarre. The threat posed by Navarre and his Protestant Huguenot allies prompted Felipe to support a French Catholic faction under Henri de Guise when France slid into yet another civil war in 1588. If that were not enough, Elizabeth's brutal execution of Catholic Mary of Scotland in February 1587 promoted Felipe's own (tenuous) claim to the English Crown through his daughter Isabella. This raised the stakes significantly. It was one thing for Spain to invade England and place Mary of Scotland on the English throne; it was another to have the throne occupied, Portuguese-style, by Felipe II.[37]

Elizabeth had declined an offer of Dutch sovereignty in May

1585, but she arranged an alliance with the United Provinces on August 20. This was just after Antwerp, England's most important offshore trading center, had fallen to Parma. The Treaty of Nonsuch called for England to send 7,350 troops to the Netherlands under Robert Dudley, the Calvinist-leaning Earl of Leicester, in return for a cost-sharing arrangement with the States General, a Dutch commitment to contribute twenty ships if England was attacked by Spain, and access to Brill and Flushing. The two Dutch ports could help to block the Spanish navy. While the Dutch had already committed 2.4 million guilders to continue the war against Parma, the added cost of the English expeditionary force provided additional protection. Johan van Oldenbarnevelt (1547–1619), the successor to William of Orange, would rule the fledgling Dutch Republic from The Hague until his controversial ouster and execution in 1618.[38]

Elizabeth expanded English privateering actions against Spain between 1585 and 1587. Drake and twenty ships were dispatched to the Caribbean to intercept American treasure ships and attack the ports of Havana, Santo Domingo, and Cartagena. England's most illustrious pirate failed to capture a single Spanish treasure ship, but he did manage to occupy Santo Domingo for one month, sailing off with a modest ransom of 25,000 ducats, and held up Cartagena for a ransom of 107,000 ducats. The underdefended Caribbean ports were easy prey for a dedicated raider despite local efforts made to refortify or patrol them. Merchants apparently preferred to pay ransoms to prevent their properties from being burned or sacked. In late April 1587, Drake directed his most audacious raid against the port of Cadiz. This was in Spain. He managed to sink, destroy, or capture twenty-four Spanish warships that had been slated to join the Invincible Armada.[39]

The Dutch were happy to have Drake as a naval ally, but England's occupation of the United Provinces was barely more tolerable than the Spanish version. After Leicester arrived in late 1585, he infuriated the leaders of Holland and Zeeland by attempting to co-opt their political authority in the States General. Holland and Zeeland responded to the English power play by elevating Prince Maurits of

Nassau, son of William of Orange, as their own military commander. Leicester's political problems mounted after he instituted a trade embargo against Spain and the southern provinces in April 1586. While this was merely a response to Felipe's order of May 1585 calling for the seizure of any English ship that dared to enter a Spanish port, the retaliatory embargo weakened an already fragile Dutch economy. Dutch shipping business was diverted to the neutral Hanseatic League. Leicester's short reign in the United Provinces was finished after his attempted military coup was thwarted in the summer of 1587—the grasping earl was sent back to England.[40]

If the English-Dutch alliance was looking somewhat shaky, so was the Invincible Armada. Felipe's invasion fleet was to be commanded by Álvaro de Bazán, the Marquis de Santa Cruz, and administered by Alonso Pérez de Guzmán el Bueno, the seventh Duke of Medina Sidonia. The skillful Santa Cruz had commanded the ninety-eight-ship Spanish fleet that secured the Azores for the unified Iberian crown in July 1583—in the process, removing sixty French ships sent by Catherine de Medici to support António's claim and providing another model for a possible amphibious invasion of England. The Invincible Armada was so important to Felipe that he assigned the Duke of Medina Sidonia to the newly created post of Purveyor-General of the Armadas and Fleets of the Indies. The purveyor-general relegated the casa factor, an officer who had outfitted the royal ships for nearly a century, to deal solely with the export of Almadén mercury to the mines of New Spain. The mining boom had converted the casa's facilities in Seville into a quicksilver warehouse.[41]

Santa Cruz was not on the same page with the Duke of Parma. The admiral estimated in October 1586 that four million ducats would be required to organize an armada of 150 ships and sixty thousand men for an eight-month-long invasion of England. The price tag could be covered easily by the spectacular metallic wealth being generated in the Americas. In contrast, the Duke of Parma's earlier plan of April 1586 had called for 30,500 Spanish troops to be loaded onto barges at Dunkirk, Nieupoort, and Gravelines and then escorted across the Strait of Dover to a landing site between

Dover and Margate. Parma's lower-budget plan alleviated the need for a massive armada if total secrecy could be preserved. Of course, this was wishful thinking.[42]

Felipe, a monarch with zero military experience, countered with an invasion plan of his own. Santa Cruz would organize an armada and landing force in Lisbon and then rendezvous the fleet with Parma's unescorted barge-borne army near the mouth of the Thames River. How the two massive invasion forces would manage to coordinate their actions, especially given the fickle winds and currents in the English Channel and North Sea, would be left to the fates. And as Antwerpers and their economic rivals knew all too well, there were few deep-water ports along the coast to accommodate the largest Spanish warships. There was one other problem. While the English army was no match for the battle-tested Army of Flanders on land, the feisty English navy had helped to pioneer the shift in naval strategy from boarding and fighting ships to sinking them. Drake and other Englishmen had demonstrated their high skill levels when it came to the deployment of heavy shipboard artillery. But no matter—if the Romans and Normans had accomplished the feat of invading England, so could Felipe II.[43]

Felipe's primary advantage in 1588 should have been in money. In addition to accelerating treasure shipments from the Americas, he could count on a variety of income sources to help defray the whopping cost of his invasion force. In 1585, Pope Sixtus V had renewed the papal cruzada for seven years to fight Protestantism in England, France, and the Netherlands and to prepare for the next "Holy War" against the now-overestimated Ottoman Empire. The Holy See of Toledo chipped in with 400,000 ducats between 1586 and 1588, encouraged by a papal agreement of July 1587 that guaranteed a one-million-ducat subsidy to Felipe if he could land Spanish troops on English soil by year-end. Rome also agreed to recognize the claim of Felipe or any other Catholic to the throne of England. Elizabeth returned the favor by sending envoys to Istanbul and Morocco. She hoped to enlist the sultans in an anti-Catholic alliance of her own.[44]

But there were never enough funds—not even with record shipments from Potosí, New Spain, and Colombia. Funding issues delayed the invasion of England for over a year. With the Netherlands under a naval blockade, Felipe was forced to rely on Hanseatic, Baltic, and even renegade Dutch shippers to supply timber, equipment, weapons, and provisions for the Invincible Armada. Morale faltered after Drake's successful raid on Cadiz harbor in April 1587. Parma was left to spend the summer of 1587 constructing a network of canals to ensure that his invasion force could be floated to the designated departure points at Sluys, Nieupoort, and Dunkirk. This took time and money. Even worse, a combination of overwork, famine, disease, warfare, and desertions reduced Parma's army to a mere seventeen thousand by March 1588. Coupled with the almost certain leakage of the invasion plan to English and Dutch spies, the lack of a deep-water port to accommodate the large escort galleons, and the difficulty in coordinating any naval escort to cover a dangerous channel crossing and landing, Parma voiced his opposition to Felipe's increasingly high-risk scheme.[45]

A staggering ten million ducats were invested in the armada during the first nine months of 1587. The delays were extremely costly. Every month of delay forced Felipe to assume an estimated 700,000 ducats in expenses to maintain the ships, troops and supplies, loans, and asientos. A six-month delay squandered 4.2 million ducats. In addition, the confiscation of Spain's finest merchant vessels for the armada had strengthened the English and Dutch naval blockades and decimated the transatlantic trade—outgoing tonnage from Seville in 1587 and 1588 were mere fractions of the 1586 level. Drake guessed that something was up after his successful raids of 1586 and 1587 had gone unpunished. But Felipe's cash flow problems did nothing to dissuade him from his mission—having been humiliated by the Protestant queen of England and her second-rate realm, Felipe was committed to restoring his reputation at all costs. The Invincible Armada would be so powerful that no English ship would dare to cross her path.[46]

Yet another wrench was thrown into Felipe's grand scheme when

Admiral Marquis de Santa Cruz died unexpectedly in Lisbon on February 9, 1588. Rather than call the whole thing off, an action that would have pleased the Duke of Parma and altered the course of history in Spain's favor, Felipe stayed the course. He immediately promoted the thirty-eight-year-old Duke of Medina Sidonia to Santa Cruz's position. The appointment was a curious one. As governor-general of Andalucia and purveyor-general of the armada, Sidonia had effectively outfitted a series of transatlantic fleets, including the armada, and had impressed Felipe with his redesign of Cadiz's defenses after Drake's raid of April 1587. However, Spain's grandest grandee had no experience at sea, even less in naval warfare, and was also opposed to the invasion plan. Sidonia was apparently chosen for his organization skills and his "glamour." Spain had plenty of experienced captains to sail the fleet, and Sidonia was smart enough to understand that the looming naval battle with England would be won with heavy guns, not with old-fashioned boarding and fighting.[47]

Sidonia had "star power." No matter that he claimed to be 900,000 ducats in debt in 1588 and had scraped up less than thirty thousand pesos for the armada campaign. Sidonia was still regarded as the wealthiest peer in Spain, supported by annual Andalusian rental income of 150,000 ducats, with a reputation that was likely to attract well-connected nobles and some of the best ships and captains from the transatlantic trade. Sidonia's command ensured that the armada would be manned by Spain's most capable crews. The duke also enjoyed excellent, long-standing relations with the wavering but badly needed Portuguese nobility. Sidonia's wife was half Portuguese, and the duke himself had entertained King Sebastião in Cadiz within weeks of the monarch's fateful departure to Morocco. The Portuguese connection bore fruit when Portugal agreed to contribute 12 heavily armed galleons, 5 other warships, 14 supply ships, 4,623 men, and 347 pieces of artillery.[48]

Sidonia had made himself very familiar with galleons and shipboard artillery. The three- and four-masted galleons adopted by most European navies by 1550 were slimmer and more maneuverable than carracks and *naos*. Every nation had its own version, but most galleons sported a fortified hull, rows of hinged gunports, trapezoidal (square) sails and top-sails on the fore and main masts, two fully planked decks, and an elevated quarter-deck and poop deck. It was the low-slung forecastle, complemented by a protruding beakhead below the bowsprit, that gave the galleon its famous crescent-shaped profile. While galleons were designed to carry the heaviest artillery, top-heavy failures like Henry VIII's one-thousand-ton *Henry Grace a Dieu* (1513) and the refurbished six-hundred-ton, seventy-eight-gun *Mary Rose* (1536) had called for a ship's heavy guns to be relocated from the bulky castles to the midsection. The *Mary Rose* had sunk off Portsmouth harbor in 1545 with over five hundred English sailors. The relocation facilitated the shift from old-fashioned boarding, fighting, and plundering to broadside shooting and sinking. In an irony of history, Felipe had encouraged his wife Mary Tudor to expand England's modest thirty-ship navy in September 1555 to increase the pressure on France.[49]

By midcentury, Spain had opted for frame-built, three-masted "man-of-war" galleons. These combined a sail plan from the nao with a tier of hinged, watertight gunports in the ship's waist. But experimentation continued. Another round of top-heavy failures, including the wreck of the *São Paulo* off Sumatra in 1561, had led to a scaling down of galleons to promote speed and maneuverability. Galleons of three hundred to six hundred tons, built mainly along the Cantabrian coast, accounted for nearly half of Spain's transatlantic tonnage by the late 1560s. Although few of these vessels were sturdy enough to complete more than four round-trip sailings, they could accommodate the latest in naval gunnery. Cannon that was breech-loaded, castle-mounted, and made of wrought iron had been found to be underwhelming in battle—their projectiles (stone, iron, or lead) were nearly harmless. These were replaced by high-quality bronze cannon or muzzle-loaded twenty-four- to fifty-pounders of cast iron that were mounted on four-wheeled truck carriages. Cast-iron guns were heavier than bronze but were only one-third as expensive. Bronze cannon were reserved for military campaigns and land-based operations. Gunners also discovered an optimal gunpowder formulation of saltpeter (75 percent), charcoal (15 percent), and sulfur (10 percent).[50]

Felipe eventually settled on midsized galleons of four hundred to five hundred tons to provide the best trade-off between speed, capacity, and protection capability. Under the direction of Cristóbal de Barros between 1562 and 1589, the Crown offered financial incentives to the primary shipyards as well as access to a twenty-thousand-ducat pool of low-interest loans. The loans weren't always repaid, but the incentives helped to revitalize the Cantabrian shipbuilding industries in Bilbao and Viscaya. Admiral Pedro Menéndez de Avilés was even allowed to experiment with hybrid sail-and-oar–powered galleons in 1567–1568. Barros eventually persuaded Felipe to commission nine new galleons—two at 400 *toneladas* and seven at 300—between November 1581 and February 1583. But with the king consumed with Portugal and the southern Netherlands, Barros rarely received construction funds in a timely fashion. He was also affected by the various naval blockades. These delayed the receipt of bronze cannon from Flanders and Germany, copper from Hungary, gunpowder and sulfur from Italy, wooden masts from the Baltic, and hemp from the Netherlands and Germany. When the nine new galleons finally left for Seville in early 1584, they sailed off without all of the specified artillery—this was dangerous during the mid-1580s. But the investment paid off. Contributions from a rival shipbuilding program under Lope de Avellaneda are murky, but Barros's dual-purpose galleons defended the treasure ships, served in the front line of the armada in 1588, and managed to return home safely after the famous ordeals off Scotland and Ireland.[51]

The 117-ship Invincible Armada that sailed into the English Channel in late July 1588 was split evenly between Iberian-built vessels and ships recruited from Italy, Flanders, and Germany. The fleet was also split into sixty-six warships armed mainly with bronze cannon rather than cast iron—Barros's nine Spanish galleons, eleven Portuguese galleons, four galleasses, twenty-nine Cantabrian naos, thirteen Mediterranean naves, and two hulks—and fifty-one supply vessels. While all but three of the galleons were sturdy enough to survive the coming battle in the channel and an unscheduled trip around the British Isles, their only flaw was a deep draft that prevented them from entering shallow coastal waters. The opposing 150-ship English fleet is somewhat of a mystery. We have details on fourteen English galleons of five hundred tons or higher—including the sluggish 900-ton, 64-gun *Elizabeth Jonas*, the 800-ton *Victory*, the 800-ton *Ark Royal*, the 600-ton, 60-gun *Lion*, and the 600-ton *Hope*—and sixteen other ships. However, we have no information on the fleet's other 120 craft and, thus, have limited means with which to substantiate the claim that most of the English galleons had been slimmed down in width, increased in sail area, reduced in crew size, and rearmed significantly.[52]

Felipe's half-baked invasion plan was based on the recent conquests of Portugal and the Azores and the likelihood that the vaunted Army of Flanders could crush English opposition on land. From departure points at Lisbon on May 30 and La Coruña on July 21, 117 members of the 141-ship Invincible Armada were instructed to rendezvous with Parma's thirty-three-thousand-man invasion force somewhere off Flanders in early August and support Parma's barge fleet across the channel. The landing site would be somewhere in Kent. With the Scheldt River blockaded by the Dutch, Parma had been forced to dig a new canal between Ghent and Nieupoort to provide a secure passageway for the 130 invasion barges that were scheduled to depart from Nieupoort and Dunkirk. Parma's multinational invasion force—comprising Spaniards, Italians, Walloons, Burgundians, Germans, Irish, and Scots—was being funded by a monthly outlay of 300,000 ducats from the Spanish treasury. A mere 4,800 troops were left to defend the entire Spanish garrison network in the Netherlands.[53]

The Dutch blockade of Flanders, effective as it had been, was uncoordinated. The two hundred merchant ships that were rearmed between November 1587 and July 1588 represented a small fraction of the 2,700-ship Dutch merchant fleet. No Dutch warship was larger than two hundred tons, nor did any carry more than sixteen small guns. In addition, the effort was plagued by age-old rivalries between Hollanders and Zeelanders. Over one hundred ships, mainly Zeelanders, had blockaded the Scheldt River since the fall of Antwerp in August 1585 with limited support from Holland. Even Justin of Nassau, the illegitimate son of William of Orange and the recently appointed lieutenant admiral of Zeeland, had been unable to persuade Holland to act more aggressively. The provincial rivalry, coupled with limited naval funds, Leicester's hasty departure, and the first of three consecutive poor grain harvests (1587–1589), had allowed Parma to seize the port of Sluys on August 5, 1587, to supply his inland headquarters at Bruges. Parma also recaptured Deventer and a number of northeastern fortresses. Holland's recalcitrant leadership finally had a change of heart when news was received that a massive Spanish Armada had left Lisbon in May 1588 and was en route to the English Channel.[54]

If no Dutch ships joined the subsequent fray between the Spanish and English fleets off Gravelines, the Dutch blockades were critical to England's victory. The expanded 135-ship blockade of the Scheldt was supplemented by a 90-ship blockade of the Flemish coast, a 100-ship blockade of Delfzijl, and a 32-ship cordon around Sluys. Justin of Nassau subsequently rearranged the blockades to monitor Parma's expected departure points at Nieupoort and Dunkirk. The action, in turn, reduced the probability that Parma's troop barges would be able to rendezvous with Sidonia's armada. Although Parma managed to conceal his final movements from the Dutch, the threat posed by the blockades must have affected Sidonia's decision-making process between August 6 and 8. Parma was forced into inaction. His flat-bottomed troop barges were incapable of sailing across the English Channel without an armed escort.[55]

One of the most anticipated naval battles of the century ensued. An Invincible Armada comprising 117 ships and over two thousand shipboard artillery pieces passed the southwestern tip of Cornwall on July 29 and sailed up the English Channel in crescent formation. Iberian naval commanders were renowned for their naval line formations. The 150-ship English fleet, under the command of Lord

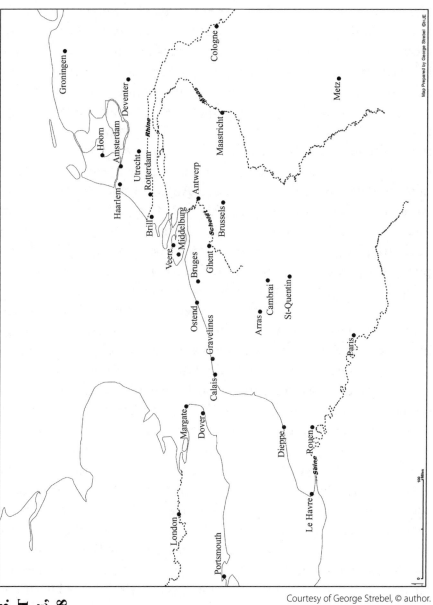

## MAP 8. NORTHWEST EUROPE, 1557–1588

Courtesy of George Strebel, © author.

Henry Seymour, Lord Charles Howard of Effingham, and Sir Francis Drake, lay to the windward of the Spanish fleet off Plymouth but refrained from firing their guns at long range. The English artillery was not heavy (or accurate) enough to hurt powerful Spanish galleons at that distance. Although four early skirmishes were inconsequential, they forced Sidonia to consume a large quantity of ammunition and placed the easterly moving armada against the wind. Sidonia subsequently lost two ships en route to Calais, a deep-water port that could accommodate eighteen of his deep-drafting galleons, and discovered on August 6 that Dutch warships were blockading Parma's weakly defended barge fleet at Nieupoort and Dunkirk. There would be no rendezvous. With the wind at their backs, the English fleet attacked Sidonia's fleet on August 7 and temporarily separated the armada off Calais. The English resumed the offensive off Gravelines on the following day.[56]

Sidonia had broken formation when he anchored eighteen of his ships off Calais on August 6. This left him vulnerable to unfavorable winds and currents, shallow waters, and well-designed sorties from eight English fireships. Sidonia's hasty cutting of the fleet's anchor cables drove his ships into the Flemish shoals to face the entire English fleet on their own. When Parma learned at Dunkirk that the armada had been scattered off Calais, he could do nothing to help. Sidonia lost only one nao and two Portuguese galleons during the only significant battle of the entire campaign, but the losses and artillery damage incurred off Gravelines were enough to scuttle the entire enterprise. Sidonia skillfully regrouped his remaining 110 ships into formation and followed the favorable winds into the North Sea. Then he shook off his English pursuers off the southeast coast of Scotland on August 11 and exploited his enemy's problems. The English were infected with typhoid fever and were running low on provisions.[57]

Sidonia's escape came at a great cost. The crippled armada somehow managed to round the northern tip of Scotland on August 20 before facing a few more weeks of violent storms, typhus, and scurvy. Twenty-eight ships, mainly the Italian, Flemish, and German

vessels, were lost off the Irish coast, and many of the eighty-two surviving vessels were damaged severely. There were thousands of
human casualties. When the survivors finally reached the ports of
Santander and Laredo between September 21 and 23, they discovered that Sidonia had salvaged forty-four of the Crown's best warships but that over twenty of these vessels were damaged beyond
repair. Sidonia's professionalism was overwhelmed by the magnitude of the affair—what should have been a minor naval setback for
Felipe II was transformed into a financial sinkhole and a possible
"changing of the guard" in the Western world.[58]

The defeat of the Invincible Armada off Gravelines on August 8
was caused mainly by a weak strategic plan, unfavorable wind conditions, and Sidonia's inability to hold the impregnable formation
that had guided the armada safely up the channel. There is no evidence that the Spaniards were undersupplied with ammunition.
Their gunners were skilled at long-distance shooting and they held
their own in the early skirmishes with Howard, Drake, Frobisher,
and Hawkins through August 3. Sidonia had kept his artillery
loaded at all times. The excavation of eight sunken warships from
the depths of the English Channel suggests that the Spaniards had
not run out of standard nine-pound iron shot between August 2 and
August 8. The twenty-two-gun *Trinidad de Escala* and the twenty-
one-gun *San Francisco* had consumed only a small fraction of their
fifty-round supply of ammunition when they went down off Gravelines. In addition, there is no evidence that the English fleet was any
faster than the armada, that an English ship could make five tacks
in the time required for a Spanish ship to make one, and that the
English could weave in and out of the Spanish line in order to fire
twice as many broadsides as the Spaniards.[59]

Nevertheless, the defeat of the Invincible Armada was big news. A
reporter for the *Fugger News Letter* service declared from England
(via Hamburg) on November 19, 1588, that a "Great Armada" of 135
Spanish ships had been defeated in early August after having failed to
rendezvous with the Duke of Parma for an invasion of England. The
reasonably accurate report (Letter #92) added that the Spanish fleet,

after losses, had been pursued for five days as far as Scotland and returned (via Ireland) to Coruña with only seventy-eight vessels. Devoted readers of the *Fugger News Letters* must have been shocked by the results. *News Letters* of 1586–1588 had covered Drake's plundering of Santo Domingo (#72), his defeat of a twenty-two-ship Spanish fleet off Cadiz (#80), and Thomas Cavendish's capture of a Mexican-bound Spanish galleon (#90). But these had been relatively small-scale affairs, tarnished by less-than-accurate claims of plunder. The events of August 1588 were on an entirely different level.[60]

## FELIPE'S MULTI-FRONT WARS

The defeat of the armada was a severe, double-barreled blow to Felipe II—psychologically and financially. While Felipe's claim to the English throne had been a reach, he had at least hoped to remove the English navy from the Netherlands. The regional situation was further complicated by the English triumph and a brewing revival of the French threat. Despite record production volumes from Potosí, the Spanish treasury was paying roughly two-thirds of its annual revenues on interest charges associated with the debt-financed armada and Army of Flanders. Taxes had already been raised to the hilt, and future state revenues had been pledged to secure existing borrowings. Yet enormous funds were needed to rebuild the armada (and Parma's army) as soon as possible, to regain the momentum in the Netherlands, and to halt Henri of Navarre's claim to the French throne. If that were not enough, Dutch and English privateers were already scheming to step up their raids in the Caribbean and along the Iberian coast.[61]

The sheer humiliation of the affair, coupled with the record output from the mines of Spanish America, pushed Felipe to restore the Spanish fleet almost as quickly as the Ottoman fleet had been restored after Lepanto. Despite the devastation inflicted against Spanish power—lives, ships, skills, money, and prestige—England was many decades away from becoming a serious global power. By 1590, Felipe had launched a total of twenty-one new galleons from the yards at Santander, Bilbao, Lisbon, Gibraltar, and Vinaroz and commissioned a fleet of smaller, faster armed frigates (*zabras*) to patrol the Caribbean. In addition, the noted military engineer Juan Bautista Antoneli was recruited to redesign the fortifications at Havana and San Juan and to build an entirely new port facility at Portobelo—Nombre de Dios was too vulnerable. The new defensive schemes worked very well. None of the two hundred or so English privateers operating in the Atlantic managed to intercept a single Spanish treasure ship between 1589 and 1591. The irrepressible Drake was repulsed off Lisbon in 1589 and off Puerto Rico and Panama in 1595. He and Hawkins succumbed to a fatal tropical disease in 1596.[62]

Record contributions from Spanish America were being spent as fast as they were received. In addition to the investments indicated above, the Invincible Armada was being rebuilt, Spain was still at war with England and the Dutch Republic, and a pent-up rebellion was brewing in Aragon. The Aragonese had wearied of their second-class status in Spain. If that were not enough, the French Protestant movement had emerged as a dangerous threat to Spain. The near destruction of the armada in 1588 had emboldened Henri III to raise the stakes significantly. In December of that year, he ordered the assassination of Felipe's principal Catholic allies—the Duke de Guise and his brother, the Cardinal of Lorraine. When Henri III, in turn, was assassinated on August 1, 1589, Protestant-leaning Henri of Navarre became next in line to seize the vacant French crown. The assorted claims against the Spanish treasury took a back seat to the brewing conflict with France.[63]

Felipe responded by declaring war on Navarre and his Huguenot allies on September 2, 1589, and relocating the Army of Flanders to the French frontier. Although the abandoned invasion of England had kept a portion of Parma's Army of Flanders intact, the pending campaign against France split this portion in two and weakened royal defenses in the Spanish Netherlands. Garrisons were placed only in the major towns. Parma even made an offer of conciliation to improve his position—Dutch rebels would be permitted to worship Calvinism on a private basis if they submitted to royal authority. The offer was rejected. Parma somehow managed to extend his military control to the northeastern Netherlands in 1590 and even established some Spanish garrisons in neighboring Germany. Garrison troops evicted Dutch forces from the northeastern areas (and would do so again in 1597). In 1591, the Army of Flanders, once again, could probably have recaptured the breakaway Dutch provinces with little difficulty. However, Felipe decided to suspend an invasion, once again, in favor of a full-scale assault against Navarre.[64]

The Army of Flanders crossed the French border in August 1590 and forced a very surprised Navarre to abandon his position in Paris. But there was a large price to be paid. With Parma fighting a two-front war in the north and south, Dutch forces managed to capture Breda—their first military triumph in twelve years. The unexpected victory spurred Maurits of Nassau, son of William of Orange, to recapture most of the northeastern territories in Parma's absence. When an extremely well-timed rebellion sprang up in Aragon, the forty-seven-year-old Parma refused to proceed any further. He had had enough, or so he thought. Felipe intervened and forced the underappreciated duke to reinvade France in December 1591.[65]

Unfortunately for Felipe, the overstretched Army of Flanders was running out of money. Outlays from Spain plunged from 18 million florins in 1590 to just 4.4 million in 1592. The shortfall would be partly remedied by recorded receipts from viceregal Peru, but record contributions from the Americas, as usual, were not enough. Two unpaid Spanish tercios mutinied in 1591, giving Parma yet another (internal) front with which to be concerned. Parma was finally allowed to retire on December 31, 1591, only after being wounded during the successful recapture of Rouen. One of the most capable military leaders of the sixteenth century eventually died of his wounds on December 3, 1592. Parma's death was yet another stroke of good fortune for the Dutch. The Count of Fuentes, Parma's successor, was unable to overcome a fresh wave of mutinies that broke out in the summer of 1593 and in 1594. His-

tory repeated itself in the Netherlands. Maurits of Nassau consolidated his victories north of the Maas River between 1591 and 1594
and recaptured Groningen in July 1594. The Dutch Republic was
restored to her former self.[66]

Felipe's fixation with France turned out to be another huge mistake. A third Spanish invasion, delayed by Parma's death until
March 1593, was stripped of its crusading zeal when Navarre
agreed to convert to Catholicism. A humiliated Felipe II saved face
(and money) by striking a six-month truce with Navarre in July
1593. Of course, the truce allowed Navarre to consolidate his recent
triumphs. After being crowned King Henri IV in February 1594,
Navarre turned the tables on Felipe by declaring war on Spain in
January 1595 and striking a peace treaty with England in May
1596. England countered Spain's capture of deep-water Calais in
April 1595 by attacking Cadiz in the summer of 1596. When the
cash-strapped Felipe attempted to return the favor in October, a violent storm sank one-third of his invasion fleet with over two thousand men off Ferrol (Spain).[67]

Felipe was heading for another bankruptcy decree, despite record
bullion contributions from Spanish America, an impressive restructuring program in viceregal Peru, and record alcabala revenues in
Spain. The familiar problem was that the costs to rebuild the Spanish
Armada, resume the offensive in the Netherlands, and invade France
had simply overwhelmed the Crown's financial position. Spanish
state revenues, after doubling during the first phase (1556–1573) of
Felipe's reign, were exploding. The quinto received from fourteen
million pesos worth of Spanish American metals was around four
times higher than that received in the 1560s and accounted for
around one-fourth of total state revenues—double the 11 percent
share recorded in 1554. However, history repeated itself. The rising
quinto and huge volumes of confiscated treasure shipments had been
either prepledged many years in advance to secure the asientos or
swapped into new layers of juros. The pending declaration of
another state bankruptcy in 1596 would essentially wipe out the 8.4
million ducats in American treasure received in that one year.[68]

No matter that the metals-driven explosion in trade had tripled the level of tax receipts from the alcabala since 1560. Annual alcabala receipts would total 3.9 million pesos in 1598 and account for 29 percent of Spain's 13.5 million pesos in state revenues. A sizable portion of the alcabala surge was derived from Felipe's fateful decision to lift the reviled trade embargo with the Netherlands in 1590. Spanish individuals and enterprises operating on both sides of the Atlantic Ocean were generating taxable private wealth that was possibly 2.5 times the value of the annual quinto. The alcabala receipts even surpassed the three million pesos contributed by Spanish America, excluding confiscations, and the contributions from the Spanish Church and the papal cruzada.[69]

Felipe had struggled with the powerful municipal councils of Spain. Whether they were controlled by transplanted Genoese or by native Spaniards, the councils had enjoyed the financial fruits of Charles V's shortsighted agreement to forego the inflationary portion of locally collected alcabala and tercia taxes. In the crisis year of 1575, Felipe had managed to triple the quota through which a rising tide of juros was secured or serviced by the municipalities of Spain. The quota was raised again after the defeat of the armada in 1588 with an amendment that called for the taxation of Spain's tax-exempt nobles and clergymen for the very first time. While the new tax schemes helped to generate a whopping eight million ducats in state revenues during the 1590–1596 period, they had failed to relieve the Crown's ever-increasing financial problems. Attention was shifted to Felipe's single most valuable possession—the Viceroyalty of Peru.[70]

The Spanish Americas contributed around one-fourth of annual state revenues during the 1590s, excluding confiscations, with bullion providing security for a rising flood of royal asientos. Roughly two-thirds of the American contribution was derived from the Viceroyalty of Peru, and an estimated 90 percent of this contribution (one million pesos ensayados in 1590) was generated by the royal quinto from Cerro de Potosí. Although Potosí had helped to build and rebuild the Invincible Armada and the Army of Flanders,

Felipe believed that she could do much better. In 1590, incoming viceroy García Hurtado de Mendoza was charged with the tasks of raising productivity at Potosí and Huancavelica, like Toledo had done successfully in the 1570s, and squeezing out additional funds to finance Felipe's next campaigns.[71]

The new viceroy had nothing to do with Potosí's production record in 1592. Instead, he focused his attention on cutting the accumulating bloat in government administration, raising taxes, collecting more tribute monies from Indian communities, and arranging a campaign fund to collect gifts and other contributions. A Royal Decree of November 1, 1591, introduced a 2 percent alcabala (sales tax) to viceregal Peru, a new tax on foreign residents, and an opportunity for prosperous mestizos to purchase royal offices. The same decree also called for an investigation of land titles within the viceroyalty (subject to confiscation). The new measures increased revenues significantly. Confiscations, foreign taxes, and the new campaign fund generated over one million pesos. The alcabala raised thirty-five thousand pesos in Lima alone, nearly seventy thousand in La Plata (Sucre), and thousands more in Quito. No matter that an armed revolt had to be crushed in Ecuador and the sale of government offices was rife with graft and corruption. The post of sheriff-bailiff (*alguacil mayor*) of the Audiencia de Charcas, a position responsible for "controlling" contraband silver, went for as much as 120,000—three times the purchase price of the position of assaying officer (*ensayado*) at Potosí. Sixteen council (*regidor*) seats were sold for around 7,500 pesos each in Potosí, followed by the sale of thirteen more at La Plata.[72]

Mendoza's new sources of nonmining revenues raised nearly 2.5 million pesos between 1592 and 1596—representing one-fourth of the Viceroyalty of Peru's total revenues during the most productive period in Potosí's history. Roughly half of the nonmining total was generated from confiscated land titles and taxes applied to foreigners. The other half was derived from the sale of offices, the alcabala, and the Indian communities. Of course, the other three-fourths of Mendoza's revenue base was realized from the spectacularly successful mining industry. The five-year period (1591–1595) included four of Potosí's strongest production years in history. This was reflected in annual tax receipts of around 1.5 million pesos and a 500,000-peso tax contribution from the royal mercury works at Huancavelica. While mercury prices remained a moving target, efforts were made to recover unpaid mercury loans. An audit in 1597 estimated that American mineros owed the Crown as much as 600,000 ducats for five thousand quintals of mercury that had been bought on credit between 1582 and 1590.[73]

Felipe received between 1.5 and 2.5 million pesos annually from the Viceroyalty of Peru during the 1592–1596 period. In 1592, a year in which Potosí produced a record 220 tons of silver, the Peruvian treasure fleet delivered 1.5 million pesos plus another 1 million pesos in gifts, loans, repaid debts, and "contributions" from the Mendoza program. The level of contributions would continue through 1596, although a portion of the 1595 funds were applied to a new Pacific fleet and local defenses. One observer, Francisco López de Caravantes, estimated that the Viceroyalty of Peru generated as much as nine million ducats during the 1591–1596 period— extremely impressive, but less than the annual interest charges on the Crown's escalating mountain of royal debt.[74]

The expense of Felipe's multifront wars was exacerbated by the inflationary effect of the American treasure imports. The cost of Spain's materials, freight, and wages were the highest in Europe. Shipbuilding costs tripled between 1580 and 1610, making it three times more expensive to rebuild the armada in Spain than in England. The Spanish textile industry had already been priced out of the market. These and other costs were embedded in the huge asientos issued to finance the latest military campaigns. The Council of Finance in Madrid had raised the ante in 1589 by issuing a single asiento of 2.4 million ecu to Agostino Spinola. This was a prelude to the whopping three- to four-million-ducat asientos issued in 1595. Investors preferred the asientos to investment opportunities in Spain's uncompetitive industries. Silver-backed asientos were repayable by the Crown partly in cash, partly in assigned treasure cargoes, or from unpledged royal revenues. Spain's core economic problem was that the asientos issued between 1590 and 1596, the world's largest financial transactions to date, left most of the American treasure in the hands of non-Spanish creditors, merchants, and manufacturers. Spain had run up a balance of trade deficit with virtually everyone in Europe.[75]

The state bankruptcy declared on November 29, 1596, was a near replay of the crises of 1557–1560 and 1575–1578—massive state expenditures had overwhelmed the state's otherwise massive borrowing capacity. The associated risk of a bankruptcy decree had prompted many lenders to raise interest rates substantially, in some cases above 50 percent, in anticipation of confiscations and forced debt conversions. The escalation in interest charges proved to be too much. Like its predecessors, the bankruptcy decree of 1596 did not cancel royal obligations. It simply forced royal creditors to convert their outstanding claims into juros that were worth between one-half and one-third of their claims. If outstanding juros totaled 100 million ducats in postbankruptcy 1598, the Crown's actual (pre-restructured) debts may have approached 200 million.[76]

The latest Spanish state bankruptcy was somewhat less severe than the debacle of 1575–1578. Despite the eighty-eight million florins that were squandered in the Spanish Netherlands during the 1590s, Potosí was still going strong, and Felipe had learned his lessons. The 1596 decree exempted the Fuggers and a few other merchant bankers who had the capability to transfer American treasure directly to the military front. By weakening the bargaining position of the Genoese and other hard-nosed creditors, the exemption helped to produce a more favorable medio generale on February 14, 1598. The security provided by American treasure persuaded the recapitalized creditors to lend the Crown a fresh 7.2 million ducats in 1598. This raised the Spanish treasury's outstanding debt to eighty-five million ducats—twice the level outstanding back in 1557, but manageable for the time being.[77]

However, the Spanish state bankruptcy of 1596–1597 further devastated a home economy that was already reeling from taxes, uncompetitiveness, and warfare. Somehow, the flow of European trade was shifting to the backwater Dutch Republic. Felipe's relaxation of the trade embargo with the Netherlands between 1590 and 1598 had raised alcabala revenues in Spain, but the suspension had inadvertently created a new commercial power. A decision by newly crowned Felipe III to reapply the embargo in late 1598 was constrained by its unenforceability. In the meantime, the governor general of the Netherlands, Archduke Albert, improved the prospects of the medio generale by brokering the Treaty of Vervins between Spain and France on May 2, 1598. The treaty freed up resources for the Army of Flanders, facilitated the recapture of territories lost during the French campaign (and hopefully more), and even launched a proposed marriage between Albert and the Infanta Isabella. Unfortunately for Felipe II, he didn't live to see these through. The seventy-one-year-old monarch succumbed to a blood disease that he had been fighting since 1595. One of the towering figures of the sixteenth century died within the walls of his beloved El Escorial on September 13, 1598.[78]

# CHAPTER TEN

# THE DUTCH ADVANCE

Spain would remain a formidable global power for centuries, but the late Felipe II had just squandered one of the greatest treasure troves in world history. The declaration of a third Spanish state bankruptcy in 1596, declared in the very middle of Potosí's peak decade of silver production, left millions of ducats in the hands of non-Spanish creditors, a mutiny-wracked Army of Flanders, and a star-crossed Spanish Armada. The vaunted, but financially draining, Army of Flanders had been in a position to crush the fledgling Dutch Republic for good in 1575, 1585, and even 1595. But in each case, a northern invasion plan had been canceled because of Felipe's endless cash flow constraints and a need to wage campaigns in "more important" locales. The so-called Invincible Armada, humiliated by the English navy in 1588 and riddled with bad luck, was being constantly rebuilt to control the increasingly powerful English and Dutch. The Portuguese navy had its own problems. Carracks not lost to age, weather, or privateering, like the English raid against the Welser-Fugger spice fleet of 1592, were being co-opted by private Portuguese interests and official corruption.

The accession of Felipe III in September 1598 did nothing to alter the fact that imperial Spain and Portugal, unified or not, had passed their peaks. Despite record inflows of American silver and a Golden Age of private-sector profits from Macao, Nagasaki, and Manila, the unified Iberian Crown was incapable of managing her spectacular wealth effectively. Felipe III (d. 1621) was headed for a fourth bankruptcy decree in 1607 thanks to the military outlays required to control the English, French, and Dutch in Europe and abroad. As early as July 1599, a *Fugger News Letter* reporter was warning about the serious Dutch threat to Iberian interests in Asia. The basic problem for the two Felipes, following in the footsteps of Holy Roman Emperors Maximilian and Charles V, was that they spent (and overspent) their metallic windfalls as fast as they were received. Soaring royal revenues were somehow always exceeded by soaring royal expenditures.

Felipe III was ready to compromise with the feisty Dutch Republic when he assumed the Spanish throne. Backed by the medio generale with his creditors and the Peace of Vervins with France, he instructed his regent, Archduke Albert of Austria, to be more tolerant of Protestantism and to consider recognizing Maurits of Nassau as Stadholder of Holland. Of course, Maurits (1567–1625) and Johan van Oldenbarnevelt had come too far to abandon their nation's hard-earned independence. There would be no compromises with Spain. The Dutch responded to Felipe's olive branch by sending a thirty-five-thousand-man army (mostly foreign mercenaries) into Flanders in the spring of 1600 and defeating Spanish forces at Nieuwpoort on June 30. Spain's tit-for-tat siege of Ostend, the last important Protestant town in Flanders, was undermined by a series of Spanish mutinies after a rocky start in June 1601. The siege of Ostend was so chaotic that Ambrosio Spinola agreed to assume command of the ground forces in September 1603 on the condition that he alone would control the siege and its financing. As a member of one of Genoa's leading merchant families, Spinola knew that he could borrow military funds at roughly half the rate received by Felipe III. The capable Spinola was promoted to second-in-command of the entire Army of Flanders after the surrender of Ostend on September 22, 1604. However, his great triumph was tarnished by a series of Dutch victories that had been achieved while the three-year-plus siege was taking place.[1]

The Spanish Netherlands maintained its reputation as a financial sinkhole. When the Army of Flanders wasn't wracked by mutinies, it drained an average of nearly five million ducats from the Spanish treasury annually between 1598 and 1604. If much of these funds were derived from huge silver shipments from Bolivia, the Bolivian cash machine was being split into pieces. An increasingly large portion of America's mineral wealth was being diverted to Spanish Manila, Brazil, and even Angola. Felipe II's financial mismanagement of the Iberian Empire had allowed Maurits to regain the offensive in the northern Netherlands and fine-tune a Dutch military machine that was regarded as the most innovative in Europe. Maurits's innovative deployment of artillery, fortifications, and siege schemes was even being adopted by creative types in Prussia and Sweden.[2]

Felipe's quagmire in the Netherlands extended into the Indian Ocean. Over 330,000 cruzados in Indian customs duties were pledged to Ottavio Centurione in 1604–1606 to repay a portion of his 2.6-million claim against the Army of Flanders. The resumption of royal control over the pepper trade in 1598 served mainly to divert Carreira profits to the Army of Flanders. As it was, Lisbon's pepper imports averaged only eighteen thousand quintals annually between 1598 and 1604, pepper prices had plunged by nearly 40 percent since 1590, and royal spice ships were being exposed to English and Dutch naval attacks. Reduced spice profits meant less funds available to repair (or replace) Portuguese warships and trading vessels. Shipwrecks rose. If that were not enough, Portugal's share of the spice business was being diverted to English and French traders in the Levant and to Dutchmen in Indonesia.[3]

One of the great ironies of the Dutch Republic's spectacular advance in global affairs was that the expansion occurred with the vaunted Army of Flanders very much engaged in the Netherlands. It would remain so until 1648. The Army of Flanders, slimmed down to just ten thousand troops in the post-Armada trauma of 1591, was gradually restored to sixty-two thousand by 1607 and coordinated with a chain of Spanish garrisons in northwestern Germany. But Felipe III's fateful decision to reinstate a trading embargo against the Dutch in 1598 after his peace offering had been rebuffed by Maurits and Oldenbarnevelt, was as counterproductive to Spanish interests as any Dutch regiment. Not only did the embargo reignite the simmering rebellion, it inadvertently fueled Amsterdam's growth by pushing

Dutch merchants into direct trading relationships in Iberian spheres of interests. Seville and Lisbon would be bypassed. In addition, the embargo failed to revive Antwerp and sabotaged the Brussels-based regency under Archduke Albert of Austria and Felipe II's daughter Isabella. A possible era of peace and prosperity in the southern provinces was sacrificed to warfare between 1600 and 1607.[4]

The Dutch struggled to overcome Spinola's effectiveness and the death of Queen Elizabeth in 1603. The accession of James I, the Catholic son of Mary of Scotland, allowed Felipe III to make peace with England in August 1604 and remove a source of Protestant support for a Dutch state that had become a bitter economic rival. In short order, Ostend surrendered and Spinola regained the offensive. Spinola's forces even managed to breach the "Great Wall," a mainly earthen wall of fortifications that defended the IJssel and Waal rivers, and might have marched deep into Holland if the Spanish treasury was not running low on cash. Falling contributions from the Royal Treasury—from 5.3 million ducats in 1605 to 3.7 million in 1607—failed to cover a series of largely inconsequential sieges and a steady expansion in the Army of Flanders. When another round of Spanish mutinies sprang up along the IJssel and Maas rivers in 1606, Felipe convened negotiations with the Dutch to staunch the financial damage. The war-weary sides finally agreed to an armistice in April 1607, a month in which a Dutch fleet had even destroyed Spanish naval defenses at Gibraltar. While the armistice was a prelude to the signing of a formal Twelve-Year Truce in April 1609, it occurred too late to avert a fourth Spanish bankruptcy decree on November 9, 1607.[5]

By 1607, there was no doubt that a collection of seven backwater Dutch provinces had assumed a lofty status in military and commercial affairs. Amsterdam had eclipsed Antwerp, Lisbon, Seville, and Venice as the most dynamic trading entrepôt in Europe. Scores of Dutch traders were penetrating some of Portugal's most attractive markets in the East. A nuisance-level revolt started in April 1566 with a petition to abolish the antiheresy laws in the Spanish Netherlands had somehow led to the Union of Utrecht

(1579), the Act of Abjuration (1581), and the Dutch Republic. The separation of the historically laggard Calvinist north from the historically prosperous Catholic south (present-day Belgium) had inadvertently launched a new global economic powerhouse. No matter that Holland, Zeeland, and the five other northern provinces squabbled endlessly and would remain a fragile political state until 1648. Yet the Dutch Republic (1581–1795) would somehow lead Europe in global trade, shipping, finance, agriculture, and military technologies for nearly two centuries.[6]

The ability of a federal nation of around 1.5 million people to accomplish these feats is as remarkable as the earlier achievements of Portugal and Castile. An additional irony is that the northern provinces had underperformed the prosperous southern provinces of Flanders and Brabant for centuries. One hidden advantage was that backwater Holland had been a relatively egalitarian state during the later Middle Ages. Holland's two hundred or so noble families and the church together owned no more than 15 percent of the province's agricultural land. This left more than two-thirds of the state's agricultural territory in the hands of the Dutch peasantry. Supported by crop specialization and innovative drainage systems, windmills, dikes, and canals, the farms of the Netherlands achieved the highest agricultural yields in Europe. Cheese, butter, and beer emerged as high-margin export products. This agricultural wealth supported an increasingly urbanized population. Roughly half of Holland's population lived in cities and towns by 1560, including 27,000 or so inhabitants in Amsterdam and Utrecht. While the leading Dutch towns were much smaller than Antwerp (85,000), they were catching up to the southern centers of Brussels (50,000), Ghent (45,000), and Bruges (35,000).[7]

The nearly constant warfare experienced in the Netherlands since the Calvinist revolt of 1566 had made Amsterdam more important than it should have been. After the fall of mighty Antwerp in 1585, Amsterdam and other northern towns were flooded with as many as 100,000 Protestant refugees having the type of capital, business skills, and trading networks that had been absent from a region centered on fishing and shipping. Few refugees placed Amsterdam on a priority list in 1585. Many of Antwerp's leading merchants relocated initially to established trading centers like Cologne, Hamburg, Bremen, and Frankfurt before making the move to modest Amsterdam. The experienced ex-Antwerpers jump-started the local economy with great effect. Amsterdam's population was set to triple to around ninety thousand inhabitants by 1600 with another thirty-nine thousand located at nearby Haarlem.[8]

Few of Antwerp's merchant class had participated in northern "bulk trades" like cheese, butter, beer, herring, and freighting. But they recognized that the Dutch were specialists. Dutch farmers focused on value-added export products, and her freighters focused on the low-cost carriage of fish, salt, grain, and timber. This latter specialty allowed enterprising Dutchmen to gain market share in the grain and timber ports of the Baltic Sea and in the salt and fish centers of the Netherlands, France, and Portugal. Dutch fisherman had trawled for herring in the North Sea since at least 1295 and had exchanged their catch for salt and other products in France, Spain, and even Italy. When the wheat and rye fields of the Baltic region became the great granary of northern Europe, it was only natural that Dutch fish merchants would diversify into the carriage of grain and timber. Efficient Dutch shippers wrested control of the grain, salt, herring, timber, and cloth markets from the Hanseatic League.[9]

The Dutch shipping industry was driven by cooperation—a function of a capital-poor economy that promoted syndicates to spread risk among merchants, small investors, captains, and employees—and innovative naval designs. At the same time that the Baltic shipyards at Danzig and Riga were constructing frame-built "hulks" of three hundred tons or higher, ships that were half the size of Iberian carracks, Dutch shipwrights were introducing even smaller vessels of two hundred tons or less that could penetrate coastlines that were too shallow to accommodate hulks and car-racks. This smaller-scale approach also cut out the middleman, leaving Dutch shippers to trade directly with the managers of local timber, grain, and salt supplies, and eliminated the need for large (costly) crews, excessive rigging, and weaponry. Taken together, the new designs allowed Dutch freighters to underprice their Hansa competitors and still make money. Lower-cost salt imported from Portugal and western France crippled the Hansa-dominated salt mines of northern Germany and Poland. When the Dutch added French wine to their product mix, a Zeeland port like Middleburg emerged as a herring center, salt refiner, and wine distributor. Amsterdam was focused on herring, grain, and timber.[10]

Amsterdam was still insignificant to Antwerp (and even Hamburg). Despite the growing clout of her shipping industry, she conducted no trade with the Mediterranean, had no wealthy merchant class, no banks or bourses, and virtually no textile industries. Lodovico Guicciardini commented on the "modest scale" of her wealth in 1561. Thomas Gresham mentioned Amsterdam only once in his regular communications with the English Crown. The quirky winds of the Zuider Zee made the port difficult to enter and exit, requiring the deployment of local "lighters" on a regular basis, and European aristocrats sniffed about the city's focus on mundane commodities. But Amsterdam freighters overwhelmed the Hanseatic League during the 1550s, gained market share in France during the French Wars of Religion, and prepared to invade the Muscovy (Russian) market. The expansion was led by cargo-specific vessels, easy access to the Rhine River, logistical support from the deeper-water ports of Texel and Vlie, and the creative shipwrights of Hoorn.[11]

Gresham should have paid more attention. Dutch merchants operating out of Danzig and Riga were so impressed with the newly formed English Muscovy Company (1558), a trail-blazing enterprise that served as a model for the later Levant Company (1581) and the East India Company (1601), that they schemed to replicate it. English traders had received privileges from Ivan IV ("the Terrible") to trade with the White Sea port of Archangel at the mouth of the Dwina River. Fourteen English Company ships arrived annually to obtain Russian furs, timber, grain, and oil that had been otherwise handled by the Hansa overland route to Novgorod and Moscow or via the daunting Norwegian sea route around the North

Cape. The English success encouraged Dutch freighters to enter the fray with their lower-cost vessels in the 1560s. The Dutch circumvented both the English and the Hansas by exchanging textiles for furs and hides at Narva on the Gulf of Finland and by trading with the rival White Sea ports of Kola and Keyor. They schemed of even new ways to penetrate the Russian market when the Ottoman-Persian conflict of 1578–1590 disrupted the traditional overland route to Muscovy.[12]

Gains in the Russian market coincided with a budding monopoly in regional freighting, the warehousing of bulk commodities, and the highest agricultural yields in Europe. But it was two factors—a massive infusion of capital and business skills from Antwerp and the suspension of the Spanish trade embargo between 1590 and 1598—that catapulted Amsterdam into the front ranks of Spanish colonial trade. The Dutch achievement was to consolidate the disjointed Antwerp-Lisbon-Seville commercial axis that had dominated global affairs since 1500 at a single location, Amsterdam, and then reconfigure this axis into something new and different. Amsterdam's fledgling financial community expanded their investments in ships, trading enterprises, government-backed debt securities, and maritime insurance. They even published weekly commodity price lists in 1585, roughly eighty years before these appeared in London. Refugees from Antwerp established a host of textile enterprises at nearby Haarlem and Leiden and diverted the sugar trade from Hamburg to Amsterdam after Hamburg had stolen it away from Antwerp. Dutch shippers even began to capture some of the luxury goods trade from Hansa and English merchants. When Felipe III reimposed the on-again, off-again colonial trade embargo against Dutch shippers in late 1598, he was too late to stop a "speeding freight train." The Dutch responded to the latest embargo by organizing massive direct investments in the Americas and in Asia. These actions targeted the century-old Spanish and Portuguese monopolies for extinction. According to one historian, no one commercial power had ever dominated the processes of global trade as completely as the Dutch between 1590 and the mid-1700s.[13]

## AMSTERDAM

Amsterdam's spectacular growth could not have been predicted. Where Antwerp sat on the deep-water Scheldt, Amsterdam was built atop wooden piles that had been sunk into reclaimed marshland. The marshland itself was located where the Amstel River met the IJsselmeer, a southwestern bay of the Zuider Zee, and was a remnant of the rising sea levels that invaded the region in the first millennium. This inhospitable body of water had also formed a part of the northern boundary line of the Roman Empire and, consequently, insulated the region from the powerful Roman influence. Farms and villages had begun to sprout up in the 1000–1300 period only after their inhabitants had learned to control the inflow of sea water with dikes, dams, and locks. Amsterdam built her first recorded dam in 1173 and added a dike across the raised banks of the Amstel River in 1180. The new dike provided a main street for the small group of thatched houses that were appearing along the riverside. The backwater village of "Amsteldam," home to a few hundred farmers and fishermen, was mentioned for the first time by Count Floris V of Holland on October 27, 1275.[14]

Amsterdam's earliest coat of arms, featuring a cog, two men, and a dog, signaled the town's dependency on the sea. However, the town was also dependent on Bruges during her early years—Bruges was the principal entrepôt of the Netherlands. The development of the northern cog between 1150 and 1200 had invigorated Bruges. The cog enabled Hollanders to carry commodities like grain, fish, salt, and timber in coastal waters, avoid many of the region's tolled inland waterways, and establish long-distance trading relationships. Scores of dams, toll booths, mills, and textile looms began to dot the countryside to support a Bruges-dominated trading system that was driven ultimately by Hansas and Italians. When primitive cog wharves were finally constructed in Amsterdam in the late 1200s, Amsterdam joined the regional expansion as well. Floris V even built a toll-free inland waterway to Dordecht to provide an alternative passage to the route dominated by Bruges.[15]

Amsterdam traders managed to carve out a measure of independence from the Bruges-Hansa network. Except for beer, Amsterdam freighters sought to bypass Hamburg and other Hansa ports by sailing around the Jutland Peninsula into the Baltic Sea. Trading with Baltic ports like Danzig and Riga enriched the coffers of the king of Denmark, who controlled the toll house at Copenhagen, but it allowed Amsterdam to develop her own business relationships in the Baltic. When Denmark blocked the Danish Sound in 1368 during a conflict with Sweden, Amsterdam contributed a single cog to the Hansa-led alliance that reopened the passageway to commercial traffic. The grateful Swedish Crown gave Amsterdamers the right to fish in Swedish herring waters for the first time. Within a few decades, over half of the ships sailing through the Danish Sound were based in Holland, and a Dutch trading colony was established south of Copenhagen.[16]

Amsterdam escaped much of the feudal, medieval processes that dominated Europe, but she was renowned as a religious center. The town was even regarded as the "Canterbury of the Low Countries" between 1350 and 1550. Church affairs, freighting, and fishing raised Amsterdam's population to nearly five thousand by 1400, prompting an expansion in land-reclamation projects, canal digging, and the replacement of fire-prone wooden structures with stone. Although taxpayers grumbled that stone foundations were extremely expensive, a war with Utrecht in 1480–1483 sped up the construction of the city walls. Stones were even listed as an acceptable method of tax payments. With nearly ten thousand inhabitants, a thriving shipping industry, and roughly twenty convents and monasteries, Amsterdam was even becoming attractive to the House of Habsburg. Future Holy Roman Emperor Maximilian allowed Amsterdam to add the imperial crown to its coat of arms in 1489—the only Dutch city to be so honored.[17]

Amsterdam's city walls, finally completed in 1508, were unnecessary when it came to religion. The real struggle for power in the Netherlands was less about Christianity than it was about the growing clash between the new merchant class and the Spanish-backed nobility. Calvinism, a Geneva-based movement that became popular in Holland during the 1550s, failed to draw an expanding population of twenty-two thousand Amsterdamers into the brewing conflict in the Spanish Netherlands. Other than a protest filed by a business group in 1564 to challenge the perceived corruption of the Catholic-dominated city council, Amsterdam remained neutral through the rebellion of 1566, the Dutch Revolt of 1572, and the subsequent five years of chaos that engulfed the Netherlands. The city fathers branded the Sea Beggars as pirates and terrorists. Some Amsterdamers provided financial support to the Army of Flanders. The balance of local power did not shift to the Protestant cause until thousands of Protestant exiles began to arrive in Amsterdam in early 1578.[18]

Amsterdam was jolted by the geopolitical events of 1578. On May 26, the Sea Beggars exploited the financial collapse of the Army of Flanders by gaining control of hold-out Amsterdam in a bloodless coup and establishing Calvinism as the official religion. While this so-called Alteration led to the imprisonment of many of Amsterdam's Catholic officials, the town's large Catholic population was unharmed and the city's modest merchant class was free to play both sides of the international conflict. Amsterdamers supplied the timber needed to rebuild Spanish ships after they had been sunk by Amsterdam-based warships and provisioned the Army of Flanders after it was contested by Protestant rebels. Investments were made in local warehouses, freighting companies, and the construction of *boyers*, *vlieboots*, and eventually *fluits*. The number of Dutch merchant ships trading with the Baltic surpassed two thousand.[19]

Amsterdam's economy surged after Antwerp's surrender to Spain in August 1585. The ability of the Sea Beggars to blockade the Scheldt River crippled Antwerp and induced tens of thousands of Antwerpers to leave town for good. Roughly half of Amsterdam's immigrants in the 1585–1600 period were from Antwerp and the southern provinces. It was these refugees who started Amsterdam's first silk finishing, sugar refining, diamond cutting, and publishing enterprises. An estimated one-third of Amsterdam's fifty thousand inhabitants spoke with an Antwerp dialect. Another new dialect arrived in the persons of Portuguese Jews and Conversos. While Amsterdam prohibited these immigrants from joining guilds or even building a synagogue (until later), the hundreds of new arrivals must have been thrilled by the separation of church and state, the absence of an Inquisition, the ability to reside in any neighborhood of their choosing, and the right to acquire property. The immigrants returned the favor by introducing Amsterdam merchants to the extended Converso families who dotted the globe.[20]

If Amsterdam was on her way to prominence, her economic foundation lay in shipping. Dutch freighters had schemed to control their construction costs and operating costs from the very start. They had already borrowed ideas from single-masted, square-sailed cogs (naos) of less than 250 tons, cargo-only ships that required a crew of less than twenty, as well as from the frame-built salt ships of Breton. After experimenting with three-masted, fully rigged round ships, Dutch shipwrights determined that predesigned, frame-first ships limited their ability to mix and match hull shapes effectively. Instead, they combined frame construction with the Viking clinker technique of old to come up with a host of specialty ships— flat-bottomed troop ships, lightly armed patrol boats, heavily armed gunboats, and even galleys. These and other specialty ships were refined by Dutch freighters, merchants, pirates, and ship captains during the course of the sixteenth century. One design in particular, the fluit, would even make economic history.[21]

The Dutch fluit was a hybrid design that married the benefits of a fully rigged sailing ship with those of herring busses, boyers, and vlieboots. Fluits were nearly flat-bottomed, shallow-drafting ships with a rounded stern, low bow, huge hold, and a full deck that was supported by pulleys and tackles to minimize crew size. They were built solely to control labor costs. Like fully rigged sailing ships, they deployed a square sail on the foremast, two square sails on the mainmast, a lateen mizzen, and a square mizzen topsail. Their unique feature, however, was a spritsail at the bowsprit. The fluit was apparently introduced in 1595 at the port of Hoorn, north of Amsterdam, but the initial design was refined through trial-and-error experimentation until at least 1610. By then, fluits had become as large as carracks. These were too slow to manage the Asia trade—their smallish masts and spars provided less canvas area than Dutch galleons (*pinnas*) and pinnaces (frigates)—but they were just right to carry bulk cargo like grain, salt, herring, and timber between the Baltic and Mediterranean. As many as eighty fluits were in operation in the early 1600s.[22]

An average-sized fluit of four hundred tons emerged as the lowest-cost bulk carrier in Europe. With a boyer-like ton-to-man ratio of 20:1, or nearly three times the typical 7:1 ratio found in the Mediterranean, the design catapulted Amsterdam to the top of Europe's shipping industry. Dutch Oostervaerders traded with the Baltic, Noortsvaerders with Scandanavia, and Straetsvaerders with the Mediterranean. A four-hundred-ton fluit of 115 feet in length and 22 feet in width sported an unusually wide length-to-width ratio of 5.2:1. While the largest fluits added square topsails to the foremast to boost speed, the focus was always on operating costs. In fact, cost considerations prompted many Dutch shipwrights to return to the "combination" method that had worked so well with other vessels. This meant applying a skeleton frame to a fluit's upper hull only after the traditional clinker-planked bottom had been completed up to the waterline. An oak-hulled fluit lasted for around twenty years, but most yards switched to pine—it was less sturdy than oak, but it was cheaper, lighter, and easier to work with.[23]

Dutch vlieboots and fluits burst on the European scene with the famines that devastated southern Europe between 1587 and 1600. The famines called for the importation of massive shipments of Baltic grain to the Mediterranean, a development that bypassed the Venetian shipping industry and provided an opportunity to demonstrate the productivity of vlieboots and fluits. Only sixteen Italian-bound grain ships left Hoorn, Amsterdam, and other Dutch ports in 1590, but the profits were so spectacular that they attracted an estimated two hundred Dutch ships to the Mediterranean in 1591. Twenty-one of the twenty-two grain ships that arrived at the Italian port of Livorno in 1592 were from Amsterdam. Their arrival signaled that Dutch freighters were capable of competing head-to-head with Genoese and Venetian shippers who had dominated the Mediterranean cargo trade for centuries. The only problem was that the profitability of the return cargoes was often disappointing. This pushed the Dutch into the regional carrying trade. Since shallow-drafting vlieboots and fluits were capable of penetrating a wider variety of harbors, river mouths, and inlets, the Dutch could gain access to smaller Italian ports that had been ignored by the major carriers. Huge volumes of Italian marble were imported in this way to fuel Amsterdam's building boom.[24]

The continuing poor harvests in southern Europe created an enormous market for Baltic wheat and rye and rewarded the industry's most efficient freighters. No matter that the Italian bonanza depleted Dutch grain stocks in 1596. The shortages extended the trading boom by catalyzing a new wave of land-clearing operations in Poland and Russia. The fluit's maneuverability allowed Dutchmen to deal directly with a wide variety of Baltic growers and merchants. The additional grain supply, in turn, raised the number of Dutch grain ships operating in the Baltic Sea and the Mediterranean. Amsterdam received a hard-to-believe six hundred grain ships from the Baltic in a single week in 1598. As many as nine hundred grain boats left Dutch ports en route to the grain depots in the Baltic Sea within a single three-day period in 1601.[25]

More Dutch capital was invested in grain than in any other commodity and demoted venerable herring from its status as the "mother trade" of Amsterdam. Over half of Amsterdam's warehouses were devoted to wheat storage. In the process, the grain trade expanded Amsterdam's inventories of salt, timber, naval supplies (hemp, tar, and pitch), and silver bullion and catalyzed investments in two new classes of warships. The galleon-like pinnace, a fast, square-sterned, fully rigged warship of fifty to one hundred tons, was the Dutch escort ship of choice during the 1590s. However, it gradually gave way to three-hundred-ton, forty-gun frigates built in Hoorn. Although pinnaces and frigates were needed in the English Channel and in the Mediterranean, they were rarely deployed in the Baltic Sea. Danzig, Konigsberg, and Riga controlled nearly all of the local trade and had no incentive to interrupt Dutch shipping—nearly two-thirds of all ships passing through the Danish Sound in 1600 were Dutch.[26]

Virtually all of the imported Baltic grain that was not sent to Italy was exchanged for salt. Modest volumes of salt had been produced in the Netherlands for centuries by boiling salt-encrusted peat until salt crystals had formed. The expansion of the salted fish industry forced the Dutch to locate larger supplies of marsh salt in warmer climes like France, Portugal, and even Venezuela. Sea water and marshes were dammed into salt ponds in which the salty brine (left to be evaporated by the sun) could be harvested on a larger scale. The Dutch and Hansas traded with the salt works located at the mouth of the Loire River and, to a lesser extent, with the Bay of Biscay— Biscay had produced salt since the days of the Roman Empire.[27]

A typical Dutch trade loop sent grain-laden vessels from Amsterdam to southern France, Biscay, or Setubal, Portugal, for salt, then north through the Danish Sound to Danzig for more grain, and then back to Amsterdam to repeat the process. The profits generated from these transactions piled up increasingly as cash (silver bullion). Prior to the Wars of Religion, French salt and wine had typically been exchanged for herring and Baltic grain at St. Malo, Dieppe, Nantes, and Bordeaux. But as Dutch freighters began to arrive with Baltic goods and even Asian merchandise, many French shippers and shipbuilders found themselves in the same position as the Baltic Hansas—out of business. Displaced Frenchmen viewed Dutch shippers as somewhat crooked, but made every effort to copy their business practices whenever possible. French salt represented roughly half of all Dutch freight carried north to the Baltic grain ports. The other principal outbound trade items were herring, cloth, French wine, spices, and lead. Exotic Asian goods began to appear in volume in the early 1600s.[28]

While Dutch shippers had gained control of the primary bulk commodities, the carriage of finished textiles and "rich trades" remained firmly in the hands of Iberians, English, Italians, and

Hansas. However, Amsterdam's ex-Antwerper community salivated over Felipe II's fateful decision to lift the Spanish trade embargo in 1590—a response to the pending reconquest of the southern provinces, the expansion at Cerro de Potosí, and the financial needs of a post-armada economy. The lifting reopened the door to the high-margin luxury trades. Lower-cost Dutch ships were shortly carrying more pepper and spice cargoes through the Danish Sound than the venerable Hansa merchants of Hamburg and Lubeck. When Iberia and Italy were wracked by poor harvests and famines, the Dutch carried massive volumes of Baltic grain and American silver into the Mediterranean and returned with even larger amounts of luxury goods. If the Dutch lagged behind the English and French when it came to trading directly with the Ottoman East, at least one Dutchman managed to exchange a silver cargo worth 100,000 ducats for spices and silks at Aleppo in 1595.[29]

Nearly constant warfare and on-and-off trading embargoes made global commerce an adventure. England was briefly allied with the Dutch after the Treaty of the Hague in 1596 but reversed course when James I signed the Treaty of London with Spain in 1604. This left the Dutch to fend for themselves against the weight of the Spanish Empire. But England's defeat of the Invincible Armada in 1588 and the ability of Dutch and English privateers to raid the West Indies and Portuguese Africa had sent a clear message to Felipe III—Spain's global empire was vulnerable. Viceregal authorities were forced to invest even more heavily in Caribbean fortifications and received only temporary relief from the death of the estimable Francis Drake off Panama in 1596. Scores of gnarly Dutchmen had taken his place. Even Portuguese Brazil was being threatened. Although Brazil's twenty thousand or so Europeans and sixty-six plantations had produced only 2,230 long tons of sugar in 1584, this was a small fraction of the sugar (and slaving) volumes to come. Brazilian sugar mills and slave-based plantations would receive an estimated eight million cruzados in investments by 1612—all dependent on ripping at least two thousand unfortunate Africans annually from their homes in Guinea, Angola, and the

Congo. More than a few Dutch smugglers began to trade unregistered Brazilian sugar, slaves, and textiles and tap into the "back door" silver that was being delivered to Buenos Aires.[30]

Spain's options were limited. Felipe II had allowed Dutch ships to deliver grain to his famine-wracked subjects in Naples and Sicily, but he was less than thrilled about the ability of Amsterdam to overwhelm his prime trading centers at Antwerp, Seville, and Lisbon. He even ordered the capture of twenty-six Dutch grain ships off Gibraltar in 1591 to send a royal message—that it could be done. While the capture turned out to be a rare exception, at least one trading pattern was becoming predictable: Dutch and European products were being carried in Spanish ships to the Spanish Americas, exchanged there for Spanish American silver bullion, and carried back to non-Spanish customers in Europe. An even more worrisome pattern for Spain was that Dutch traders, smugglers, and privateers were bypassing official Spanish channels to conduct business directly with American merchants. The Americans were happy to receive merchandise at steep discounts from fixed monopoly prices. The risk was that a Spanish patrol ship could confiscate almost any Dutch trading vessel at will—if they could be found.[31]

## DUTCH ENTERPRISE

The reinstatement of the Spanish embargo of 1598 was a call to action in Amsterdam. Although Dutch traders had been beaten to Asia by Portugal, Spain, and even England, they had their own ideas about doing business in the east. Prior to around 1595, Dutch trading activities with Asia were almost completely dependent on intermediaries from Portugal and Aleppo. Adventurers like Jan Huygen van Linschoten had reached Goa in 1583 only by sailing in the service of the Portuguese Crown. Linschoten spent five educational years under Vicente de Fonseca, the archbishop of Goa, and subsequently published his famous *Itinerario* in 1596 while he was en route to the Russian Arctic. *Itinerario* included a wealth of firsthand and secondhand information about business and political conditions in India, Sri Lanka, and Indonesia, but the book's greatest contribution (to interested Dutch parties) was its implication that the overextended Portuguese commercial empire in Asia could be attacked by energetic businessmen.[32]

When Dutch merchants were denied official access to Iberian markets in 1598, they expanded their direct trading activities with the Americas and schemed to break the tenuous Portuguese presence in Asia. Fortunately for the Dutch, Felipe II had failed to follow through on his ample opportunities to crush the fledgling republic between 1580 and 1598. This eighteen-year "window of opportunity" was exploited by enterprising Dutchmen and thousands of profit-seeking refugees from Antwerp and Portugal. Decades of war and innovation had produced an integrated Dutch military-naval-business machine that was arguably the finest in Europe. The Dutch leadership prepared to redirect this three-pronged machine against Portugal, the weaker partner of the unified Iberian Crown. Portugal's population was equivalent in size to the Dutch Republic, and Lisbon was struggling to manage a global empire that was overstretched, hollowed out, privatized, and vulnerable.

If anyone cared about legalities, the Protestant Dutch had no reason to accept the papal bulls that had given Portugal exclusive control of Africa, India, and East Asia. Dutch legal scholars like Hugo Grotius argued that the Portuguese had not "discovered" any of the Eastern lands, since they had been inhabited for millennia, and that the pope had no authority over their non-Christian native populations. Of course, Portuguese (and Spanish) scholars responded in kind, the Italians were too weak to interfere, and the English and French had already demonstrated their own opinions on the subject. In fact, it had been England's defeat of the Invincible Armada in 1588 and the relative success of globe-trotting English privateers that had encouraged the laggard Dutch to move forward in global affairs.[33]

Dutch traders invaded Portuguese West Africa in force between 1593 and 1607. The plan was to extend Amsterdam's growing sugar- and salt-trading activities with Lisbon, Brazil, and Venezuela to Portugal's trading colonies in West Africa. West Africa was not unfamiliar territory. Merchants from Bruges had followed Portuguese traders to the Guinea coast as early as 1475, followed in 1500 by other Flemings who became active in the sugar (and slave) trade with the Canaries. Flemings held slave-trading asientos with the Genoese prior to 1540, and a few merchants, like Balthasar de Moucheron, built sizable fortunes during the 1570s. Dutch visits to West Africa were limited to São Tome (1562) and the Canaries (1571) until trading activity accelerated in the 1590s. Over two hundred Dutch ships visited the region between 1593 and 1607 in search of the same sugar, gold, ivory, and salt that had driven Portugal's activities since 1440—more ships (and a focus on slaving) were on the way.[34]

Trade was generally peaceful in West Africa until 1598. Led by ex-Antwerpers, three or four Dutch merchant ships sailed south to collect gold, ivory, and sugar from a string of Portuguese factories en route to São Tome. Founded as a sugar and slaving center in around 1485, São Tome had gradually superseded venerable São Jorge da Mina as the most profitable enterprise in West Africa. But the slave trade remained firmly in Portuguese hands until Felipe III reinstituted the Spanish trade embargo in 1598. The embargo had the opposite effect—it accelerated Dutch activity in West Africa. In 1599, eight Holland-based companies even attempted to assert their own monopoly on the West African trade, two decades prior to the creation of the Dutch West Indies Company in 1621, and might have succeeded had they not excluded the rival Zeelanders. The Dutch sent around twenty ships per year to West Africa between 1600 and 1608.[35]

In the Spanish Caribbean, Dutch privateers followed the hit-and-run model that had been pioneered by Hawkins, Drake, and other Englishmen. Many of them replaced the Portuguese entrepreneurs who had traded cheaper contraband goods (and slaves) with grateful Spanish American colonists. Although Dutchmen had traded with Panama since 1572 and Brazil since 1587, they focused on Venezuela in 1596. Venezuela's pearl and salt deposits made the port of

Cumana a notorious haven for Dutch smugglers and helped to bolster the gnarly reputation of Dutch pirates and slavers in the Caribbean. The discovery of a ten-square-mile salt deposit near Punta de Araya unleashed a Venezuelan salt boom between 1599 and 1605—the site drew nearly one hundred Dutch salt ships annually, mainly out of Hoorn, and raised the status of a collection of erstwhile pirates and privateers. A host of Dutch trading enterprises, a few of which were arranged through the Portuguese Converso network, were built on the delivery of embargoed Caribbean salt, sugar, hides, and tobacco to Amsterdam's fledgling refining industries. The salt marshes of Punta de Araya encouraged some Dutchmen to establish a chain of fortified trading posts along the neighboring "wild coast" of Guyana. These provided staging areas for possible incursions into the sugar plantations of northeastern Brazil. By 1605, a Spanish fleet of eighteen ships was required to evict the pesky Dutch from these nominally held Iberian territories.[36]

Amsterdam's commercial interest in the Americas, Brazil, West Africa, Russia, and the Mediterranean was dwarfed by her obsession with the Far East. England had once again shown the way. The establishment of the English Levant Company in 1581, a four-merchant enterprise modeled on the English Muscovy Company, had been catalyzed by Drake's achievements and a growing sense that the spectacular riches of the overstretched Iberian Empire were vulnerable to predators. Merchant adventurers John Newbery and Ralph Fitch left London in February 1583 on behalf of the Levant Company to find a possible overland route between Persia and China. The unfortunate Newbery died en route, but Fitch reached Aleppo intact, evaded capture by Portuguese authorities in Hormuz, "the driest island in the world," and noted the impressive orchards and gardens of the "fine city" of Goa. From there, Fitch documented his remarkable travels through East Asia.[37]

After meandering his way through the bazaars of India, Burma, Thailand, and Malaysia, Fitch reached his final destination, Malacca, on February 2, 1588, and stayed there for two months. He was impressed with the multinational array of trading vessels, the "great stores" of spices, drugs, diamonds, and jewels, and the fact that the entire enterprise was licensed by the captain of Malacca. While Fitch would have been even more impressed had he visited Malacca in 1518, he also learned that the merchants of Portuguese Macao were sending silk, gold, musk, and porcelains to Nagasaki and were returning with "nothing but silver." He estimated that the great carrack "bringest from [Japan] every year about 600,000 crusadoes [of silver] and 200,000 crusadoes more in silver [brought] yearly out of India, [the Portuguese] employ to their greatest advantage in China."[38]

Fitch returned to England in April 1591 with illuminating, if somewhat sketchy, reports about a part of the world that was still a mystery. The Portuguese, as well as the Fuggers and Jesuits, had managed to cloak their long-standing trading relationships in secrecy. Charts, maps, sailing directions, and local market conditions were rarely disclosed to the general public. This left would-be European competitors to form their own ideas about the Eastern world. As early as 1510, an intrepid Venetian named Ludovico di Varthema had published a "traveler's report" (*Itinerario*) of the Asian marketplace. This was one of the few available reports concerning Portuguese Asia until the publication of Tome Pires's *Suma Oriental* (in Venice) in 1563, Linschoten's *Itinerario* of 1596, and Richard Hakluyt's *Principal Navigations* of 1599. While Fitch's vague journal was included in Hakluyt's landmark work, we must assume that his most valuable observations were reserved for either the directors of the Levant Company or for Queen Elizabeth herself. It was Linschoten's more detailed accounts from 1583–1589, with input from on-the-ground characters like Dirck Gerritszoon Pomp ("Dirck China"), that almost single-handedly catalyzed the formation of the English and Dutch East India Companies.[39]

In the meantime, market intelligence reports from Fitch and Linschoten—circulated word-of-mouth long before they were published in book form—prompted a London-based syndicate to finance an expedition to the Far East in 1591. Three ships under James Lancaster and George Raymond left Plymouth in early April only to face hardship after hardship en route to the spice markets of Indonesia. After losing Raymond and two ships en route to Zanzibar, the expedition party was reduced to Lancaster and thirty-four sailors by the time it reached the Malaysian port of Penang (north of Malacca) in June 1592. The Englishmen subsequently raided the Straits of Malacca with some success before retracing their route back to the Cape of Good Hope. From there, Lancaster steered directly to the West Indies to avoid Iberian warships. This turned out to be a bad idea. Lancaster evaded Iberian vessels successfully, but fifteen of his sailors decided to wage a mutiny and leave the captain and nineteen loyalists for dead in the Caribbean. However, Lancaster and the survivors returned heroically to England, after discovering a new method to control scurvy (lemon juice), and provided Englishmen and Dutchmen alike with a fresh source of business intelligence. An opportunistic syndicate of nine Amsterdam merchants responded almost immediately. They organized the Compagnie van Verre (Company of Far Lands) in March 1594 to trade directly with the Far East.[40]

Having made the bulk of their fortunes from trading (and privateering) in the Baltic, Russia, and the Caribbean, Amsterdam's merchant community were just as experienced as the English when it came to long-distance trade. If Portugal's position in Indonesia was shaky, the Dutch were not about to let the English seize the opportunity. The Compagnie van Verre pooled 290,000 guilders (128,000 ducats)—an amount roughly equivalent to the value of sixty large (five-thousand-guilder) homes in Amsterdam—to send Cornelius de Houtman and a four-ship fleet to Java in April 1595. Houtman arrived at the Javanese pepper port of Bantam in June 1596 with a heavily armed fleet, 249 men, and 64 cannon. This was more than enough firepower to arrange a commercial treaty with the local sultan and exploit an Achinese-controlled passageway through the underutilized Sunda Strait between Sumatra and Java. However, Houtman's triumph was marred by Dutch involvement in the murder of a rival Javanese sultan and the loss of three ships and 160 men during the return trip. Although the pepper, nutmeg, and mace that returned to Amsterdam with Houtman's single ship in August 1597 were insufficient to offset the heavy losses, no one questioned the business potential. At least seven separate Dutch fleets were sent back to Java in 1598, including two from Zeeland and Rotterdam, to defy Felipe III's reinstated trade embargo as well as the Portuguese warships that patrolled the West African coastline.[41]

Houtman's humiliation had been softened by the publication of Linschoten's *Itinerario* in 1596—it raised commercial interest in the Far East to a fever pitch. Linschoten's experiences in the service of the Portuguese (and even the House of Fugger) highlighted Portugal's weak points in the Indian Ocean and suggested ample business opportunities for persons with energy and capital. Houtman's fleet had actually been welcomed in Bantam. He had sold out his Dutch cargo on the spot and was invited by the local sultan to reload the ships with pepper, spices, and other exotica. Java's merchant community preferred almost anyone to the dreaded Portuguese, and there were hundreds of potential trading partners in town—Javanese, Malays, Chinese, Indians, and private Portuguese. If that were not enough, Houtman had avoided contact with any Portuguese warships en route.[42]

The seven Dutch companies organized for a second expedition to Indonesia were reduced to six after the cost-conscious Compagnie van Verre merged with a rival firm. The merged enterprise, the so-called Old Company, subsequently raised 768,466 guilders to dispatch an eight-ship fleet under Jacob Corneliszoon van Neck in the spring of 1598. Four other companies sent a total of fourteen ships to Indonesia in that same year, each crammed with silver bullion, arms and ammunition, linens, velvets, and glass. Most of the investors were rewarded. Virtually all of the vessels returned safely, including Van Neck after a fifteen-month voyage, and delivered profits of 100 to 400 percent. The one failure, a nine-ship fleet under Olivier van Noort, was undone by Van Noort's ambitious attempt to make a westerly crossing to Asia through the Magellan Straits. This one ended like most of the others. It was the unimpeded cape route, blazed by da Gama a century earlier, that delivered the profits in 1599 and induced an even larger Dutch fleet of sixty-five ships to make the voyage in 1601. In this case, greed triumphed over common sense. While Van Neck bombarded Macao unsuccessfully (for show) and destroyed a Portuguese galleon, he and his competitors paid too much (double) for spice purchases in Indonesia and returned with too much supply. Indonesians were thrilled, but Dutch profit margins were squeezed on both sides of the globe.[43]

If profit margins were under pressure, the Dutch invasion of Indonesia had attracted attention. *Fugger News Letter* (#201) reported from Amsterdam on July 24, 1599, that four Dutch ships had returned from "India" with three hundred loads of pepper (around six hundred tons) and large quantities of cloves, nutmeg, and cinnamon while four other Dutch ships were en route to the Moluccas. The reporter warned that "if the King of Spain does not beware and put a stop to the [Dutch], in time great harm will befall the Kingdom of Portugal and the Venetians." Letter #204 from Antwerp on October 22, 1599, added that "Dutch navigation to the Indies is becoming even greater, which will cause the Portuguese heavy damage in their trade." Letter #214 from Antwerp on July 2, 1600, reported that two (of nine) Dutch ships had arrived from the Moluccas with 620,000 pounds of nutmeg and 100,700 pounds of mace, cloves, and pepper. The reporter observed that the Dutch invasion of East India was "very harmful to the Spaniards, because it has been decided in Holland and Zeeland that this journey should be made through Portuguese and Spanish waters."[44]

The excessive price competition encountered in 1601 prompted the Dutch States General to nationalize the entire Asian enterprise in 1602. There was no good reason to have Dutch merchants fighting among one another for spices when the process destroyed profit margins for all and enriched the Indonesian sultans. In addition, a series of uncoordinated expeditions to Siam and even Japan had little hope for success on their own. A Dutch merchant syndicate had sent Jacob Quackernaeck on a voyage to Japan in April 1600 and had not heard from him since. The thinking was that if the Iberians could manage global monopolies and the English could establish a Muscovy Company, a Levant Company, and a brand-new East India Company, then the Dutch could establish a new and improved state-backed monopoly of their own. The English Crown had joined with Thomas Smythe, a director of the Levant Company, and other London merchants to found the English East India Company in 1601. The plan was to respond aggressively to Amsterdam's Compagnie van Verre and to exploit global capabilities that had been established by Drake's circumnavigation, the defeat of the Spanish Armada, and Lancaster's hair-raising adventures.[45]

The English East India Company, chartered on December 31, 1600, was structured like the Merchant Adventurers of London, the Muscovy Company, and the Levant Company. All of these English enterprises were chartered, long-term–oriented partnerships that received a renewable trading monopoly from the English Crown. While having a company-owned fleet of ships added a measure of permanence to the voyage-by-voyage partnerships that had typified long-distance trade for centuries, the English Company operated with no common capital, no central administration, and left each voyage to be financed and conducted independently. Three separate English firms would be forced to compete against each other in Java. When a true joint-stock company was finally organized in 1612, it was undercapitalized and dependent on London for policy guidance, dispute resolution, and military assistance. Private interests overwhelmed those of the English Company as a "company" until 1660. It was then that the English Company made the leap from renewable terms to the type of permanent capital structure that defines a modern corporation.[46]

Despite these concerns, the English Company's maiden four-ship fleet under Lancaster and John Davis was warmly received by the sultan of Aceh in June 1602. The sultan was especially pleased by their ability to capture a Portuguese carrack en route to the pepper marts of Bantam. Assisted by letters and presents from Queen Elizabeth herself, Lancaster managed to establish an English Company factory at Bantam before returning home in September 1603 with Portuguese plunder and 515 tons of pepper. That was the peak. The company followed up Elizabeth's death in 1603 with two more expeditions in 1606–1607—Henry Middleton's four-ship expedition of 1606; William Keeling and William Hawkins in 1607—but essentially struggled thereafter.[47]

## THE VOC

The Dutch responded in 1602 with an East India Company of their own. The Verenigde Oostindische Compagnie (VOC) was chartered by the States General on March 20 and capitalized at a whopping 6.5 million guilders—twenty-two times the initial equity of the Compagnie van Verre and equivalent to 2.3 million cruzados. The VOC was pushed through by Oldenbarnevelt and Maurits when it was learned that Lancaster and Davis were en route to Indonesia. The latest Dutch enterprise may have been created on its own, but the English Company had provided a call to action. The 6.5 million guilders were raised almost immediately—3.7 million from Amsterdam, 1.3 million from Zeeland, and the rest from Delft, Enkhuisen, Rotterdam, and Hoorn. VOC shareholders were broad-based. The Amsterdam chamber received at least 5,000 guilders from 184 different investors, highlighted by 50,000 guilder-plus investments from ex-Antwerpers, Iberian Conversos, and Jews. Refugees from Antwerp and the southern provinces accounted for 301 of the VOC's 1,086-person investor base in Holland. Some 69 Zeelanders chipped in with investments of at least 5,000 guilders.[48]

Backed by a renewable twenty-one-year commercial monopoly, the VOC was intended to reorganize fourteen separate Dutch fleets—comprising sixty-five ships, seven provinces, and six regional chambers—into a single, state-backed enterprise. The States General empowered the VOC board of directors, the Heeren XVII, to finance a fleet of ships, hire employees, establish military garrisons, conduct foreign relations, wage "defensive" wars, and even sign treaties. Headquartered within the confines of the present-day University of Amsterdam, eight of the VOC's seventeen directorships went to the powerful Amsterdam chamber, four went to Zeeland, and the other five were allocated to the other four chambers. While each chamber managed its own capital position and accounts, the VOC held all of the company's capital in common and determined that no shareholder was to receive a single dividend payment until trading profits exceeded 5 percent of the company's initial capitalization. Unfortunately for economic historians, the VOC was never really tested in economic combat against the English Company. Elizabeth's death in 1603 deprived England of her leading globalist. The English Company never threatened the VOC (or the Portuguese) in the Far East and eventually shifted its attention to Persia and India.[49]

Elizabeth's death allowed the Dutch to focus exclusively on the Iberians. Like the Portuguese model established by Afonso de Albuquerque a century earlier, the VOC was authorized to establish a monopoly on all Dutch trade east of the Cape of Good Hope, construct a chain of fortified naval bases to support the armed VOC fleets, and apply whatever policies were deemed necessary under its twenty-one-year charter. These authorizations made the VOC a virtual state within a state. The shift to a military posture also meant that a century of relatively peaceful Dutch trading was coming to an end. The type of people who were employed by the VOC were not necessarily the most mannered of Dutchmen. Cutthroat Dutch sailors, smugglers, and pirates had already gained a reputation in the Caribbean for their gnarliness. The VOC charter was very explicit about the geopolitical objectives—the company was encouraged to attack the economic power and prestige of the Iberian Crowns throughout Asia.[50]

With the Portuguese well entrenched in Goa, Malacca, Colombo, Macao, and Nagasaki, the newly chartered VOC decided to concentrate initially on the Spice Islands of Indonesia. Of course, the pending Dutch assault would require some improvisation to avoid an inevitable military response from Malacca or Manila. Consequently, the VOC established its regional headquarters at the second-tier Javanese port at Bantam. Thanks to Houtman's diplomatic work, Dutch vessels had been authorized to travel directly across the Indian Ocean to Java through the Sunda Straits. This allowed the VOC to avoid the well-defended Straits of Malacca and to partially control the monsoons that had plagued Portuguese sailings for nearly a century. The Dutch had reason to be confident. They had engineered the highest quality ships in the world, stalemated the vaunted Army of Flanders, and, thanks to the capital markets of Amsterdam, had access to much greater financial resources than either Portugal, Spain, or England. The Dutch also had a killer commercial drive. They would outdo even the detested Portuguese in their attempts to destroy all local Asian competitors in the Spice Islands.[51]

The VOC's maiden ten-ship fleet of 1603 was commanded by Steven van der Hagen. Backed by the nine-hundred-ton *Dordecht*, a ship equipped with six 24-pound guns and eighteen 8–9 pounders, Van der Hagen was empowered to apply military force against Portuguese ships and facilities and any local ruler who refused to sign a spice contract with the VOC. He was extremely successful. He raised 1.3 million cruzados in privateering "prizes" in 1603 by somehow capturing a fully loaded carrack in Macao harbor and seizing an anchored vessel at Johore. Van der Hagen's shelling of Macao in 1601 had been ineffective, but Dutch warships had demonstrated the value of their shipboard artillery, their flexible sail plans, speed, and shall draft. The royal shipyards in Iberia, the Caribbean, Goa, and Manila had not sat still after the armada's defeat in 1588, but Portuguese warships were sluggish, undergunned, undermaintained, and undermanned. Portugal's polyglot crews of underpaid Asian sailors were rarely as skilled, disciplined, or motivated as Dutchmen. In addition, Portuguese naval commanders, mainly well-connected *fidalgos*, were less capable at sea than they were on land. In short, the Portuguese had finally met their match.[52]

The Dutch gradually wore down the Portuguese with numbers. Dutch galleons, cannon, and longer-range culverins were no more advanced technologically than the Iberian varieties, but naval battles were being determined by exchanges at close range (two hundred yards or less), gunnery skills, speed, and an ability to sail close to the wind. Those with the most ships and guns usually won. The Dutch warships (pinnas) that protected the fluits and took on the Iberians benefited from a more efficient construction process and a shift to inexpensive cast-iron gunnery. Excavations of vessels sunk off Malacca in 1606 reveal that the Dutch held their own with lower-quality iron cannons at close range—twelve of the sixteen guns aboard the *Nassau* were made of iron (four of bronze) while fourteen of the sixteen aboard Portugal's *São Simão* were made of expensive bronze (two of iron). Heavily armed Portuguese galleons were too expensive, too few in number, and too inflexible in battle. The Portuguese managed to defeat the Dutch (or English) on only five occasions between 1588 and 1658—drawing thirteen times, suffering ten defeats, and losing nearly twice as many warships (ninety-six) as the Dutch (fifty-four). The 120-ship Dutch war fleet of 1626 was twice the size of the Iberian fleets combined.[53]

The Dutch advance was supported by a home economy that was more vibrant, more entrepreneurial, and less medieval than the Iberian model. Dutch traders had ready access to exportable trade merchandise and bullion. The great irony was that the mineral-poor Dutch Republic was building a huge stockpile of precious metals by taking transatlantic trading profits away from the Iberians. The fabulous mines of the Spanish Americas were somehow being redirected to the gnomes of Amsterdam. It was left to the state-sponsored VOC to reinvest these Western profits in the high-risk/high-reward markets of the Far East. Following yet another Iberian model, the stepping stone one, the VOC plan was to focus initially on Indonesia from a single headquarters (initially Bantam) and then chip away at the entire Portuguese Empire in the Far East. The objective was the establishment of a Lisbon-styled long-distance trading monopoly centered by Amsterdam. While the VOC would have to deal with Portugal's private sector—a powerful array of Portuguese merchant families who were committing two million cruzados annually to the Asian trade and leaving most of the military costs to the beleaguered Crown—there would be no viceroyalties. There would also be no royal, aristocratic, or patronage-based appointments, no religious orders, no opportunities for VOC employees to line their own pockets, and no qualms about applying military force.[54]

Starting with Van der Hagen's maiden voyage in 1603, the VOC operated as a ruthlessly professional business enterprise. When the position of governor-general was established at the VOC's first headquarters at Bantam (Java) in 1610, it became clear that businessmen (not nobles) would be running things. The Dutch factories placed eventually at Bantam, Batavia, Ternate, Tidore, Ambon, and the Banda Islands were managed by experienced merchants or military officers—not by bureaucrats. And unlike the Iberians, little effort was expended to convert natives to Christianity. The Dutch were happy to conduct business with Hindus, Muslims, Jews, or fellow Christians as long as there were profits in it. They had observed that the Jesuits and the other religious orders had probably alienated as many Asians as they had enlightened. The Dutch focused solely on business. A one-sided treaty with the sultan of Bantam delivered the sultan's entire pepper crop to the VOC at cost. Slave labor was deployed routinely on the Bandas and at other spice plantations that were faced with labor shortages.[55]

The Dutch gained market share in pepper and spices at the expense of the Portuguese. Despite their past successes, Portugal had never dominated the Asian spice trade completely and always had to compete with traditional Muslim merchants. These traditional merchants operated with lower overhead and required profit margins of only 10–15 percent—versus the 40–70 percent margins demanded by Europeans to offset transportation costs, risk, and shareholder expectations. The cape route was expensive and dangerous. This left the Ottoman-controlled Red Sea and the overland caravan routes in great demand for many decades. While the Dutch were stuck with the same business environment as the Portuguese, the threat of Dutch military force and the superiority of private Dutch shipping companies made up for many of the logistical problems.[56]

The VOC also focused on Indonesia initially because the islands were still viewed as a gateway to China, the Portuguese were reviled there, the Sunda Straits provided an alternate trade route across the Indian Ocean, and Macao and Nagasaki, backed by the Spanish fleet at Manila, were considered impregnable. While Spanish officials in Manila attempted to recover regional "hearts and minds" by cleaning up some of Portugal's most blatant corruption practices, they were too late. Another wrinkle was that the Islamization of Java was essentially complete. By 1600, the last of Java's powerful Hindu-Buddhist states were gone and the Sultanate of Mataram had conquered the Malacca-friendly rice port of Demak. The Muslim conquests pressured Malacca's always-tenuous food problems and promoted the VOC's own agenda against the Portuguese. Indonesian sultanates were quick to sign commercial treaties with the Dutch more because of their disdain for the Portuguese than because of the Dutch arms pointed at their heads. Mataram eventually swung back to the Portuguese after Batavia (Jakarta) was captured and refortified by the VOC in 1619—the sultan recognized that Batavia gave the Dutch an excellent harbor, control of the Sunda Straits, a platform for regional brutality, and an expanding spice center.[57]

The VOC's modest successes in Java were followed by an attack on Ambon in 1605. Focused on the one-third of the island's spice crop that was reserved for the Portuguese Crown, the Dutch captured Ambon with assistance from the local Hituese. They also won local favor by evicting the island's Portuguese Catholic missionaries. While displaced Portuguese traders managed to relocate to the friendlier confines of Grise, Timor, and Macassar, the VOC's conquests on Java and Ambon were the start of a gradual and irregular invasion of Portugal's eastern trading posts. In addition to pepper and spices, the initial VOC headquarters at Bantam imported indigo and gold from Sumatra, tin from Malaysia, cottons from India and, later, coffee and tea. Portugal's senior partner, Spain, did her part. The Spanish fleet at Manila helped to secure Tidore, Ternate, and Timor from the Dutch in 1606 and helped to thwart a series of brazen Dutch attacks on Malacca in that same year. These actions freed up Portuguese ships to defend Malacca (and Goa) from attempted blockades. A Dutch fleet was even defeated off Mozambique in 1607.[58]

If a Dutch conquest of the Portuguese spice empire was not quite inevitable, the attacks on Malacca in 1606 had sent a powerful message. Admiral Cornelius Matelief besieged Malacca in the summer of 1606 with eleven VOC warships, ranging in size from 220 to 700 tons and armed with the latest (mainly iron) artillery, plus 1,279 Dutch sailors, and an allied fleet of galleys and seven hundred troops from Johore. Malacca, still impregnable to a frontal assault, barely survived a series of naval skirmishes and a prolonged siege. Starvation and disease were taking their toll when Viceroy Martim Afonso de Castro sailed into the straits with a thirty-ship rescue fleet from Goa in mid-August. Afonso de Castro broke the three-month siege and stalemated the retreating Dutch fleet off nearby Cape Rachado. Losses were heavy on both sides. Matelief returned to fight two more inconclusive battles off Malacca in late October before withdrawing with just four surviving warships. Malacca had hung on, despite Portuguese losses that exceeded those of the VOC.[59]

The Portuguese had not been completely asleep. Lisbon had warned Goa back in March 1595 about Houtman's pending expedition to Asia, but Goa's overextended navy had been powerless to stop it. After Dutch privateers had scared away prospective investors in the five-year Asia Contract of 1597, incoming viceroy Cristóvão de Moura (1600–1603) declared facetiously that only three things were necessary to defend Portuguese India—money, men, and ships. Afonso de Castro was fortunate to receive fifteen large carracks between 1604 and 1605 thanks to a seven-year program (1604–1610) to extort contributions from Portuguese Conversos. The new vessels helped to break up the siege of Malacca in 1606. Promises for an even larger fleet were interrupted by a brazen (but temporary) Dutch blockade of Lisbon harbor—the blockade delayed Goa's receipt of seven new ships until 1607 and nineteen more until 1608 and 1609.[60]

The looming Spanish state bankruptcy of 1607 and the subsequent signing of a Twelve Years' Truce with the Dutch in 1609 prevented the Iberians from following through. They failed to counterattack the Dutch in naval battle and declined to combine the formidable Spanish and Portuguese fleets in either the Atlantic or the Far East. Neither the funds nor the political will were there. The union of the two Iberian Crowns was underutilized. Seville's powerful merchant community blocked repeated attempts by Portuguese businessmen to gain greater access to the royal court in Madrid (until the 1620s), and direct trade was prohibited between Manila and Macao (until 1619). Protectionism ruled. The increasingly overextended Iberian Empire was devolving into a patchwork of side deals. Felipe III confiscated 500,000 cruzados in pepper profits to shore up the Army of Flanders in 1602 in exchange for a deal through which the Duke of Braganza received commercial rights to the Goa-Lisbon trade. There were not enough state funds to mount a serious counterattack against the VOC—this despite the extortion of 1.8 million cruzados from Iberian Conversos in exchange for the Crown's confirmation of a 1604 papal decree that pardoned all suspected Iberian heretics from the Inquisition. The terms of the pardon were predictably revoked in 1614.[61]

Macao, Nagasaki, and Manila were still impregnable. Iberian naval defenses had been stepped up after the VOC captured three Portuguese ships and 1.5 million cruzados worth of cargo off Macao in 1603 and nearly took Malacca in 1606. Heavily armed Dutch "Indiamen" stayed away from Malacca, even with offers of assistance from the always-willing Achinese, and stayed away from Manila. No attacks were waged against Macao until after the expiration of the Twelve Years' Truce in 1621. Even then, half-hearted Dutch assaults in 1622 and 1629 were complete failures. The VOC may have been wary about interfering in the Macao-based affairs of their prospective business partners in Japan. Malacca was a different matter. Dutch intelligence reports indicated that the great entrepôt was in decline. This signaled an opportunity to penetrate the vibrant textile markets of the eastern Indian Ocean, a sphere that had been transformed by private Portuguese initiative, and cut even further into Malacca's business. Malacca's traditional spice markets had already been carved up among the Javanese, private Portuguese, and Dutchmen.[62]

The overstretched Portuguese Empire did not go down easily. However, a series of budget deficits, the VOC, and bad weather were all working against it. Over one-third of the Portuguese ships sent to Goa between 1604 and 1608 never made it—some were wrecked, others were captured, burned, or forced to return to Lisbon. Hormuz barely escaped blockades in 1602, 1607–1608, and 1614–1615. With a multinational population approaching forty thousand, including only two hundred Portuguese households and a modest five-hundred-man garrison, Hormuz was faring only slightly better than Malacca. The barren island was a convenient site for the exchange of American silver for Persian silks, Arabian horses, and Indian trade goods, but not much more. All of Hormuz's food supplies and most of her water had to be imported, most of her troops were engaged in private business ventures, and non-Portuguese traders—Persians, Armenians, Hindus, and Jews—were extorted regularly. The 12–15 percent royal customs duty had raised 200,000 cruzados annually during the 1590s, but it would have raised as much as 500,000 if not for official corruption, untaxed silver, and a very modest tax on textiles. It was no accident that the three-year captaincy of Hormuz sold for an impressive 125,000 cruzados in 1615, one-fourth of the Estado da India's office sales in that year, despite competition from the revived caravan trade (through Kandahar) and the silk markets at Basra and Shiraz. Portuguese ships should have been more efficient than the thirteen thousand camels who were employed annually along the overland route from Lahore.[63]

Things were looking increasingly shaky for the Estado da India when Felipe III finalized a temporary truce with the Dutch Republic in 1609. What turned out to be a twelve-year truce outraged the merchant communities of Lisbon, Goa, and even Amsterdam—private Portuguese merchants feared losing more of their hard-earned markets to the Dutch; the VOC feared that the truce would stall its momentum in South Asia. The VOC was mistaken. As the private Portuguese had predicted, the Dutch improved their position in Asia during the 1609–1621 period and cashed in on the opening of legal trade with Iberia and the Spanish Americas. The truce solidified the Dutch advance. Another winner was England. English privateers, on their second attempt, defeated a Goa-based fleet of Portuguese galleons off Surat in 1614. This opened a commercial door to the Gulf of Cambay and the Persians who traded there. England exploited the expiration of the Twelve Years' Truce by supporting Safavid Persia's successful seizure of Portuguese Hormuz in 1622.[64]

Portugal hung on in the Far East. The VOC had focused initially on breaking the tenuous Portuguese spice monopoly in Indonesia because the monopoly was crumbling anyway. A few attempts to take Macao, the new crown jewel of Portuguese Asia, had failed miserably. Van Neck's bombardment of Macao in 1601 had been inconsequential, and his destruction of a Portuguese galleon was a one-time event. But armed with increasingly large quantities of American silver bullion, earned from the transatlantic trade, the Dutch aspired to replicate the Portuguese business model in the Far East. Since the Portuguese were handling the exchange of Chinese silk, porcelain, and gold for increasingly large volumes of Japanese silver, it was easy for the Dutch to come to the same conclusion as the Spaniards—American silver provided the very same business opportunity. Silver bullion was also supporting European trading activities with the Indian textile merchants of Surat, Coromandel, and Bengal.[65]

Blocked in Macao, Nagasaki, and Manila, the VOC schemed of ways to penetrate the dynamic markets of Japan and China on her own. It was inadvertent contacts between Japanese traders and Dutchmen in Patani, a Siamese vassal port on the eastern Malay Peninsula, that promoted the first Dutch expedition to Japan in 1600. The expedition was paved by Dirck Gerritszoon Pomp's earlier visit to Japan in 1584. The adventurous Pomp had gained notoriety as "Dirck China" for his firsthand experiences in the Orient. In fact, Pomp had been the primary source for many of Linschoten's secondhand observations in *Itinerario*. Pomp's observations, coupled with market intelligence pried out of Japanese traders at Patani, prompted a Dutch merchant syndicate to send Jacob Quackernaeck on a maiden voyage to Japan in April 1600. This was nearly three years prior to the VOC's first expedition to Indonesia and within months of Tokugawa Ieyasu's epic triumph at Sekigahara. Quackernaeck benefited from his shipwreck off the northeastern coast of Kyushu on April 29 with Melchior van Santvoort, twenty-three other survivors, and some salvaged cargo. The Dutchmen were presented to Tokugawa Ieyasu at Edo (Tokyo) and were thrilled to receive a cache of gold and some trading privileges. The would-be shogun figured that the entrenched Portuguese needed some local competition now that Ieyasu's prospectors had just discovered record deposits of silver and gold on Sado Island. The small catch was that Quackernaeck and his colleagues were forbidden to leave Japan until 1605.[66]

A trading junk owned by the daimyo of Hirado carried Quackernaeck, Santvoort, and crew back to Patani in December 1605 with a shogunate letter addressed to the "king of the Netherlands." Ieyasu issued a trading pass to the VOC's factor at Patani in late 1606 in anticipation of an official reply from Prince Maurits. Unfortunately for Santvoort, who had returned to Japan promptly, no one knew that an armistice was being negotiated between the Dutch and the Iberians. It would lead to the signing of a formal Twelve Years' Truce between the two powers in 1609 and raise fears that the Portuguese monopoly in Japan would be frozen. A follow-up expedition to Japan was delayed until 1609. Two VOC trade envoys, Abraham van den Broeck and Nicolaes Puijck, were dispatched from Bantam in May with Maurits's belated reply and instructions to link up with Santvoort at Hirado. After arriving at Hirado on July 1, the threesome were allowed to meet with the shogun at Sumpu on August 12. The meeting was a success. The VOC was issued a coveted Red Seal license (*shuinjo*), a permit to establish a factory at Hirado (in direct competition with the Portuguese at Nagasaki), and another letter from the shogun. Ieyasu stated that Maurits's long-awaited reply had been received "as if it had come from a close friend."[67]

Ieyasu (1542–1616) had defeated his rivals with guile and skill. Following Hideyoshi's death in 1598, he had agreed to serve on a five-member daimyo council that pledged its support to Hideyoshi's five-year-old son Hideyori. But Ieyasu apparently had personal scores to settle. He had been hostaged by his own family, the Matsudaira clan, to a rival daimyo between the ages of six and eighteen. The painful memory prompted Ieyasu to withdraw his support of Hideyori and somehow gain the allegiance of the "warrior faction" of Hideyoshi's own army. The so-called political faction, led by Hideyoshi's Toyotomi clan and other western daimyos, responded by declaring war against Ieyasu's rebel army. When the two opposing armies of roughly one hundred thousand met on October 20, 1600, at Sekigahara in central Honshu, victory was won only when the Mori clan—the same clan who had introduced the *nambambuki* smelting process to the mines of Iwami in 1581—switched abruptly to Ieyasu's side. Tensions eased after the defeated Toyotomi clan was allowed to retain three provinces and a generous 650,000 *koku* (325,000 cruzados) in annual revenues.[68]

The unification of Japan was finally complete. Ieyasu received the title of shogun from the imperial court in Kyoto in 1603 and headquartered his Tokugawa Bakufu (government) at the refortified castle town of Edo (Tokyo). While succession concerns prompted Ieyasu to transfer the position of shogun to his son Hidetada (1579–1632) in 1605, Ieyasu served as Japan's primary ruler until his death in 1616. He dealt ruthlessly with Hideyoshi's son Hideyori. In early 1614, Ieyasu sent his army to besiege Osaka Castle—a fortress that was considered as impregnable as Ieyasu's own Edo Castle—and crafted a well-designed ruse to breach the castle walls. Not only was the Toyotomi claim eliminated for good, a message was sent to Japan's western daimyos. Ieyasu rebuilt Osaka Castle into an even mightier fortress. He forced some fifty-eight western daimyos to contribute a staggering sixteen million *koku* (eight million cruzados) in cash, labor, and materials to rebuild the complex between 1620 and 1629. The new five-story central keep was surrounded by 4.8 miles of walls, two moats, royal living quarters, turrets, gates, warehouses, and other facilities.[69]

The funds to solidify the Tokugawa triumph were derived from the mines of Iwami, Ikuno, and Sado. Sado Island was placed under

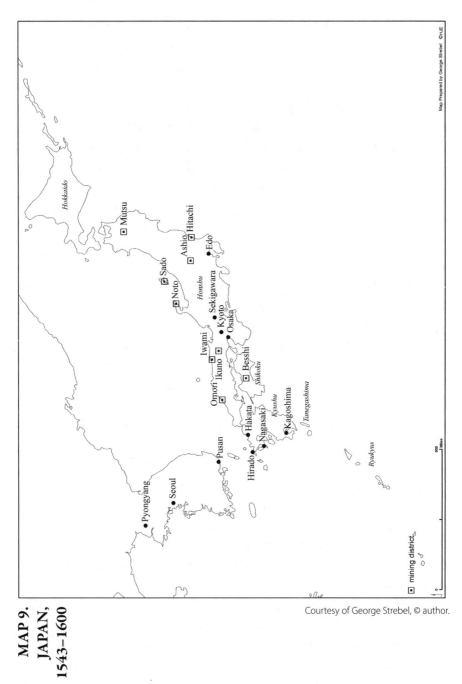

MAP 9.
JAPAN,
1543–1600

Courtesy of George Strebel, © author.

direct Tokugawa control shortly after a huge strike was made at Aikawa in 1601—an estimated thirty-eight tons of silver were produced in 1602 alone. This dwarfed the fourteen tons of silver produced in Iwami in 1601, the ten tons produced from the Ikuno mines of Tajima in 1598, and another three tons realized from Inabu, Chugoku, and Nakase. The gradual introduction of the American amalgamation process may have raised total Japanese volumes to one hundred tons or more by the 1630s. While estimates that Sado produced as much as 174 tons of silver plus 200,000 ryo (3.6 tons) of gold annually between 1613 and 1648 are very hard to believe, so are rival estimates that Sado's silver and gold output averaged only twenty-one tons. More reasonable estimates have pegged average annual Japanese silver exports at sixty-five tons during the entire 1604–1639 period, backed by Portuguese claims that they alone were shipping over eighty tons of Japanese silver to Macao annually during the 1630s.[70]

The doubling of Japanese silver production to one hundred tons or more was probably achieved by amalgamation. One can assume that Portuguese merchants or Jesuits introduced the process to Japan as early as 1580 to treat the Potosí-like deep-seated ores and waste slag that predominated at Iwami, Ikuno, and eventually Sado. While the Portuguese shipped modest amounts of Chinese mercury to Nagasaki prior to 1600, most of the 8,400 or so quintals of mercury shipped to Japan between 1598 and 1638 were probably concentrated in the 1630s. The mercury was apparently purchased by Portuguese traders in Guangzhou for forty cruzados and resold to Japanese mining magnates for ninety-one. Based on the ratio of mercury to amalgamated output at Potosí, two thousand quintals of Chinese mercury would have produced eighty tons of Japanese silver. China's mercury works, wherever they were, even invited interest from the mercury-deprived mineros of Mexico. In 1609, the viceroy of New Spain entered into a contract to import four thousand quintals of Chinese mercury into Mexico at a price (fifty pesos per quintal) that was half the cost of the Almadén variety. If China was capable of producing that much mercury, the record is murky.

Mexico received a grand total of two hundred quintals from China in 1612 and much less after that.[71]

Sado's metallic output, whatever it was, was shipped to the Tokugawa offices in Edo for registration and minting. The Tokugawa *bakafu* (shogunate) applied an imperial monopoly over Japan's mining industry in 1601 and provided financial incentives to encourage prospecting activities and the adoption of the *nambambuki* refining process (and possibly amalgamation). Ieyasu also created Japan's first unified currency system. In the case of Sado, silver was directed to Edo's silver mint (*ginza*) starting in 1612 and gold to the gold mint (*kinza*) until Sado received her first mint in 1622. The oval-shaped *koban*, weighing 17.86 grams and initially 84.29 percent pure gold, was minted from the gold deposits of Sado and the two to three tons of gold that were received collectively from Ani, Hodatsuyama (Noto), Oguchi, Osarizawa, Rikuchu, and Satsuma. As for copper, Japan finally got around to minting her own copper coins after large deposits were discovered at Ashio in the north and at the Sumitomo-controlled Besshi mine on Shikoku Island in the south.[72]

Most of Japan's silver exports were exchanged for raw Chinese silk. Nearly two-thirds of the 256,362 cruzados worth of Chinese goods aboard a typical Portuguese vessel in 1600 was represented by raw silk, with the remainder comprised of gold, tin, and mercury. Raw silk was finished into ceremonial fabrics at the Nishijin silk-weaving district in Kyoto. Fabrics made from prized white raw silk were reserved for the shogun and high government (bakufu) officials. Ieyasu expanded the silver-for–silk exchanges. Backed by the 2.5 million cruzados worth of precious metals held at Edo Castle and the recent strikes on Sado, he established a Red Seal trading system in 1602 and organized the silk merchants of Sakai, Kyoto, and Nagasaki into a trading cartel in 1604. A short list of "Red Seal" merchant ships, with seals stamped by the shogun himself, were licensed to supply the *Itowappu-nakama* cartel with product. The cartel was authorized to fix prices and facilitate bulk purchases through the Portuguese and other licensed freighters. Within ten years, Japan was importing 6,300 piculs of raw Chinese silk annually and splitting the business evenly among the Red Seal fleets. Silk was received either from Macao, designated Chinese ports, or through a complex trading circuit that included stops at Taiwan, the Ryukyu Islands, Korea, Vietnam, and Manila. Japanese Red Seals carried roughly 41 percent of Japanese silver exports during the 1604–1635 period, followed by the Portuguese (32 percent), Chinese (17 percent), and the late-arriving Dutch (11 percent).[73]

The great surprise was that China's bimetallic ratio was unaffected by the rising flood of silver that supported the silk trade. As Japan's bimetallic ratio crept up (with her silver output) to around 12:1, it widened the profit opportunity with a Chinese ratio that fluctuated between 5.5:1 and 8:1 through 1630. Whether the silver was imported from Japan, Manila, or the cape route, the world's most attractive arbitrage opportunity remained intact through the mid-1630s. High-quality pesos de ocho received in Goa were either melted down and recast as Mogul rupees, arbitraging the Moguls' 9:1 bimetallic ratio in the process, or sent on to Macao to be melted down into bars and ingots for China. Dutch records at Fujian indicate that China's bimetallic ratio stood at a very inviting 8:1 through the 1620s and remained at an attractive 10:1 as late as 1635. No wonder that China imported as much as 127 tons of silver annually between 1600 and 1650—2.5 times the fifty-ton average received in 1550–1600. The profit potential in China more than offset the cost of shipping American silver across the Pacific or around the Cape of Good Hope. Would-be arbitrageurs like the VOC benefited from a relatively open Chinese market under Emperor Chongzen (1628–1644). By the time that China's bimetallic ratio finally converged with the revised global standard of 13:1 in 1637, the VOC was matching the two million cruzados in annual trade enjoyed by Portuguese Nagasaki.[74]

If the VOC intended to tap into Japan's silver trove, the Japanese were underwhelmed by the VOC's slim variety of Chinese silks and Indonesian pepper at Hirado. Portuguese Nagasaki carried merchandise that was more varied, higher quality, and protected by a

Hirado-based magistrate who was willing to interfere in Dutch affairs. The efforts of incoming VOC manager Jacques Specx (1609–1621), a man noted more for his diplomatic skills than his record keeping, were critical. Specx jump-started the Hirado factory by renting a local junk in 1610 and scraping up a hard-to-believe twelve thousand quintals of pepper (from Johore), fifteen thousand guldens worth of raw Chinese silk (from Patani), and one ton of lead. Otherwise, Hirado was largely ignored by the VOC. This left Specx to deal with the meddling magistrate, unofficial Chinese freighters, the shogun's silk cartel at Nagasaki, and a brief invasion by the English East India Company in 1613—an Englishman, William Adams, had been at the helm of Quackernaek's *Liefde* in 1600. But Specx was resilient. By 1615, he had parlayed a modest 57,000 guilders worth of imported merchandise into 195,000 guilders of Japanese trade goods. The latter figure included 15,000 from a captured Portuguese vessel.[75]

Dutch Hirado began to offer a much wider selection of goods after Ieyasu removed the local magistrate from Hirado's affairs. Business was also supported by the renewal of the VOC's Red Seal license in 1611, a focus on privateering, and the promotion of a nonthreatening brand of Christianity—the Dutch were happy to remind the later Iemitsu regime (1623–1651) that the agnostic VOC was a very credible alternative to the Catholic Portuguese. But if the Tokugawas were using the Dutch as a wedge against the Portuguese, they were wary about evicting the Portuguese from Japan entirely. The Portuguese were critical components of the Japanese economy. They had helped to unify Japan, catalyze foreign trade, and expand the nation's mining industry. In addition, the powerful Jesuits had been contained by a revival of Buddhist nationalism. The Jesuits were placed further on the defensive in 1614 when Shogun Hidetada decided to enforce a largely ignored ban against Catholicism. The move led to the persecution of an estimated 280,000 Japanese Christians between 1614 and 1635.[76]

Self-serving Dutchmen were behind some of the anti-Catholic (anti-Portuguese) campaigns between 1614 and 1635. But even though VOC exports of Japanese silver rose to around 400,000 guilders annually during the 1622–1627 period, the Dutch were a second-tier player in Japan until the late 1630s. The Portuguese were hurt less by the Dutch than by the Japanese raw silk cartel and the religious movement. Portuguese merchants enjoyed a final hurrah with a switch to silk fabrics, which circumvented the cartel, and the shogun's prohibition of Japanese foreign trade in 1635. The ban left the silk merchants of Sakai, Kyoto, Osaka, and Edo to rely on Portuguese, Chinese, and Dutch ships for their carriage needs. The patient Dutch were in the right place at the right time. A nine-ship VOC fleet left Hirado with nearly 1.5 million guilders worth of Japanese silver in 1636 and returned with an equivalent amount of Chinese goods. The VOC's largest haul to date trailed the 2.1 million cruzados in Japanese trade handled by the Portuguese in both 1636 and 1637, but the Portuguese were on borrowed time thanks to a few high-profile Portuguese bankruptcies and the massacre of thirty-seven thousand Japanese Christians in Shimabara (Kyushu) in 1637–1638. Iemitsu evicted the Catholic Portuguese from Japan in 1638 and replaced them with the Protestant Dutch (and selected Chinese). VOC trading volume in Japan soared to a staggering 6.3 million guilders in 1640.[77]

No matter that the evicted Portuguese had helped to launch Japan's process of unification when they stumbled onto Tane-gashima Island in 1543. Nearly one hundred years later, it was the upstart Dutch who were instructed to transfer their Hirado operations to a new factory on Deshima Island in Nagasaki harbor. Macao was decimated by the loss of her Japanese trading partners, Malacca fell in 1641, and the fruits of the Portuguese Empire were either lost to the Dutch or enjoyed by Portuguese entrepreneurs, captains, and self-dealing officials. The viceroy in Goa was left to administer the declining fortunes of the Estado da India. Most of the important names were still there—Muscat, Sofala, Mozambique, and Mombasa in East Africa; the Gulf of Cambay ports of Diu and Daman; the Malabar posts at Goa, Mangalore, Cannanore, Cranganore, and Cochin; two posts on Coromandel; plus Colombo

(Sri Lanka), the Moluccas (Tidore, Solor, and Ambon), and, of course, the crown jewel at Macao—but not for long. The newly independent Kingdom of Portugal (1640) would be fortunate to hang on to Brazil, Angola, Mozambique, Goa, and Macao.[78]

While the Dutch triumph in Japan was undermined by their inability to penetrate the huge Chinese market, they defeated the Portuguese in the Far East. Coming on the heels of their commercial dominance in Europe, their stockpiling of Spanish American precious metals, and their near-monopoly of transatlantic shipping, the Dutch had already established themselves as the most powerful commercial nation on earth. Not powerful enough to invade the Spanish Americas militarily or gain more than a few footholds in India, Persia, and the Levant, but strong enough to take Portuguese Africa and possibly even Brazil. Amsterdam had emerged as the premier global marketplace for grain, pepper, spices, salt, silks, precious metals, ivory, brazilwood, timber, hides, cochineal, cacao, tobacco, and sugar. But not yet human slaves—while unorganized Dutch fleets had imported West African gold and ivory since 1598, control of the more lucrative slave trade was still in the hands of the Portuguese. As for the so-called Spanish embargo, one royal official estimated that the value of Spanish products carried illegally by the Dutch (and other foreign flags) was six times that of Spanish products carried in authorized Spanish ships.[79]

Dutch privateering activities in the South Atlantic had overshadowed the slow-but-steady progress in the Far East. The Dutch were after the same slaves, sugar, gold, and ivory that had driven Portugal's West African operations since 1440 and had launched the whole global enterprise. No matter that "Dutch" privateers were more likely to be refugees from Antwerp, Brabant, Flanders, and even Lisbon than from the northern provinces. Privateering had become so profitable that an entrepreneur like Willem Usselinx, a native Dutchman, attempted to organize a single Dutch West Indies & West Africa Company in 1607 to circumvent a possible truce with Spain and control the profitless prosperity that had affected the early years of the Indonesian spice trade—too many Dutch trading companies and not enough demand. Since the entire system had been largely the creation of ex-Antwerpers, Usselinx also feared that a treaty with Spain—signed, no less, in Antwerp in 1609—carried the potential to shift the hard-earned Amsterdam-centric system back to Spanish-controlled Antwerp. The reopening of the Scheldt River to global commerce might restore Antwerp to her former economic status and return Amsterdam to mediocrity.[80]

Usselinx should not have worried. Antwerp was through in the same way that Bruges had been through in 1500 and that Venice, Genoa, and Florence were nearly through. The Twelve Years' Truce (1609–1621) actually enhanced Amsterdam's position in a way that twelve more years of war was unlikely to achieve. The activist Amsterdam faction eventually gained the upper hand, converted Usselinx's proposal of 1607 into the Dutch West Indies Company (WIC) of 1621, and seeded it with 2.2 million cruzados in equity capital. By then, Oldenbarnevelt's peaceful approach to foreign policy had cost him his job and his life. Prince Maurits conspired to have Oldenbarnevelt arrested and executed for treason in 1619. This, in turn, transformed the VOC from an armed trading company to an increasingly ruthless war machine and stepped up plans to establish the privateering- and slaving-oriented WIC in 1621.[81]

# CHAPTER ELEVEN

# CAPITAL OF THE WORLD

The spectacular global expansion launched by Columbus, da Gama, and access to unimaginable supplies of precious metals created a new geography during the sixteenth century. This new geography was one in which a handful of seaborne European powers, capped off by the combinatorial Dutch Republic, had come to dominate the lives of millions of native Americans, Africans, Muslims, and Asians and integrate their economic affairs into a rudimentary global system. Most of these subject peoples had been less miserable under the old order of things. Scores of medieval kingdoms, sultanates, and urban centers were rearranged in the process and forced to grapple with a global trading system that was anchored in far-off places like Cerro de Potosí, Zacatecas, and Sado Island. Few people, including the remarkable Houses of Habsburg, Fugger, and Genoa, had been aware of these mining sites extraordinaire as recently as 1540.

The sea-based power pioneered by the Portuguese and raised to new heights by the Dutch and English would be a fact of global life for the next four centuries. In the meantime, market forces had finished Antwerp, reduced Venice to second-tier status, and placed an erstwhile global titan, unified Spain and Portugal, on the defensive. If Antwerp was destroyed by warfare, Venice was crippled by old-fashioned competition. The Venetians had survived the Portuguese challenge and an uneasy eighty-year partnership with Istanbul, but the easterly invasion of English, French, and Dutch merchant ships was too much. The famous Venetian colony at Aleppo was being overrun by Westerners, the timber-challenged Venetian Arsenal was losing business to Dutch shipyards, and at least one Venetian merchant was ordering five 720-ton Dutch fluits to carry barrels of Cretan wine. Small consolation to Venice that unified Spain and Portugal had squandered huge fortunes and neglected their home economies. Spanish American treasure continued to be sunk into inconsequential military campaigns, balance of trade deficits, and non-Spanish enterprises. Portugal's middleman profit machine was eroded by Asians, Dutchmen, and energetic merchant families having little regard for Lisbon's medieval mindset. Two other candidates for global supremacy, England and France, had problems of their own. The death of Henri IV in 1610 interrupted France's nascent recovery. England was sinking into a full-blown civil war.[1]

The emerging global power was the Dutch Republic. While the seven Dutch provinces were as mineral poor as Portugal, France, and England, the ascendant Dutch were trading their way into the control of massive stocks of precious metals and recycling these metals into a global array of business enterprises. Spain's escalating balance of payment deficits and Portugal's fragmented empire were transferring huge volumes of bullion into the arms of entrepreneurial Dutch merchants who knew what to do with them. Amsterdam's growing metallic hoard was beginning to rival the mountain of bullion that had established the great Italian city-states and had made imperial Istanbul the envy of the world. The Italian, Ottoman, and Iberian empires were all on the defensive after 1600, and the unexpected Dutch would outperform all comers during the course of seventeenth century.

Amsterdam presided over a Europe that was a very different place from the downtrodden demographic entity of the fifteenth century. Europe's population had risen by nearly 27 percent between 1500 and 1600, from 61.6 million to 78 million, despite the loss of millions to warfare and emigration. Religious refugees fueled a huge jump in population in the tiny Dutch Republic to 1.5 million, while the somewhat larger Spanish Netherlands (Belgium) increased to only 1.6 million. The population of the entire Netherlands exceeded that of Portugal (1.1 million) but trailed England (4.4) and Austria-

Bohemia (4.3). Western Europe's most heavily populated nations were France (19 million), the fragmented states of Germany (16), Italy (13.1), and Spain (8.1). The dramatic growth of Lisbon, Seville, and Amsterdam during the second half of the sixteenth century had been offset by the spectacular rise and fall of Antwerp—from 40,000 in 1496 to over 100,000 in 1565 and back to 46,000 in 1591.[2]

The bleak demographic entity of the sixteenth century was Native America. While instant metropolises like Mexico City, Zacatecas, and Potosí compared with Europe's largest urban centers, their existence masked the near destruction of native American populations by European disease, overwork, and warfare. The American devastation was many times worse than that experienced in post–Black Death Europe between 1350 and 1450. Possibly 500,000 native Americans perished in the West Indies between 1492 and the first few decades of the sixteenth century, followed by the subsequent demographic carnage in pre-Cortés Mexico and pre-Pizzaro Peru. Smallpox, measles, plague, typhus, influenza, diphtheria, and mercury poisoning killed off as many as twenty-four million native Mexicans between 1519 and 1630 and as many as eight million native Peruvians between 1533 and 1630.[3]

The details of this demographic carnage are shocking. The preconquest (1519) population of Central Mexico—estimated at around 25 million, or nearly three times the preconquest level of Inca Peru—was reduced by nearly 8 million (to 17 million) between 1519 and 1523 alone. From there, the demographic collapse got only worse—falling to 6 million in 1548, 3 million in 1568, 2 million in 1580, and to a horrific low of 750,000 native Mexicans in 1630. The arrival of an estimated 250,000 or so Spaniards (and other Europeans) during the entire sixteenth century filled a very small fraction of the demographic void. New Spain's population would recover to around 1.2 million in 1650, but the growth was driven by Spanish immigrants, black slaves, and other non-Indians. Even if the preconquest estimates are inflated, there is no doubt that Mexico's native population was virtually destroyed between 1519 and 1630.[4]

The population of the preconquest Inca Empire was probably around 9 million, near the low end of an estimated range of 6 to 37 million. Based on the depopulation ratios of Mexico and the availability of relatively precise census figures for the 1570–1630 period, Inca Peru's native population plummeted from 9 million in 1520 to 1.3 million in 1570. As in Mexico, the demographic collapse worsened with age. By 1630, the number of native Americans living within the Viceroyalty of Peru was reduced to 600,000, with most of the survivors confined to the highland regions between Cuzco and Potosí. The native inhabitants of Cuzco, down to around 13,000 by 1561, and other highland towns fared much better than those in the lowlands—the coastal communities of Peru and Ecuador were wiped out almost entirely. Native Peruvians represented only 8 percent of Lima's 25,000 inhabitants in 1614—blacks (freemen and slaves) accounted for 42 percent of the total, followed by Spaniards (39 percent) and other groups (11 percent).[5]

As the Portuguese and Genoese (and Dutch) knew full well, there was a direct relationship between the near destruction of native American peoples and the acceleration of the West African slave trade. Most Iberian immigrants declined to perform manual labor, and someone had to excavate the mines, process the metals, harvest the sugar cane, and work the farms that made the entire economic system function. No matter that sugar was a minor industry in New Spain and Peru and the mining industries of Mexico and Bolivia had been revolutionized by the labor-saving amalgamation process. Charles V's prohibition of American slavery in 1542 was ignored. The ongoing demographic collapse induced sugar magnates in the West Indies and Brazil to scale up a century-old Genoese business model that had been pioneered in the sugar fields of the Canaries and Madeiras. A sizable share of the Portuguese slave trade between West Africa and Brazil was financed with "back-door" Bolivian silver and managed under multiyear asientos issued by the unified Iberian Crown. Well over thirteen thousand slaves were exported to the Americas annually between 1580 and 1600, with many more to come. The return cargo, raw sugar, was freighted in Dutch ships to the sugar refineries in Amsterdam.[6]

The WIC liked the model so much that it worked tirelessly to capture Portugal's West African slave trade. Investors had been underwhelmed by the one ton or so of West African gold and one hundred tons of ivory that was received annually. The WIC eventually added an array of new (or captured) slaving stations in West Africa—Fort Nassau (1611), More (1611), Goeree (1617), Rufisque (1617), Arguin (1634), São Jorge da Mina (1637), Cabo de Corco (1637), São Tome (1641), and Axim (1642)—and shipped the human cargo across the sea to a central slaving center on Curaçao (1634) in the Antilles. Curaçao handled many of the seventy thousand African slaves who were sent to the Caribbean and Brazil between 1630 and 1674. The Portuguese were left with second-tier slaving stations at Principe and Ano Bom until Portuguese Brazilians, operating without any assistance from Lisbon, recovered São Paulo de Loanda (Angola) from the Dutch in 1648.[7]

The WIC, try as it might, failed to capture Portuguese Brazil. Dutch interest in the region had followed the successful invasion of the salt marshes of Venezuela in 1587, regular visits to the Margarita salt pan between 1599 and 1605, expeditions to the "wild coast" of Guiana, and the ability of Dutch freighters to capture as much as two-thirds of the Brazilian sugar trade. While Dutch privateers prowled Brazil regularly in search of sugar, silver, and smuggling profits, a full-scale assault on Brazil was delayed until the expiration of the Twelve Years' Truce, the establishment of the WIC, and the receipt of an unexpected windfall in 1628. However, Portuguese Brazilians managed to withstand over thirty years of warfare, culminating in the WIC's formal surrender at Pernambuco in January 1654, with only occasional naval assistance from Lisbon. Jaws must have dropped in Amsterdam when huge deposits of gold were struck in southeastern Minas Gerais in the 1690s—Minas Gerais would produce eleven tons of gold per year during the eighteenth century.[8]

The expanding slave trade was a direct consequence of the demographic collapse in Native America and a new world order. Another aspect of this order was the emergence (and mapping) of towns that few people could name or locate in 1540. Cities like

Mexico City, Zacatecas, Lima, and Potosí in the West and Goa, Manila, Macao, Beijing, Kyoto, Edo, and Nagasaki in the East were now as recognizable as the great trading centers of the Mediterranean. The flip side was that scores of formerly great trading centers had fallen on hard times. Antwerp, the mighty creation of the Fuggers, Portuguese, and Habsburgs, had followed the downward path of medieval Bruges, Lyon, and Medina del Campo. Augsburg declined with the Central European mining industry. The great Italian city-states—as well as Istanbul, Cairo, Hormuz, Calicut, and Malacca—faced a myriad of challenges. The future appeared to belong to London, Paris, and, of course, Amsterdam. Amsterdam had somehow fused together global trading networks that had otherwise been controlled by Antwerp, Lisbon, Seville, and the Italians. Amsterdam had emerged as the business capital of the entire world.

The Amsterdam from which the Dutch East India Company (VOC) was launched in March 1602 had come a long way from her days of fishing, salting, and regional freighting. The city's population had soared from around 30,000 prior to the fall of Antwerp in 1585 to 90,000 in 1600 and would swell to 105,000 by 1622. No matter that Amsterdam's physical expansion was dependent on draining marshlands, damming or diking lakes, constructing concentric rings of canals, and sinking thousands of wooden piles into reclaimed "lands" to support building foundations. These were expensive. Amsterdam's tax-paying citizenry was faced with round after round of public works projects. These included a major new canal, the future Herengracht in 1591, the massive 1.5-million-guilder "Beemster" drainage project in 1608, and the relocation of the local shipyard from Lastage to the lower-rent Zaanstreek district. Otherwise, Amsterdam's massive shipbuilding needs were filled mainly by the yards at Rotterdam and Hoorn. If taxpayers grumbled that the 1.3 guilders per day earned by local construction workers was nearly the double the wages paid in Cologne, alternative remedies were unclear. Coupled with the mud, filth, and poverty that was a regular feature of urban life in the late Middle Ages, work-in-progress Amsterdam must have been a sight (and smell) during the building boom. Yet at least one visitor, the Duc de Rohan, claimed in 1600 that "Amsterdam had no equal in Europe for wealth and beauty save for Venice."[9]

Amsterdam's reputation for religious tolerance and a defensible military position on the Zuider Zee attracted thousands of refugees—whether they were Protestants, Catholics, Conversos, or Jews. While the intense land reclamation projects lagged the city's ability to house the new arrivals, leaving three thousand squatters to inhabit an outlying slum in 1609, the immigrants made huge contributions to local affairs. Protestant refugees from Antwerp and the southern provinces launched a number of industries and persuaded the city fathers to spend money to make money. The Portuguese Rodrigues de Vega received 120,000 guilders prior to 1600 to establish Amsterdam's first silk business with over four hundred employees. A similar arrangement was struck with one of Venice's finest makers of crystal glass. Southerners started Amsterdam's sugar-refining industry, a fifty-plant enterprise by 1662, and even the world's first daily newspaper—the commercially oriented *Courante* commenced operations in 1618. In addition to entrepreneurs, Amsterdam drew some of the greatest thinkers and artisans of their (or any other) day. These included John Locke, René Descartes, Baruch de Spinoza, and Rembrandt van Rijn. Artistry had its own ups and downs. Rembrandt, the son of a Leiden miller, started a modest portrait-painting business in the 1630s that made him famous and wealthy by 1642, bankrupt in 1656, and so destitute that he was buried in a pauper's grave in 1669.[10]

With few medieval guilds to block their way, many immigrants launched business ventures in textiles, printing, silk, leather, glass, publishing, tobacco, diamonds, and goldsmithing. Transplanted Iberian Conversos and Jews, possibly one thousand strong by 1610, became so successful in Amsterdam that Felipe III attempted to restrict their brethren from leaving the Iberian Empire. When that failed, the king averted a wholesale exodus of Converso capital to Amsterdam by easing the claws of the Inquisition—the number of autos-da-fé proceedings in Goa was halved between 1600 and 1620.

Count-Duke of Olivares (Gaspar de Guzmán), who served as regent after Felipe's death in 1621, was even more liberal. He invited wealthy Portuguese Conversos to the Royal Court at Madrid for the first time and rewarded their financial support with full trading and residential privileges in Spanish America. That is, until the general Spanish economy declined during the 1632–1636 period. Spanish nobles decided to mount a scapegoating campaign against the Conversos and persuaded Felipe IV to revive the tribunals. These and other actions catalyzed a second wave of Converso immigration to Amsterdam and other enlightened locales.[11]

Immigrants had little to do with the vaunted Dutch shipbuilding industry. Dutchmen built ships for Felipe II's refurbished navy in 1596–1597; the English Muscovy Company in 1597–1598, Henri IV's blockade of La Rochelle, a massive 1,400-ton, sixty-gun galleon (the *St. Louis*) for Cardinal Richelieu in 1626, and a 1,210-ton, sixty-four-gun vessel (the *Vasa*) for Gustavus Adolphus in 1628. Unfortunately, the unstable *Vasa* sank in Stockholm harbor during her maiden voyage in August 1628. The Swedish king had better luck with transplanted munitions suppliers like Gerard and Louis de Geer of Liege. The Geers helped to develop Sweden's rich deposits of copper, expand the kingdom's modest arms industry, and diversify the Dutch arms industry. Amsterdam had already supplanted Antwerp and Hamburg as Europe's principal source for arms and ammunition. Products that were not manufactured locally were imported from the great munitions works at Aachen and Solingen, the heavy gun makers of England, the Baltic (gunpowder), and Italy (sulfur). Thanks mainly to the Geers, Dutch-owned munitions plants in Sweden were manufacturing saltpeter, gunpowder, shot, cannon, and muskets by 1615. These and locally produced products—lead, iron, guns, brass, steel, tin, wire, armor, pikes, and matches—were either consumed by Adolphus's military campaigns or sent back to Amsterdam. The Swedish city of Goteborg was laid out with Dutch-styled canals and water systems.[12]

Amsterdam seemed to be at the center of everything. In the words of historian C. R. Boxer, the Dutch Republic was "indisputably the greatest trading nation in the world" at the conclusion of the Eighty Years' War with Spain in 1648. Like the trailblazing Portuguese, the modest territorial conquests of the Dutch masked a commercial empire of enormous clout—highlighted by the world's largest stocks of precious metals and grain. No matter that the surrounding countryside was mineral poor, wheat poor, and was engaged with the venerable Army of Flanders until 1648. Like the Venetians and Portuguese before them, the Dutch metallic hoard was captured mainly in trade. Profits earned from the 100,000–plus tons of Baltic grain imported annually were typically received in the form of silver bullion, sent on, Portuguese-style, to the merchants of the Indian Ocean, Indonesia, and the Far East to exploit the available arbitrage opportunities, and then reinvested in the Caribbean and West Africa. As late as 1666, an estimated three-fourths of the capital invested on the Amsterdam Bourse was associated with Baltic grain contracts and related enterprises. The Dutch also dominated trading in Norwegian timber, Swedish copper, Portuguese and French salt, European textiles, American and Brazilian sugar, and African slaves. A powerful English response was delayed until England had resolved her own Wars of Religion, fittingly with the assistance of William III of Orange in 1688.[13]

American bullion was critical to the Dutch triumph. The VOC's first board of directors (the Heeren XVII) stipulated in 1602 that only Spanish American silver (pieces of eight) was authorized for exchange in the Far East. No wonder that the thirty thousand reales that went down with the 320-ton *Nassau* off Malacca in 1606 had been minted at Potosí, Mexico City, Seville, Toledo, and Segovia. Some were even stamped with "Philip by the Grace of God, King of Spain and the Indies" on one side and the coat of arms of the Habsburg Netherlands on the other. The Dutch focus on (and co-option of) Spanish silver was reflected in a doubling in the value of VOC precious metal exports to Asia—to 9.7 million guilders (mainly 114 tons of silver) during the 1610–1620 period—and arbitrage opportunities tied to a higher-than-average bimetallic ratio of over 12:1 in Amsterdam. Bullion accounted for nearly all of the VOC's eastbound cargo in 1615, the start of an easterly metallic flow that would reach twenty-nine million guilders between 1690 and 1700.[14]

Amsterdam eclipsed Seville and Genoa as Europe's premier silver-trading center—garnering as much as one-fifth of the total volume of American silver that was registered in Seville between 1595 and 1630. The silver was exchanged for grain, salt, sugar, and manufactured European products or sent on to the four corners of the globe in search of trading opportunities. Local real estate opportunities were so slim that wealthy Amsterdamers gravitated to investments in business enterprises, company shares, and government bonds. Amsterdam became a town of speculators. René Descartes, rationalist philosopher extraordinaire, declared in 1635 that "there is nobody [in Amsterdam] who does not trade in something." While investors in the VOC and WIC suffered through a regular string of losses during the early years, shareholders were occasionally rewarded by sheer luck. Piet Heyn reversed seven years of WIC losses in one fell swoop when he hijacked nine weather-beaten Spanish treasure ships off Matanzas, Cuba, in September 1628. The heist added a staggering fifteen million guilders worth of silver (eighty-nine tons), jewels, and other goods to the WIC account and raised Amsterdam's bullion reserves to a level that nearly matched that held by the rest of Europe combined.[15]

Heyn's heist was not to be confused with the huge volumes of precious metals earned from trading, freighting, and financial services. Amsterdam became a financial colossus. The open market in commodities trading, inherited from Antwerp, prompted wild speculation in metals and in the bills of exchange that were secured by them. The issuance of bills of exchange between Amsterdam and London had surged during the brief English occupation and catalyzed the introduction of a number of creative (if sometimes risky) financial instruments and services. Unsophisticated investors struggled with debentures, unsecured interest-bearing notes issued by security-challenged borrowers, and discovered the hard way that debentures were far more risky than "merchant notes" secured by bullion. Otherwise, investors had benefited from the publication of daily transactions between Amsterdam and ten other cities, the printing of marine insurance policies (1592), and the founding of the Chamber of Insurance (1598) to regulate them. These innovations were followed by the establishment of the Amsterdam Stock Exchange (1609) to monitor trading in commodities and stocks and then the quasi-public Wisselbank (City Loan Bank) in 1611 to provide low-interest business loans (initially 2.5–6.5 percent). Naval warfare, privateering, and pirating activities may not have been always good for stocks, but they were great for the insurance business—434,700 guilders in annual insurance premiums were registered in 1635.[16]

The Amsterdam Stock Exchange, modeled on the Antwerp Bourse, was incapable of reining in the type of freewheeling trading that had made Antwerp famous. However, it ensured that nearly three hundred local brokers were licensed and prohibited by a 1581 law from participating in brokered transactions. Otherwise, speculative trading in company shares, futures, and options exploded with Amsterdam's global expansion. It was extended from grain to other commodities like timber, herring, spices, and whale oil. VOC shares more than doubled in value after a modest slice of the Spice Islands had been seized in 1605. A separate Grain Exchange was established in 1617 in an attempt to prevent powerful grain merchants from manipulating the grain market in the style of the Fuggers and Genoese. Iberian Conversos and Jews, people who knew a few things about the trading business, helped to improve the transparency of the issuance, syndication, distribution, and trading of company shares and government revenue bonds. The improved transparency helped to subscribe a one-million-guilder, 3 percent public-works bond in just two days in 1664. Amsterdam served as the financial capital of Europe until London seized the title in the 1690s.[17]

Investors required patience when it came to the VOC (and the later WIC). The investment of fifteen million guilders in 246 heavily armed ships and thirteen fortified Asian factories produced a string of losses between 1602 and 1619. The 1.8 million cruzados in revenues earned during the 1613–1621 period, one-third of which was nonrecurring privateering "prizes," were less than half the Portuguese volume in a single year. Small consolation that the one million cruzados worth of pepper, gold, and spices delivered by three "Indiamen" vessels in July 1618 dwarfed anything received to date in London. While ongoing losses through 1625 reduced the VOC's equity position to 1.6 million cruzados and raised outstanding debt to 2.8 million, VOC shareholders were bolstered by the firm's ability to borrow low-cost funds from the Wisselbank. They also recognized that the losses reflected heavy front-end investments that would hopefully pay off later. Long-suffering VOC shareholders were rewarded by a steadily rising stream of profits between 1626 and 1640. The stream reflected an average of two million cruzados of Asian merchandise received annually, an ability to limit shipwreck and other losses to 5 percent, and a more ruthless emphasis on privateering. The VOC earned an impressive 11.9 million guilders during the heady 1631–1640 period.[18]

The VOC's early trials were cushioned by Amsterdam's vast array of business enterprises. The city's wealthiest citizen in 1585 had been Dirck Janszonn Graaf, a burgomaster and iron trader who was worth around 140,000 guilders. This may have been petty cash to the Fuggers and Welsers, but no more than five other Amsterdamers owned fortunes greater than 100,000 guilders in that year. By 1631, however, over one hundred Amsterdamers exceeded that total and formed part of the one thousand local-shareholder base that had invested in the overseas expedition companies. Vincent van Bronckhost, the son of a grain merchant who had invested in Houtman's expedition of 1595, helped to establish the VOC. Coenraad Bury, a former envoy to Russia, administered the WIC's New Netherland colony prior to the arrival of Peter Stuyvesant. Dirck Spieghel, the grandson of a herring merchant and the son of a soap boiler, became a WIC director during in the 1660s.[19]

These and other Amsterdamers scoured the world for new business opportunities. Emboldened by a Dutchman's successful expedition to Japan in 1600, a host of Dutch navigators were commissioned to explore some of the last uncharted regions of the globe.

Profit opportunities in the Russian Arctic had drawn scores of fur-trading fleets to the north prior to the establishment of a single Amsterdam syndicate in 1594. The syndicate sent over twenty trading expeditions to Russia (via Archangel) between 1594 and 1600 and eventually established the White Sea Company in 1619. In the meantime, the energetic Cornelius de Houtman entered the fray by establishing the Northern Company in around 1600. A shaky reputation earned from the Indonesian expedition of 1595–1597 failed to prevent Houtman from raising funds to explore the Spitsbergen region for whales (oil), polar bears (pelts), and walrus (tusks). But Houtman was again ahead of his time. It would be a successor to the Northern Company, the Noordse of Groenlandie, that would exploit the region successfully after 1614 and establish the Dutch whaling village of Smeerenburg. Known also as "Blubbertown," Smeerenburg would have as many as one thousand inhabitants during the boom years of 1617–1624.[20]

The Dutch and their competitors failed to find a Northwest Passage to Asia, an achievement that awaited the Norwegian explorer Roald Amundsen in 1905. In the meantime, efforts had been made to find one. Estêvão Gomes, a Portuguese who was one of the few survivors of the Magellan expedition, explored the entire eastern coastline between Florida and Labrador (for Spain) in 1525 in search of a possible northwest passage. Not only were his efforts ignored, the overstretched Iberians essentially abandoned the uncharted (mineral-poor) northern territories of North America to all comers. Giovanni de Verrazano, sailing for France in 1524, beat Gomes to the mouth of the Hudson River and laid the groundwork for a subsequent expedition by Jacques Cartier in 1534. But Cartier's exploration of the Saint Lawrence River failed to uncover a passage and demonstated only that the region was rich in fur-bearing animals—it took the efforts of Cardinal Richelieu to organize the Cent Associés de la Nouvelle France in 1628. Similarly, Englishmen Martin Frobisher and John Davis impressed few European investors after rounding the northwestern tip of Quebec in 1576 and approaching the straits to the future Hudson Bay.[21]

It was not until 1608 that the English Muscovy Company hired Henry Hudson to follow through on the earlier efforts of Frobisher and Davis. By then, the English had lost most of the Russian market to the Dutch and were hoping to beat the relentless Dutch to the next business opportunity. Hudson was encouraged to follow the same Novaya Zemlya route that had been traced (and abandoned) by the remarkable Jan Huygen van Linschoten between 1594 and 1597. The difficult-to-handle Hudson fared no better than his predecessors. He was blocked by ice and forced to return by a mutinous crew. Hudson must have impressed someone because he was recruited immediately by the VOC to try again on behalf of the Dutch Republic. The VOC provided him with the *Half Moon* ship, a crew, and enough provisions to conduct a hopefully successful arctic voyage to Asia in the spring of 1609. While Hudson was assisted by John Davis's quadrant invention of 1604, an instrument that was superior to the typical cross-staffs of the sixteenth century, he encountered problems with his sailors once again. En route to Novaya Zemlya, the crew refused to sail any farther north and demanded that Hudson chart a new course to the warmer climes of northeastern America. No matter that the demand violated strict VOC policy.[22]

Hudson was probably aware of a series of English expeditions that had been made to "Virginia" and of John Smith's recent establishment of Jamestown Colony in May 1607. He was certainly unaware of Samuel Champlain's expedition through upstate New York and Vermont in that same summer of 1609. According to a journal kept by Hudson's first mate, Robert Juet, the *Half Moon* reached as far south as Penobscot Bay, Maine, by mid-July, stocked up on local codfish and lobster, and met with tribes who had had plenty of experience with French traders. The natives advised Hudson that the surrounding region held tantalizing quantities of gold, silver, and copper. Hudson was somehow induced to plunder a local village (on July 25) as the *Half Moon* made its way south to Cape Cod. Named by Çaptain Bartholomew Gosnald in 1602, the cape and a nearby landing site at Plymouth would be visited by seafaring Pilgrim colonists in 1620. In the meantime, Hudson and his men were greeted warmly on August 3 and introduced to the pleasures of green tobacco.[23]

After rounding Cape Cod, Hudson struggled through bad weather as he sailed southward. He passed a number of river mouths, including New York, Delaware, and Chesapeake bays, but declined to make landfall on possible English territory—he was, after all, on the Dutch payroll. Hudson returned north to New York Bay on September 2 under the assumption that this uncharted strip of coastline was unclaimed by either England or France. The *Half Moon* anchored in New York harbor on September 3, leaving the crew to take soundings, observe the oak-covered landscape, and fish for salmon, mullets, and rays. The following day, some deerskin-clad inhabitants of this "pleasant land" welcomed their visitors with green tobacco and offered up red copper pipes, yellow copper, maize, bread, furs, and hemp in trade. These and other items were purchased for European knives, hatchets, beads, and "trifles."[24]

An otherwise friendly encounter was rocked on September 6. Canoe-borne Indians ambushed a sounding party and killed one of Hudson's men, John Colman, during the skirmish. A more wary Hudson broke camp on September 11 and entered the river that would bear his name. When the winds were unfavorable, the shallow-drafting *Half Moon* drifted upriver with the incoming tide, emphasizing the spritsail, and anchored when the tide worked against it. Juet noted that the river was one mile wide in places, bordered by high land on both sides, and within reach of nearby mountains. The white-green cliffs opposite "Manna-hata" (Manhattan) and the barren hills below Newburgh were both thought to contain deposits of copper or silver. The crew traded cautiously with local Indians for corn, pumpkins, tobacco, oysters, grapes, and beaver skins as the ship headed north. On September 22, having reached the river's treacherous confluence with the Mohawk River opposite present-day Troy, Hudson decided to turn back to New York harbor.[25]

Hudson's achievement was underwhelming to the VOC. The frontier territories claimed on behalf of the Dutch in 1609 were viewed as insignificant compared with slaving and privateering opportunities in the West Indies and Brazil. When Hudson returned to the English fold in early 1610 to find an alternative northwest passage to Asia, the expedition ended in failure. The courageous but extremely unpopular Hudson was stranded in Hudson Bay by another mutinous crew and died there of exposure in 1611. Hudson's efforts were ignored until 1614. In March of that year, the Dutch States General offered a trading monopoly to any company willing to finance four expeditions to the unclaimed territories situated between New France (French Canada) and English Virginia within a three-year period. The lands, laying between 40 and 45 degrees latitude, were promoted as a possible source of fur-trading profits and agricultural goods. A four-merchant syndicate from Amsterdam and Hoorn organized the New Netherlands Company in 1614 but failed to follow through. The States General rejected a subsequent proposal to establish a Puritan colony in New Netherland in 1620. Dutch investors were unimpressed by the English experience in Virginia and preferred to focus on business opportunities located in other geographic regions.[26]

# NOTES

## INTRODUCTION

1. John Maynard Keynes, *A Treatise on Money*, vol. 2 (London: Macmillan & Co., 1950), pp. 156, 157.

## CHAPTER ONE: THE PORTUGUESE HEAD EAST

1. K. S. Mathew, *Indo-Portuguese Trade and the Fuggers of Germany* (New Delhi: Manohar, 1997), p. 101.

2. Ibid., p. 102; Joseph Gies and Frances Gies, *Merchants and Moneymen* (New York: Crowell, 1972), p. 275.

3. Mathew, *Indo-Portuguese Trade*, p. 104; Gies and Gies, *Merchants and Moneymen*, p. 275.

4. Pierre Vilar, *A History of Gold and Money, 1450–1920* (Atlantic Highlands, NJ: Humanities Press, 1976), pp. 71, 72; John Munro, "Bullion Flows and Monetary Contraction in Late-Medieval England and the Low Countries," in *Precious Metals in the Later Medieval and Early Modern Worlds*, ed. J. F. Richards (Durham, NC: Carolina Academic Press, 1983), p. 119.

5. Richard Ehrenberg, *Capital and Finance in the Age of the Renaissance* (New York: Harcourt, 1928), pp. 66, 133, 134; Gies and Gies, *Mer-*

*chants and Moneymen*, p. 274; Jacob Strieder, *Jacob Fugger the Rich*, trans. Mildred L. Hartsough (New York: Adelphi Company, 1931), pp. 28, 40, 41, 57; Mathew, *Indo-Portuguese Trade*, p. 121; John U. Nef, "Silver Production in Central Europe," *Journal of Political Economy* 49, no. 4 (1941): 576, 578; Vilar, *Gold and Money*, p. 72; Cedric E. Gregory, *A Concise History of Mining* (New York: Pergamon Press, 1980), pp. 95, 96.

6. Gies and Gies, *Merchants and Moneymen*, p. 275; Ehrenberg, *Capital and Finance*, pp. 64–66; Strieder, *Jacob Fugger*, pp. 52, 54, 56; Victor von Klarwill, ed., *The Fugger News-Letters* (New York: G. P. Putnam, 1924), p. xxii; Mathew, *Indo-Portuguese Trade*, pp. 103, 104, 120, 144.

7. Ehrenberg, *Capital and Finance*, pp. 66, 133, 134; Gies and Gies, *Merchants and Moneymen*, p. 274; Strieder, *Jacob Fugger*, pp. 28, 40, 41, 57; Mathew, *Indo-Portuguese Trade*, p. 121.

8. Nef, "Silver Production," pp. 576, 586.

9. Strieder, *Jacob Fugger*, pp. 63–65; Mathew, *Indo-Portuguese Trade*, p. 119.

10. Ehrenberg, *Capital and Finance*, p. 67; Mathew, *Indo-Portuguese Trade*, pp. 106, 122.

11. Hermann Kellenbenz, "Final Remarks," in *Precious Metals in the Age of Expansion*, ed. Hermann Kellenbenz (Stuttgart: Klett-Cotta, 1981), p. 310; Ian Blanchard, "England and the International Bullion Crisis," in Kellenbenz, *Precious Metals*, p. 110; Ekkeland Westermann, "Tendency in the European Copper Market," in Kellenbenz, *Precious Metals*, p. 71.

12. Kellenbenz, "Final Remarks," p. 310; Gies and Gies, *Merchants and Moneymen*, pp. 276, 278–80; Oszhar Paulinyi, "The Crown Monopoly," in Kellenbenz, *Precious Metals*, p. 38.

13. Gies and Gies, *Merchants and Moneymen*, p. 281; Strieder, *Jacob Fugger*, pp. 120–23; Eugenia W. Herbert, "The West African Copper Trade," in Kellenbenz, *Precious Metals*, pp. 119, 120, 128.

14. Hermann Kellenbenz, *The Rise of the European Economy* (London: Weidenfeld & Nicolson, 1976), pp. 5, 80; Strieder, *Jacob Fugger*, pp. 124, 126; Ehrenberg, *Capital and Finance*, p. 67; Mathew, *Indo-Portuguese Trade*, pp. 107, 123.

15. Jervis Wegg, *Antwerp* (London: Methuen, 1916), pp. 126, 174; Mathew, *Indo-Portuguese Trade*, p. 101.

16. Vilar, *Gold and Money*, pp. 71, 72; Munro, "Bullion Flows," pp. 100–102.

17. J. H. Parry, *Age of Reconnaissance* (New York: New American Library, 1963), pp. 9, 10; Lyle McAlister, *Spain and Portugal in the New World* (Minneapolis: University of Minnesota Press, 1984), p. 41.

18. Parry, *Age of Reconnaissance*, pp. 5, 11–13.

19. C. R. Boxer, *The Portuguese Seaborne Empire* (New York: Knopf, 1969), pp. 16, 17, 19, 28; J. D. Fage, *An Introduction to the History of West Africa* (Cambridge: Cambridge University Press, 1955), p. 42.

20. Boxer, *Portuguese Seaborne Empire*, pp. 17, 19; Fage, *History of West Africa*, pp. 43, 44.

21. Boxer, *Portuguese Seaborne Empire*, pp. 24–26; Fage, *History of West Africa*, p. 44; António Henrique de Oliveira Marques, *History of Portugal* (New York: Columbia University Press, 1972), pp. 142, 144–50.

22. Boxer, *Portuguese Seaborne Empire*, pp. 24, 27; McAlister, *Spain and Portugal*, pp. 50, 51, 55; J. H. Parry, *The Spanish Seaborne Empire* (New York: Knopf, 1966), p. 40; Parry, *Age of Reconnaissance*, pp. 20, 21, 131–33, 147.

23. Marques, *History of Portugal*, pp. 151, 221.

24. Richard W. Unger, *The Ship in the Medieval Economy* (London: Croom Helm; Montreal: McGill-Queen's University Press, 1980), pp. 212–14, 231; Martin Elbe, "The Caravel and the Galleon," in *Cogs, Caravels and Galleons*, ed. Robert J. Gardiner (London: Conway Maritime Press, 1994), pp. 91–96; Boxer, *Portuguese Seaborne Empire*, pp. 24–26, 30; McAlister, *Spain and Portugal*, pp. 48, 49; Parry, *Age of Reconnaissance*, p. 131; Vilar, *Gold and Money*, p. 51; Fage, *History of West Africa*, pp. 2, 3, 44; Kellenbenz, "Final Remarks," p. 311; Marques, *History of Portugal*, p. 150.

25. Vilar, *Gold and Money*, pp. 51, 52; Fage, *History of West Africa*, p. 53.

26. John Day, "The Great Bullion Famine of the Fifteenth Century," *Past & Present* 79 (May 1978): 35–40; Kellenbenz, "Final Remarks," p. 311; Andrew M. Watson, "Back to Gold and Silver," *Economic History Review* 20, no. 1 (1967): 19, 20.

27. Boxer, *Portuguese Seaborne Empire*, pp. 20, 29; Parry, *Spanish Seaborne Empire*, p. 42; Marques, *History of Portugal*, p. 217.

28. Boxer, *Portuguese Seaborne Empire*, pp. 28, 31; Vilar, *Gold and Money*, p. 53; Parry, *Age of Reconnaissance*, pp. 133, 135, 136; Fage, *History of West Africa*, pp. 47–49, 57; Marques, *History of Portugal*, pp. 218, 246.

29. Boxer, *Portuguese Seaborne Empire*, pp. 24–26, 30; Vilar, *Gold and Money*, pp. 50, 56; McAlister, *Spain and Portugal*, p. 48; Herbert, "West African Copper Trade," pp. 121, 122.

30. Boxer, *Portuguese Seaborne Empire*, pp. 29, 30; Parry, *Age of Reconnaissance*, p. 136.

31. Marques, *History of Portugal*, pp. 172, 173, 207; James C. Boyagian, *Portuguese Trade in Asia Under the Habsburgs, 1580–1640* (Baltimore: Johns Hopkins University Press, 1993), p. 3.

32. Marques, *History of Portugal*, pp. 165–67, 265.

33. Parry, *Age of Reconnaissance*, pp. 136, 137; Marques, *History of Portugal*, p. 219.

34. Parry, *Age of Reconnaissance*, pp. 137, 138.

35. Henry Kamen, *Spain, 1469–1714* (London; New York: Longman, 1983), p. 3; Boxer, *Portuguese Seaborne Empire*, p. 33; Parry, *Age of Reconnaissance*, pp. 138, 139; Mathew, *Indo-Portuguese Trade*, p. 3; Marques, *History of Portugal*, pp. 220, 221.

36. Parry, *Age of Reconnaissance*, pp. 139, 140, Marques, *History of Portugal*, p. 227.

37. Parry, *Age of Reconnaissance*, pp. 140, 141; Ernest S. Dodge, *Islands and Empires* (Minneapolis: University of Minnesota Press, 1976), p. 222; Sanjay Subrahmanyam, *The Career and Legend of Vasco da Gama* (Cambridge; New York: Cambridge University Press, 1997), pp. 79–87, 121, 128; Marques, *History of Portugal*, p. 223.

38. Mathew, *Indo-Portuguese Trade*, pp. 107, 112–19, 123.

39. Boxer, *Portuguese Seaborne Empire*, pp. 40, 41; Spiros Vryonis Jr., "The Question of the Byzantine Mines," *Speculum* 37, no. 1 (1962): 1, 3.

40. Parry, *Age of Reconnaissance*, pp. 141, 142; Subrahmanyam, *Vasco da Gama*, pp. 128, 130, 136–38.

41. Subrahmanyam, *Vasco da Gama*, pp. 151, 175; Marques, *History of Portugal*, p. 226.

42. Subrahmanyam, *Vasco da Gama*, pp. 175, 178–81.

43. Ibid., pp. 181, 182.

44. Ibid., pp. 182, 183, 185.

45. Richard Reusch, *History of East Africa* (New York: F. Ungar, 1961), pp. 228, 230, 231.

46. Subrahmanyam, *Vasco da Gama*, pp. 190, 195, 200–206; John F. Guilmartin Jr., *Galleons and Galleys* (London: Cassell, 2002), p. 82.

47. Subrahmanyam, *Vasco da Gama*, pp. 212–17; Geoffrey Parker, *The Military Revolution* (Cambridge; New York: Cambridge University Press, 1996), p. 104.

48. A. R. Disney, *Twilight of the Pepper Empire* (Cambridge, MA: Harvard University Press, 1978), pp. 32–37.

49. Subrahmanyam, *Vasco da Gama*, pp. 221–23.

50. Ibid., pp. 224–29, 232.

51. Ibid., pp. 237, 238; Mathew, *Indo-Portuguese Trade*, p. 149.

52. Mathew, *Indo-Portuguese Trade*, p. 149; Boxer, *Portuguese Seaborne Empire*, p. 61; Parry, *Age of Reconnaissance*, p. 50; Ehrenberg, *Capital and Finance*, pp. 51, 53.

53. Mathew, *Indo-Portuguese Trade*, pp. 4–7.

54. Wegg, *Antwerp*, p. 68; Ehrenberg, *Capital and Finance*, p. 68; Herman Van Der Wee, *The Growth of the Antwerp Market and the European Economy* (Louvain: Biblioteque de l'Universite, 1963), p. 127.

55. Mathew, *Indo-Portuguese Trade*, pp. 7–9; K. S. Mathew, *Portuguese Trade with India in the Sixteenth Century* (New Delhi: Manohar, 1983), pp. 155–58; Marques, *History of Portugal*, p. 213; Parry, *Age of Reconnaissance*, p. 50.

56. Parry, *Age of Reconnaissance*, pp. 50, 51.

57. Mathew, *Indo-Portuguese Trade*, p. 4.

58. G. S. P. Freeman-Grenville, "The Coast, 1498–1840," in *History of East Africa*, ed. Roland Oliver and Gervase Mathew (Oxford: Clarendon Press, 1963), p. 134; Reusch, *History of East Africa*, pp. 231, 232; R. Coupland, *East Africa and Its Invaders* (Oxford: Clarendon Press, 1938), p. 44; Subrahmanyam, *Vasco da Gama*, pp. 239, 240.

59. Mathew, *Indo-Portuguese Trade*, p. 113; Freeman-Greenville, "The Coast," p. 135; Reusch, *History of East Africa*, pp. 230, 233; Marques, *History of Portugal*, p. 249.

60. Vilar, *Gold and Money*, pp. 56, 57, 92; Marques, *History of Portugal*, p. 260; Kellenbenz, "Final Remarks," pp. 311, 312.

61. Parry, *Age of Reconnaissance*, pp. 142, 143; Mathew, *Indo-Portuguese Trade*, pp. 69, 70.

62. Mathew, *Indo-Portuguese Trade*, pp. 74, 75, 77.

63. Subrahmanyam, *Vasco da Gama*, p. 256; Mathew, *Indo-Portuguese Trade*, pp. 76–78.

64. Subrahmanyam, *Vasco da Gama*, p. 258.

65. Ibid., pp. 258, 259.

66. M. N. Pearson, *Coastal Western India* (New Delhi: Concept, 1981), pp. 69, 70; Mathew, *Indo-Portuguese Trade*, p. 13.

67. Pearson, *Coastal Western India*, pp. 72, 73; Boxer, *Portuguese Seaborne Empire*, pp. 46, 48; Parry, *Age of Reconnaissance*, pp. 144, 145; Dodge, *Islands and Empires*, pp. 225–27; Marques, *History of Portugal*, p. 270.

68. M. A. P. Meilink-Roelofsky, *Asian Trade and European Influence in the Indonesian Archepelago, 1500–1630* (The Hague: Nijhoff, 1962), pp. 27, 28, 31, 36; M. C. Ricklefs, *A History of Modern Indonesia* (Basingstoke, UK: Palgrave, 2001), pp. 18, 19.

69. Meilink-Roelofsky, *Asian Trade*, pp. 27, 28, 31, 36; Ricklefs, *Modern Indonesia*, pp. 18, 19.

70. Meilink-Roelofsky, *Asian Trade*, pp. 37–40; Ricklefs, *Modern Indonesia*, pp. 18, 19; Tome Pires, *The Suma Oriental of Tome Pires*, trans. Armando Cortesão (London: Hakluyt Society, 1944), p. 269.

71. Ricklefs, *Modern Indonesia*, p. 21.

72. Meilink-Roelofsky, *Asian Trade*, pp. 122, 123.

73. Parker, *Military Revolution*, p. 83; William S. Atwell, "International Bullion Flows and the Chinese Economy Circa 1530–1650," *Past and Present* 95 (May 1982): 79; Richard Von Glahn, *Fountain of Fortune: Money and Monetary Policy in China* (Berkeley: University of California Press, 1996), p. 114; Paul Kennedy, *The Rise and Fall of Great Powers* (New York: Random House, 1987), pp. 4–7; Edward Farmer et al., *Ming History: An Introductory Guide to Research* (Minneapolis: History Department, University of Minnesota, 1994), p. 20.

74. Steve Shipp, *Macau China* (Jefferson, NC: McFarland & Co., 1997), p. 13; Clive Willis, ed., *China and Macau* (Aldershot, UK: Ashgate, 2002), p. xvi; Farmer, *Ming History*, p. 20.

75. Ina E. Slamet-Velsink, *Emerging Hierarchies* (Leiden: KITLV Press, 1995), pp. 19, 23, 25, 246–48; B. R. Chatterji, *History of Indonesia* (Meerut: Meenakshi Prakashan, 1967), pp. 5, 6; Leonard Y. Andaya, *The World of Maluku* (Honolulu: University of Hawaii Press, 1993), pp. 1, 2, 104; J. Spruyt and J. B. Robertson, *History of Indonesia* (Melbourne: Macmillan, 1973), pp. 6, 8, 10.

76. Bernard H. M. Vlekke, *Nusantara* (The Hague: W. van Hoeve, 1959), p. 75.

77. Meilink-Roelofsky, *Asian Trade*, pp. 93–100.

78. Andaya, *World of Maluku*, p. 115.

79. Ibid., pp. 116, 117.

80. Pires, *Suma Oriental*, pp. 286, 287.

81. Mathew, *Indo-Portuguese Trade*, pp. 164–67, 169.

## CHAPTER TWO: THE SPANISH HEAD WEST

1. J. H. Parry, *Age of Reconnaissance* (New York: New American Library, 1963), p. 134.

2. Henry Kamen, *Spain, 1469–1714* (London; New York: Longman, 1983), pp. 1, 3, 7; John Lynch, *Spain, 1516–1598* (Oxford: Blackwell, 1991), p. 41.

3. Kamen, *Spain*, pp. 17, 19.

4. Pierre Vilar, *Spain: A Brief History* (Oxford; New York: Pergamon Press, 1977), p. 22; Jaime Vicens Vives, *Approaches to the History of Spain* (Berkeley: University of California Press, 1967), p. 56; Kamen, *Spain*, pp. 13, 32; Lynch, *Spain 1516–1598*, pp. 3, 39.

5. J. H. Parry, *The Spanish Seaborne Empire* (New York: Knopf, 1966), p. 43; Parry, *Age of Reconnaissance*, p. 149; António Henrique de Oliveira Marques, *History of Portugal* (New York: Columbia University Press, 1972), p. 222.

6. Marques, *History of Portugal*, p. 222.

7. Kamen, *Spain*, p. 54; Vilar, *Gold and Money*, pp. 61, 63; I. A. Wright, *The Early History of Cuba* (New York: Macmillan, 1916), pp. 6, 7.

8. Carl O. Sauer, *The Early Spanish Main* (Berkeley: University of California Press, 1966), pp. 23–26.

9. Sauer, *Early Spanish Main*, pp. 25–29; Henry Petitjean Roget, "The Taino Vision," in *The Indigenous People of the Caribbean*, ed. Samuel M. Wilson (Gainesville: University of Florida Press, 1997), pp. 171, 173; Irving Rouse, *The Tainos* (New Haven, CT: Yale University Press, 1992), pp. 144, 147.

10. Jaime Vicens Vives, *An Economic History of Spain*, trans. Frances M. Lopez Morellas (Princeton, NJ: Princeton University Press, 1969), pp. 48, 51, 52, 58; Cedric E. Gregory, *A Concise History of Mining* (New York: Pergamon Press, 1980), p. 82; John C. Allan, *Considerations on the*

*Antiquity of Mining in the Iberian Peninsula* (London: Royal Anthropological Institute, 1970), pp. 2, 11, 17–19, 25–35; J. S. Richardson, *The Romans in Spain* (Cambridge, MA: Blackwell, 1996), pp. 3, 4, 42, 73, 128; Leonard A. Curchin, *Roman Spain* (New York: Routledge, 1991), pp. 20, 21, 138; Clair C. Patterson, "Silver Stocks and Losses in Ancient and Medieval Times," *Economic History Review* 25 (May 1972): 231, 232.

11.  Kamen, *Spain*, p. 54; Parry, *Spanish Seaborne Empire*, p. 45; Marques, *History of Portugal*, p. 222.

12.  Kamen, *Spain*, p. 54; Marques, *History of Portugal*, p. 222.

13.  Parry, *Age of Reconnaissance*, pp. 158, 159.

14.  Parry, *Spanish Seaborne Empire*, pp. 47, 48; J. H. Elliott, "The Spanish Conquest and Settlement of America," in *The Cambridge History of Latin America*, vol. 1, ed. Leslie Bethell (Cambridge: Cambridge University Press, 1984), p. 161; Parry, *Age of Reconnaissance*, p. 148.

15.  Clarence Henry Haring, *Trade and Navigation between Spain and the Indies in the Time of the Hapsburgs* (Gloucester, MA: P. Smith, 1964), p. 4.

16.  Vilar, *Gold and Money*, p. 63; Elliott, "The Spanish Conquest," p. 162.

17.  Samuel M. Wilson, "Introduction," in Samuel M. Wilson, *Indigenous People of the Caribbean*, pp. 4–7.

18.  Sauer, *Early Spanish Main*, pp. 62, 77, 78, 81, 197.

19.  Ibid., pp. 62, 91, 92, 198.

20.  Rouse, *The Tainos*, pp. 149, 150; Jose Carlos Mariategui, *Seven Interpretative Essays on Peruvian Reality*, trans. Marjory Urquidi (Austin: University of Texas Press, 1971), p. 39.

21.  Kamen, *Spain*, pp. 55, 56; Parry, *Spanish Seaborne Empire*, pp. 54, 55; Sauer, *Early Spanish Main*, pp. 92, 105, 106.

22.  Parry, *Spanish Seaborne Empire*, pp. 48, 53; Parry, *Age of Reconnaissance*, pp. 142, 153–55; Carlos Prieto, *Mining in the New World* (New York: McGraw-Hill, 1973), pp. 20, 39, 153.

23.  Parry, *Age of Reconnaissance*, p. 155.

24.  Samuel Eliot Morison, *The European Discovery of America—The Southern Voyages, A.D. 1491–1616* (New York: Oxford University Press, 1974), pp. 276–84.

25.  Parry, *Age of Reconnaissance*, p. 157; Morison, *European Discovery*, pp. 289, 292, 293.

26. Rouse, *The Tainos*, pp. 153, 154; Elliott, "The Spanish Conquest," p. 165.

27. Haring, *Trade and Navigation*, pp. 4, 5, 28.

28. Ruth Pike, *Enterprise and Adventure* (Ithaca, NY: Cornell University Press, 1966), pp. 3, 13, 99–101, 147.

29. Parry, *Age of Reconnaissance*, p. 51; Carla Rahn Phillips, "Time and Duration," *American History Review* 92, no. 3 (1987): 536, 545; Haring, *Trade and Navigation*, pp. 102–104.

30. Haring, *Trade and Navigation*, pp. 6–8, 13, 25; John Lynch, *Spain under the Habsburgs* (New York: New York University Press, 1981), p. 161.

31. Haring, *Trade and Navigation*, pp. 23, 24; Lynch, *Spain*, pp. 157–59.

32. Parry, *Spanish Seaborne Empire*, p. 55; Vives, *Economic History*, pp. 243, 268; Pike, *Enterprise and Adventure*, pp. 1–3, 153.

33. Lynch, *Spain*, pp. 161–64.

34. Parry, *Spanish Seaborne Empire*, pp. 56–60, 63; Elliott, "The Spanish Conquest," p. 166.

35. Louis A. Perez Jr., *Cuba* (New York: Oxford University Press, 1988), pp. 15–22; Philip S. Foner, *A History of Cuba* (New York: Integrated Publishing, 1962), p. 19.

36. Perez, *Cuba*, pp. 23–27; Foner, *History of Cuba*, pp. 25, 32; Wright, *Cuba*, pp. 18, 19, 22, 26, 59; Pike, *Enterprise and Adventure*, p. 103; Sauer, *Early Spanish Main*, p. 186.

37. Sauer, *Early Spanish Main*, pp. 25, 154, 158, 186; Perez, *Cuba*, pp. 27, 28, 31; Wright, *Cuba*, pp. 69, 81, 82, 202, 203.

38. Pike, *Enterprise and Adventure*, pp. 128–33; Haring, *Trade and Navigation*, p. 125; Eric Williams, *From Columbus to Castro* (New York: Harper & Row, 1971), pp. 26, 27, 29.

39. J. H. Parry and P. M. Sherlock, *A Short History of the West Indies* (New York: St. Martin's Press, 1987), pp. 10, 16; Parry, *Spanish Seaborne Empire*, pp. 49, 50, 61; Rouse, *The Tainos*, p. 155; Bartolomé de las Casas, *The Devastation of the Indies* (New York: Seabury Press, 1974), p. 21; Elliott, "The Spanish Conquest," p. 164.

40. Parry and Sherlock, *History of the West Indies*, pp. 11, 22, 23; Kamen, *Spain*, pp. 55, 56; Parry, *Spanish Seaborne Empire*, pp. 48, 49; Rouse, *The Tainos*, pp. 152, 156, 157; Las Casas, *Devastation*, pp. 24, 27,

29; Elliott, "The Spanish Conquest," pp. 167, 168; Pike, *Enterprise and Adventure*, pp. 56–58, 90, 169–71.

41. Williams, *From Columbus*, p. 25, Pierre Vilar, *A History of Gold and Money, 1450–1920* (Atlantic Highlands, NJ: Humanities Press, 1976), p. 67.

42. Vives, *Economic History*, p. 316; Leslie Bethell, "A Note on the Native American Population on the Eve of the European Invasions," in Bethell, *Cambridge History of Latin America*, vol. 1, pp. 145, 146.

43. Parry, *Spanish Seaborne Empire*, pp. 50, 51, 61; Vilar, *Gold and Money*, p. 111; Parry, *Age of Reconnaissance*, p. 163; Robert C. West, *Colonial Placer Mining in Colombia* (Baton Rouge: Louisiana State University Press, 1952), pp. 3–6.

44. J. H. Elliott, "Cortés, Velasquez and Charles V," in *Hernán Cortés, Letters from Mexico*, trans. and ed. Anthony Pagden (New Haven, CT: Yale University Press, 1986), p. xiv.

45. Salvador de Madariaga, *Hernán Cortés* (Westport, CT: Greenwood Press, 1979), pp. 19, 22, 25, 29.

46. Ibid., pp. 46, 47, 57–61.

47. Ibid., pp. 66–69, 99; Anthony Pagden, "Translator's Introduction," in Cortés, *Letters from Mexico*, p. l.

48. Madariaga, *Hernán Cortés*, pp. 72–76.

49. Ibid., pp. 76–78.

50. Linda Schele and David Freidel, *A Forest of Kings* (New York: Harper Perennial, 1990), p. 57; Madariaga, *Hernán Cortés*, pp. 79–81.

51. Madariaga, *Hernán Cortés*, pp. 82, 83; Cortés, *Letters from Mexico*, p. 460.

52. Madariaga, *Hernán Cortés*, pp. 88, 89, 91.

53. Ibid., pp. 90, 92, 94–96; Pagden, "Translator's Introduction," p. lii.

54. Madariaga, *Hernán Cortés*, pp. 83, 84; Lawrence Anderson, *The Art of the Silversmith in Mexico, 1519–1936* (New York: Oxford University Press, 1941), pp. 16–18; Wright, *Cuba*, pp. 72–76.

55. Wright, *Cuba*, pp. 77, 85; Morison, *European Discovery*, pp. 517, 518.

56. Madariaga, *Hernán Cortés*, pp. 99, 100.

57. Samuel Salinas Alvarez, *Historia de los Caminos de Mexico* (Mexico: Banco Nacional de Obras y Servicios Publicos, 1994), pp. 132, 138, 140; Bernal Diaz de Castillo, *The Discovery and Conquest of*

*Mexico*, ed. Irving A. Leonard (New York: Grove Press, 1956), p. xxvi; Anderson, *Art of the Silversmith*, pp. 18–20; Wright, *Cuba*, p. 85; Madariaga, *Hernán Cortés*, pp. 99, 100, 113; Elliott, "Cortés," p. xvii.

58. Hugh Thomas, *The Conquest of Mexico* (London: Hutchinson, 1993), pp. 57, 162–65, 171, 172.

59. Elliott, "Cortés," pp. xv–xix; Pagden, "Translator's Introduction," p. xl.

60. Elliott, "Cortés," pp. xix–xx; Pagden, "Translator's Introduction," p. xl; Anderson, *Art of the Silversmith*, pp. 21–27.

61. Elliott, "Cortés," p. xxii; Cortés, *Letters from Mexico*, p. 50.

62. Salinas, *Historia*, pp. 144–52; Cyclone Covey, ed. and trans., *Cabeza de Vaca's Adventures in the Unknown Interior of America* (Albuquerque: University of New Mexico Press, 1983), p. 9; Cortés, *Letters from Mexico*, p. 68.

63. Pagden, *Letters*, pp. 460, 471; Cortés, *Letters from Mexico*, p. 109; Schele and Freidel, *Forest of Kings*, p. 57.

64. Salinas, *Historia*, pp. 76, 84; Pagden, *Letters*, p. 460.

65. Bernal Diaz del Castillo, *The Memoirs of the Conquistador, Containing a True and Full Account of the Discovery and Conquest of Mexico and New Spain*, trans. John Ingram Lockhart (London: Hatchard, 1844), pp. 235–37.

66. Robert C. West, "Aboriginal Metallurgy and Metalworking in Spanish America," in *In Quest of Mineral Wealth*, ed. Alan K. Craig and Robert C. West (Baton Rouge: Geoscience Publications, Louisiana State University, 1994), pp. 9, 10, 12–14; Dorothy Hosler, *The Sounds and Colors of Power* (Cambridge, MA: MIT Press, 1994), pp. 10, 12, 13.

67. West, "Aboriginal Metallurgy," p. 10, Hosler, *Sounds and Colors*, p. 12.

68. Cortés, *Letters from Mexico*, pp. 99, 100; Pagden, *Letters*, p. 470; Anderson, *Art of the Silversmith*, pp. 35–38.

69. Elliott, "Cortés," pp. xxii–xxvi; Wright, *Cuba*, pp. 87, 88; Anderson, *Art of the Silversmith*, pp. 21–27.

70. Salinas, *Historia*, pp. 154–56; Prieto, *Mining in the New World*, p. 154; Madariaga, *Hernán Cortés*, pp. 325, 526; Elliott, "Cortés," p. xxvi; Pagden, *Letters*, p. 475.

71. Anderson, *Art of the Silversmith*, pp. 39, 44, 45; Pagden, *Letters*, p. 478.

72. Bernal Diaz del Castillo, *Discovery and Conquest*, p. xxx; Salinas,

*Historia*, p. 156; Geoffrey Parker, *The Military Revolution* (Cambridge University Press, 1996), p. 103; Elliott, "Cortés," p. xxix; Pagden, *Letters*, pp. 466, 484, 491; Cortés, *Letters from Mexico*, pp. 206–208.

73. Thomas, *Conquest*, p. 568; Salinas, *Historia*, p. 168; Anderson, *Art of the Silversmith*, pp. 47, 54, 55; Pagden, *Letters*, pp. 493, 510.

74. Parry, *Age of Reconnaissance*, pp. 125–27; Pagden, *Letters*, pp. 462, 463.

75. Elliott, "Cortés," pp. xxxi, xxxii; Thomas, *Conquest*, pp. 579–81, 622.

76. Elliott, "Cortés," pp. xxxi–xxxiv; Pagden, *Letters*, pp. 498, 504, 509; Cortés, *Letters from Mexico*, pp. 292, 330–32.

77. Elliott, "Cortés," p. xxxiv; Salinas, *Historia*, pp. 170, 178, 182, 208, 216; Cortés, *Letters from Mexico*, pp. 329, 361, 412; Pagden, *Letters*, p. 511.

78. Luis Weckmann, *The Medieval Heritage of Mexico* (New York: Fordham University Press, 1992), p. 427; Thomas, *Conquest*, p. 560; Bethell, *History of Latin America*, p. 413; Pagden, *Letters*, pp. 495, 496, 522.


## CHAPTER THREE: CONVERGENCE WITH THE HOUSE OF HABSBURG

1. J. H. Parry, *The Spanish Seaborne Empire* (New York: Knopf, 1966), p. 49.

2. Herman Van Der Wee, *The Growth of the Antwerp Market and the European Economy* (Louvain: Biblioteque de l'Universite, 1963), pp. 11, 18, 19; Richard Ehrenberg, *Capital and Finance in the Age of the Renaissance* (New York: Harcourt, 1928), p. 56; Jervis Wegg, *Antwerp* (London: Methuen, 1916), p. 57.

3. Van Der Wee, *Antwerp Market*, pp. 336, 342.

4. Wegg, *Antwerp*, pp. 59, 60, 169; Ehrenberg, *Capital and Finance*, pp. 236–38; Van Der Wee, *Antwerp Market*, p. 123.

5. Van Der Wee, *Antwerp Market*, p. 125.

6. Ibid., pp. 127, 128.

7. K. S. Mathew, *Indo-Portuguese Trade and the Fuggers of Germany* (New Delhi: Manohar, 1997), pp. 6, 7.

8. Ibid., pp. 158–63.

9. Ibid., pp. 165, 166.

10. Ibid., p. 166.

11. Frederic C. Lane, "Venetian Shipping during the Commercial Revolution," in *Crisis and Change in the Venetian Economy in the Sixteenth and Seventeenth Centuries*, ed. Brian Pullan (London: Methuen, 1968), pp. 33, 34.

12. Ehrenberg, *Capital and Finance*, pp. 68, 71, 138, 140, 229; Mathew, *Indo-Portuguese Trade*, pp. 123–25; Jacob Strieder, *Jacob Fugger the Rich*, trans. Mildred L. Hartsough (New York: Adelphi Company, 1931), pp. 62–67, 86, 90, 93.

13. Hermann Kellenbenz, *The Rise of the European Economy* (London: Weidenfeld & Nicolson, 1976), p. 172; Joseph Gies and Frances Gies, *Merchants and Moneymen* (New York: Crowell, 1972), pp. 275, 291; Strieder, *Jacob Fugger*, pp. 42, 162, 165; Victor von Klarwill, ed., *The Fugger News-Letters* (New York: G. P. Putnam, 1924), p. xxi; Mathew, *Indo-Portuguese Trade*, pp. 107, 124, 134, 135.

14. Ehrenberg, *Capital and Finance*, p. 69.

15. Ibid., pp. 70, 71, 74; Gies and Gies, *Merchants and Moneymen*, pp. 282, 286; Mathew, *Indo-Portuguese Trade*, p. 126.

16. John Lynch, *Spain, 1516–1598* (Oxford: Blackwell, 1991), pp. 49, 50.

17. Ehrenberg, *Capital and Finance*, pp. 73, 75.

18. Ibid., pp. 74, 75.

19. Ibid., pp. 75, 76.

20. Ibid., pp. 281, 289, 290; Klarwill, *Fugger News-Letters*, p. 252.

21. Ehrenberg, *Capital and Finance*, pp. 77, 78, 207; Gies and Gies, *Merchants and Moneymen*, p. 287; Strieder, *Jacob Fugger*, pp. 150–52.

22. Lynch, *Spain*, pp. 50, 56–59; James D. Tracy, *Emperor Charles V, Impresario of War* (Cambridge: Cambridge University Press, 2002), p. 99.

23. M. C. Ricklefs, *A History of Modern Indonesia* (Basingstoke, UK: Palgrave, 2001), pp. 21, 22.

24. Stanley Karnow, *In Our Image: America's Empire in the Philippines* (New York: Random House, 1989), pp. 26, 30, 31.

25. J. H. Parry, *Age of Reconnaissance* (New York: New American Library, 1963), p. 159.

26. Karnow, *In Our Image*, pp. 32, 33.

27. Ibid., pp. 33, 34.

28. Ibid., pp. 34, 36.

29. Ibid., p. 37; Parry, *Age of Reconnaissance*, p. 160; Leonard Y. Andaya, *The World of Maluku* (Honolulu: University of Hawaii Press, 1993), pp. 116–18.

30. Karnow, *In Our Image*, p. 37; Parry, *Age of Reconnaissance*, p. 160; Ernest S. Dodge, *Islands and Empires* (Minneapolis: University of Minnesota Press, 1976), p. 233.

31. William L. Schurz, *The Manila Galleon* (New York: E. P. Dutton, 1959), p. 19; Carlos Prieto, *Mining in the New World* (New York: McGraw-Hill, 1973), p. 155; Pedro A. Gagelonia, *The Filipino Historian* (Manila: Feucci, 1970), pp. 31, 32; Salvador de Madariaga, *Hernán Cortés* (Westport, CT: Greenwood Press, 1979), pp. 467–69; Anthony Pagden, trans. and ed., *Hernán Cortés* (New Haven, CT: Yale University Press, 1979), p. 525; António Henrique de Oliveira Marques, *History of Portugal* (New York: Columbia University Press, 1972), p. 261; Mathew, *Indo-Portuguese Trade*, p. 149; Cortés, *Letters from Mexico*, pp. 266, 267.

32. Schurz, *Manila Galleon*, p. 19; Prieto, *Mining in the New World*, p. 155; Gagelonia, *Filipino Historian*, pp. 31, 32; Madariaga, *Hernán Cortés*, pp. 467–69; Pagden, *Letters*, p. 525.

33. Andaya, *World of Maluku*, pp. 116–18; M. A. P. Meilink-Roelofsky, *Asian Trade and European Influence in the Indonesian Arche-pelago, 1500–1630* (The Hague: Nijhoff, 1962), p. 164.

34. Meilink-Roelofsky, *Asian Trade*, pp. 157, 158.

35. Van Der Wee, *Antwerp Market*, pp. 154, 155.

36. Meilink-Roelofsky, *Asian Trade*, pp. 155–57.

37. Hermann Kellenbenz, "Final Remarks," in *Precious Metals in the Age of Expansion*, ed. Hermann Kellenbenz (Stuttgart: Klett-Cotta, 1981), pp. 318, 319.

38. Ibid., p. 317; John U. Nef, "Silver Production in Central Europe," *Journal of Political Economy* 49, no. 4 (1941): 576, 586.

39. Nef, "Silver Production," p. 584; Kellenbenz, *European Economy*, pp. 85, 86, 107, 108.

40. Kellenbenz, *European Economy*, pp. 182–85.

41. Tracy, *Emperor Charles V*, pp. 88, 92, 100.

42. Ibid., pp. 23–25, 39.

43. Hayward Keniston, *Francisco de los Cobos* (Pittsburgh: Univer-

sity of Pittsburgh Press, 1960), pp. 5, 99–104, 117, 142; Rafael Varon Gabar, *Francisco Pizarro and His Brothers* (Norman: University of Oklahoma Press, 1997), p. 37.

44. Tracy, *Emperor Charles V*, p. 92.

45. Ibid., pp. 81–86.

46. Ibid., pp. 64–66, 88, 104; Lynch, *Spain 1516–1598*, pp. 68–72, 78.

47. Tracy, *Emperor Charles V*, p. 92.

48. Kellenbenz, *European Economy*, pp. 182–85; Tracy, *Emperor Charles V*, pp. 64–66, 88, 104, 114, 290.

49. Ehrenberg, *Capital and Finance*, pp. 79–82; Strieder, *Jacob Fugger*, p. 9; Klarwill, *Fugger News-Letters*, p. xxiv.

50. Ehrenberg, *Capital and Finance*, pp. 79–82; Strieder, *Jacob Fugger*, p. 9; Klarwill, *Fugger News-Letters*, p. xxiv.

51. Ehrenberg, *Capital and Finance*, p. 83; Gies and Gies, *Merchants and Moneymen*, p. 290; Strieder, *Jacob Fugger*, p. 127; Mathew, *Indo-Portuguese Trade*, p. 129.

52. Ehrenberg, *Capital and Finance*, pp. 84, 85, 97; Gies and Gies, *Merchants and Moneymen*, p. 296; Strieder, *Jacob Fugger*, p. 130; Mathew, *Indo-Portuguese Trade*, pp. 19, 111.

53. Mathew, *Indo-Portuguese Trade*, pp. 17, 18, 112.

54. Ehrenberg, *Capital and Finance*, pp. 88, 89; Strieder, *Jacob Fugger*, p. 160.

55. Tracy, *Emperor Charles V*, pp. 134, 135; Marques, *History of Portugal*, p. 214.

56. Halil Inalcik, *The Ottoman Empire*, trans. Norman Itzkowitz and Colin Imber (New Rochelle, NY: Orpheus Publishing, 1973), p. 128; Ann Williams, "Mediterranean Conflict," in *Suleyman the Magnificent and His Age*, ed. Metin Kunt and Christine Woodhead (London; New York: Longman, 1995), pp. 44–46; Lynch, *Spain*, p. 95.

57. Tracy, *Emperor Charles V*, pp. 4, 276; Lynch, *Spain*, p. 87.

58. Wegg, *Antwerp*, p. 185; Ehrenberg, *Capital and Finance*, pp. 81, 89, 90; Van Der Wee, *Antwerp Market*, p. 199.

59. Tracy, *Emperor Charles V*, pp. 6, 22, 45, 47; Lynch, *Spain*, pp. 86, 88.

60. Tracy, *Emperor Charles V*, pp. 47, 48.

61. Ibid., pp. 114–17, 124.

62. Ibid., pp. 117–19.

63. Spiros Vryonis Jr., "The Question of the Byzantine Mines," *Speculum* 37, no. 1 (1962): 15, 16; Kellenbenz, "Final Remarks," p. 321; Sevket Pamuk, *A Monetary History of the Ottoman Empire* (Cambridge: Cambridge University Press, 2000), pp. 37, 38; Rhoads Murphey, "Silver Production in Rumelia According to an Official Ottoman Report Circa 1600," *Sudost-Forschungen* 39 (1980): 86, 93; Sina Cirkovic, "The Production of Gold, Silver and Copper," in Kellenbenz, *Precious Metals*, pp. 59, 60.

64. Inalcik, *Ottoman Empire*, p. 35; Williams, "Mediterranean Conflict," pp. 44–46; Kunt and Woodhead, *Suleyman the Magnificent*, p. 24; Stanford J. Shaw, *History of the Ottoman Empire and Modern Turkey* (Cambridge: Cambridge University Press, 1976–77), map; Lynch, *Spain*, p. 95; Geza David, "Administration in Ottoman Europe"; Kunt and Woodhead, *Suleyman the Magnificent*, pp. 74, 76, 78, 80, 81, 84.

65. Williams, "Mediterranean Conflict," pp. 44, 46; Lynch, *Spain*, p. 92.

66. Inalcik, *Ottoman Empire*, pp. 116–18, 128; Halil Sahillioglu, "The Role of International Monetary Movements in Ottoman Monetary History, 1300–1750," in *Precious Metals in the Later Medieval and Early Modern Worlds*, ed. J. F. Richards (Durham, NC: Carolina Academic Press, 1983), p. 275.

67. Tracy, *Emperor Charles V*, pp. 7, 9, 10, 17, 37, 93, 94, 120; Lynch, *Spain*, p. 80; Inalcik, *Ottoman Empire*, p. 36.

68. Tracy, *Emperor Charles V*, pp. 120–22, 275, 278.

69. Ibid., pp. 126–32.

70. Lynch, *Spain*, p. 85; Tracy, *Emperor Charles V*, pp. 290–93.

71. Tracy, *Emperor Charles V*, p. 139.

72. Ibid., pp. 139, 141, 151, 152.

## CHAPTER FOUR: THE GREAT AMERICAN TREASURE HUNTS

1. António Henrique de Oliveira Marques, *History of Portugal* (New York: Columbia University Press, 1972), pp. 252, 253; Samuel Eliot Morison, *The European Discovery of America—The Southern Voyages, A.D. 1491–1616* (New York: Oxford University Press, 1974), p. 299.

2. Morison, *European Discovery*, pp. 287, 288, 300–302; Charles E.

Nowell, "Aleixo Garcia and the White King," *Hispanic American Historical Review* 26, no. 4 (1946): 454.

3. Nowell, "Aleixo Garcia," pp. 455, 456.

4. Ibid., pp. 456–59.

5. Ibid., pp. 455, 459, 460.

6. Ruth Pike, *Enterprise and Adventure* (Ithaca, NY: Cornell University Press, 1966), p. 104.

7. Ibid., pp. 105–109; Morison, *European Discovery*, p. 302.

8. Pike, *Enterprise and Adventure*, pp. 110–13; Nowell, "Aleixo Garcia," pp. 460, 462.

9. Pike, *Enterprise and Adventure*, pp. 113, 116, 117.

10. Robert C. West, "Early Silver Mining in New Spain," in *In Quest of Mineral Wealth*, ed. Alan K. Craig and Robert C. West (Baton Rouge: Geoscience Publications, Louisiana State University, 1994), pp. 120–22; Carlos Prieto, *Mining in the New World* (New York: McGraw-Hill, 1973), p. 21.

11. West, "Early Silver Mining," pp. 122, 123.

12. Harry E. Cross, "South American Bullion Production and Export," in *Precious Metals in the Later Medieval and Early Modern Worlds*, ed. J. F. Richards (Durham, NC: Carolina Academic Press, 1983), p. 410; Murdo J. Macleod, "Spain and America: The Atlantic Trade, 1492–1720," in *The Cambridge History of Latin America*, vol. 1, ed. Leslie Bethell (Cambridge: Cambridge University Press, 1984), pp. 359, 360; Adam Szaszdi, "Preliminary Estimate of Gold and Silver Production in America," in Herman Kellenbenz, ed., *Precious Metals in the Age of Expansion* (Stuttgart: Klett-Cotta, 1981), pp. 175, 176.

13. Henry Kamen, *Spain, 1469–1714* (London; New York: Longman, 1983), p. 56; Lyle McAlister, *Spain and Portugal in the New World* (Minneapolis: University of Minnesota Press, 1984), p. 59; J. H. Parry, *Age of Reconnaissance* (New York: New American Library, 1963), pp. 51, 52, 60; Anthony Pagden, trans. and ed., *Hernán Cortés* (New Haven, CT: Yale University Press, 1979), p. 513.

14. James M. Lockhart, *The Men of Cajamarca* (Austin: Institute of Latin American Studies, University of Texas Press, 1972), pp. 4, 5, 122, 124, 125.

15. Rafael Varon Gabar, *Francisco Pizarro and His Brothers* (Norman: University of Oklahoma Press, 1997), pp. 3–11; Lockhart, *Men of Cajamarca*, pp. 127, 137–42.

16. Gabar, *Francisco Pizarro*, pp. 12–19, 24, 36; Lockhart, *Men of Cajamarca*, pp. 5, 6, 143, 144; Parry, *Age of Reconnaissance*, p. 163.

17. Lockhart, *Men of Cajamarca*, pp. 6, 7; Parry, *Spanish Seaborne Empire*, p. 89; Parry, *Age of Reconnaissance*, p. 171; Prieto, *Mining in the New World*, pp. 27, 154; Gabar, *Francisco Pizarro*, pp. 24, 36.

18. Gabar, *Francisco Pizarro*, p. 37.

19. Ibid., pp. 38, 48, 52.

20. J. H. Elliott, "Cortés, Velasquez and Charles V," in *Hernán Cortés, Letters from Mexico*, trans. and ed. Anthony Pagden (New Haven, CT: Yale University Press, 1986), pp. xxxiv, xxxvi; Pagden, "Translator's Introduction," pp. xli–xliii; Pagden, *Letters*, pp. 436, 490, 523, 524, 525.

21. Salvador de Madariaga, *Hernán Cortés* (Westport, CT: Greenwood Press, 1979), pp. 450–55, 466; Elliott, "Cortés," p. xxxvi; Pagden, "Translator's Introduction," pp. lvii, lix, Pagden, *Letters*, p. 526.

22. Madariaga, *Hernán Cortés*, pp. 456–62.

23. Cortés, *Letters from Mexico*, p. 277.

24. Gabar, *Francisco Pizarro*, pp. 22, 23, 40, 42, 45; Lockhart, *Men of Cajamarca*, pp. 190–95; Parry, *Spanish Seaborne Empire*, p. 89; Parry, *Age of Reconnaissance*, p. 171; Prieto, *Mining in the New World*, pp. 27, 154; Pedro Querezaza and Elizabeth Ferrer, *Potosí* (New York: Americas Society Art Gallery in Association with Fundación BHN, La Paz, 1997), p. 8; Lockhart, *Men of Cajamarca*, pp. 6–8.

25. Lockhart, *Men of Cajamarca*, pp. 9, 10; National Geographic Society, *The Inca* (Washington, May 2002), map and supplement.

26. Geoffrey Hindley, *A History of Roads* (London: P. Davies, 1971), pp. 22, 23; Simon Collier et al., eds., *Cambridge Encyclopedia of Latin America* (Cambridge: Cambridge University Press, 1992), pp. 183, 187.

27. Cedric E. Gregory, *A Concise History of Mining* (New York: Pergamon Press, 1980), pp. 85, 86; Peter L. Bernstein, *The Power of Gold* (New York: John Wiley, 2000), p. 78; Parry, *Spanish Seaborne Empire*, pp. 74–81; National Geographic Society, *The Inca*, map.

28. Izumi Shimada, "Pre-Hispanic Metallurgy and Mining in the Andes," in *In Quest of Mineral Wealth*, ed. Alan K. Craig and Robert C. West (Baton Rouge: Geoscience Publications, Louisiana State University, 1994), pp. 39–44, 46.

29. Robert C. West, "Aboriginal Metallurgy and Metalworking in Spanish America," in Craig and West, *In Quest of Mineral Wealth*, pp. 5,

6; Heather Lechtman, "Issues in Andean Metallurgy," in *Pre-Columbian Metallurgy of South America*, ed. Elizabeth P. Benson (Washington, DC: Dumbarton Oaks Research Library and Collections Trustees for Harvard University, 1979), pp. 28, 29; Dorothy Hosler, *The Sounds and Colors of Power* (Cambridge, MA: MIT Press, 1994), p. 16; Heather Lechtman, "Tradition and Styles in Central Andean Metalworking," in *The Beginning of the Use of Metal Alloys*, ed. Robert Maddin (Cambridge, MA: MIT Press, 1988), pp. 344, 348, 353–55; Shimada, "Pre-Hispanic Metallurgy," pp. 48, 49, 54.

30. West, "Aboriginal Metallurgy," p. 7; Lechtman, "Issues in Andean Metallurgy," pp. 6, 18, 26, 30; Shimada, "Pre-Hispanic Metallurgy," pp. 57–62.

31. Shimada, "Pre-Hispanic Metallurgy," p. 68; Lechtman, "Issues in Andean Metallurgy," pp. 4–18, 25–30; West, "Aboriginal Metallurgy," p. 7.

32. Herbert S. Klein, *Bolivia* (New York: Oxford University Press, 1982), pp. 3–9.

33. Ibid., p. 10; Clair C. Patterson, "Native Copper, Silver and Gold," *American Antiquity* 36, no. 3 (1971): 316.

34. James D. Tracy, *Emperor Charles V, Impresario of War* (Cambridge: Cambridge University Press, 2002), pp. 152, 153.

35. Lockhart, *Men of Cajamarca*, pp. 145–51.

36. Ibid., pp. 9, 10.

37. Ibid., pp. 10, 11.

38. Ibid., p. 11.

39. Ibid., pp. 12, 161.

40. Ibid., pp. 13, 135.

41. Ibid., pp. 13, 14; Parry, *Spanish Seaborne Empire*, p. 90; Gabar, *Francisco Pizarro*, p. 44.

42. Lockhart, *Men of Cajamarca*, p. 14; Parry, *Spanish Seaborne Empire*, p. 90; Gabar, *Francisco Pizarro*, p. 44.

43. Earl J. Hamilton, *American Treasure and the Price Revolution in Spain, 1501–1650* (Cambridge, MA: Harvard University Press, 1934), p. 34.

44. Gabar, *Francisco Pizarro*, pp. 72, 73; Vilar, *Gold and Money*, p. 110; Bailey Diffie, "Estimates of Potosí Production 1545–1555," *Hispanic American Historical Review* 20 (May 1940): 107, 108.

45. Tracy, *Emperor Charles V*, pp. 155, 156.

46. Hayward Keniston, *Francisco de los Cobos* (Pittsburgh: University of Pittsburgh Press, 1960), pp. 105, 106, 109–11, 117, 144.

47. Ibid., pp. 157, 161, 166.

48. National Geographic Society, *The Inca*, map.

49. Gabar, *Francisco Pizarro*, pp. 46, 74; Lockhart, *Men of Cajamarca*, pp. 153, 157–61; Clarence Henry Haring, *Trade and Navigation between Spain and the Indies in the Time of the Hapsburgs* (Gloucester, MA: P. Smith, 1964), pp. 98–101.

50. Klein, *Bolivia*, pp. 23–25; Noble David Cook, *Demographic Collapse: Indian Peru, 1520–1620* (Cambridge: Cambridge University Press, 1981), pp. 54, 70, 114; Sergio Villalobos, *A Short History of Chile* (Santiago de Chile: Editorial Universitaria, 1996), pp. 33, 34; National Geographic Society, *The Inca*, map.

51. Parry, *Spanish Seaborne Empire*, pp. 90–92; Parry, *Age of Reconnaissance*, p. 174; Gabar, *Francisco Pizarro*, pp. 226, 268; Lockhart, *Men of Cajamarca*, pp. 15, 16, 158, 168–72.

52. Lockhart, *Men of Cajamarca*, pp. 175–80.

53. Klein, *Bolivia*, pp. 20–22, 33, 34.

54. Gabar, *Francisco Pizarro*, pp. 226, 229–30; Lockhart, *Men of Cajamarca*, pp. 161, 268–72; Pedro Querezaza and Elizabeth Ferrer, *Potosí* (New York: Americas Society Art Gallery in Association with Fundación BHN, La Paz, 1997), pp. 8, 9; Prieto, *Mining in the New World*, p. 30.

55. Lockhart, *Men of Cajamarca*, pp. 161–65, 180.

56. Felipe Ainsworth Means, "Gonzalo Pizarro and Francisco de Orellana," *Hispanic American Historical Review* 14, no. 3 (1934): 278–81.

57. Ibid., pp. 280–82, 290.

58. Lockhart, *Men of Cajamarca*, pp. 180, 181.

59. Gabar, *Francisco Pizarro*, pp. 48, 70, 84, 94, 98, 99; Lockhart, *Men of Cajamarca*, pp. 16, 181.

60. Jose Ignacio Avellaneda Navas, *The Conquerors of the New Kingdom of Granada* (Albuquerque: University of New Mexico Press, 1995), p. 3.

61. West, "Aboriginal Metallurgy"; Lechtman, "Issues in Andean Metallurgy," pp. 30, 31; Hosler, *Sounds and Colors*, p. 17; Shimada, "Pre-Hispanic Metallurgy," pp. 39, 68.

62. Avellaneda Navas, *The Conquerors*, pp. 4–6.

63. Ibid., pp. 15, 16; Vilar, *Gold and Money*, pp. 107–109.

64. Haring, *Trade and Navigation*, pp. 98–101.

65. Avellaneda Navas, *The Conquerors*, pp. 16–19, 36.

66. Ibid., pp. 6–14, 16.

67. Ibid., pp. 32–36.

68. Querezaza and Ferrer, *Potosí*, p. 8; Parry, *Age of Reconnaissance*, p. 174; Prieto, *Mining in the New World*, pp. 28, 155; National Geographic Society, *The Inca*, map; Robert C. West, *Colonial Placer Mining in Columbia* (Baton Rouge: Louisiana State University Press, 1952), pp. 6, 7; Avellaneda Navas, *The Conquerors*, pp. 37, 38.

69. Lockhart, *Men of Cajamarca*, pp. 126, 127; Pagden, *Letters*, p. 473; Avellaneda Navas, *The Conquerors*, pp. 20, 21.

70. Avellaneda Navas, *The Conquerors*, pp. 39, 40.

71. Ibid., pp. 40, 41.

72. Ibid., pp. 42, 43, 45.

73. Ibid., pp. 46, 47.

74. Marques, *History of Portugal*, pp. 253, 254.

75. Prieto, *Mining in the New World*, pp. 39, 40, 155–57; Marques, *History of Portugal*, pp. 232, 256, 364–68.

76. Prieto, *Mining in the New World*, pp. 26, 27; Lewis Hanke, *The Imperial City of Potosí* (The Hague: Nijhoff, 1956), p. 17.

77. Parry, *Age of Reconnaissance*, p. 173.

78. John Upton Terrell, *Journey into Darkness* (New York: Morrow, 1962), pp. x, xi, 10, 270.

79. David A. Howard, *Conquistador in Chains* (Tuscaloosa: University of Alabama Press, 1997), pp. 39–47.

80. Ibid., pp. 47, 51; Nowell, "Aleixo Garcia," p. 456.

81. Howard, *Conquistador*, pp. 85–88, 110, 111, 115, 121–27, 145–53.

82. Frederick W. Hodge, ed., *Spanish Explorers in the Southern United States, 1528–1543* (New York: Scribner, 1907), p. 285; Salinas, *Historia*, pp. 174, 184, 188; Parry, *Spanish Seaborne Empire*, pp. 88, 89.

83. Parry, *Spanish Seaborne Empire*, p. 88; Samuel Salinas Alvarez, *Historia de los Caminos de Mexico* (Mexico: Banco Nacional de Obras y Servicios Publicos, 1994), p. 188; Hodge, *Spanish Explorers*, p. 285.

84. West, "Early Silver Mining," pp. 119, 120, 123, 124; Cortés, *Letters from Mexico*, pp. 324, 445.

85. Peter J. Bakewell, *Silver Mining and Society in Colonial Mexico* (Cambridge: Cambridge University Press, 1971), p. 1; Cortés, *Letters from Mexico*, p. 446.

86. Madariaga, *Hernán Cortés*, pp. 464, 465, 469, 470.

87. Madariaga, *Hernán Cortés*, pp. 470–73.

88. Hodge, *Spanish Explorers*, p. 275.

89. Fr. Angelico Chavez, *Coronado's Friars* (Washington, DC: Academy of American Franciscan Historians, 1968), p. 34.

90. Hodge, *Spanish Explorers*, pp. 311–37; Chavez, *Coronado's Friars*, p. 9; Richard Flint and Shirley C. Flint, eds., *The Coronado Expedition to Tierra Nueva* (Niwot: University Press of Colorado, 1997), pp. 2, 8, 293, 297, 302.

91. Pike, *Enterprise and Adventure*, pp. 117, 118.

92. Ibid., pp. 123–26.

## CHAPTER FIVE: AN UNLIKELY BANKRUPTCY

1. Ann Williams, "Mediterranean Conflict," in *Suleyman the Magnificent and His Age*, ed. Metin Kunt and Christine Woodhead (London; New York: Longman, 1995), p. 47; Salih Ozbaran, "Ottoman Naval Policy in the South," in *Suleyman the Magnificent and His Age*, ed. Metin Kunt and Christine Woodhead (London; New York: Longman, 1995), pp. 60, 61; James D. Tracy, *Emperor Charles V, Impresario of War* (Cambridge: Cambridge University Press, 2002), pp. 7, 37; Halil Inalcik, *The Ottoman Empire*, trans. Norman Itzkowitz and Colin Imber (New Rochelle, NY: Orpheus Publishing, 1973), p. 38.

2. Tracy, *Emperor Charles V*, pp. 141, 144, 145.

3. Ibid., pp. 146–54, 157; John Lynch, *Spain, 1516–1598* (Oxford: Blackwell, 1991), p. 96.

4. Tracy, *Emperor Charles V*, pp. 152–55; Hayward Keniston, *Francisco de los Cobos* (Pittsburgh: University of Pittsburgh Press, 1960), pp. 166, 168, 171, 173–76, 181.

5. Tracy, *Emperor Charles V*, pp. 177, 78; Keniston, *Francisco de los Cobos*, pp. 185, 186.

6. Richard Ehrenberg, *Capital and Finance in the Age of the Renaissance* (New York: Harcourt, 1928), pp. 95, 96; Herman Van Der Wee, *The*

*Growth of the Antwerp Market and the European Economy* (Louvain: Biblioteque de l'Universite, 1963), pp. 178, 201.

7. Williams, "Mediterranean Conflict," p. 48; John F. Guilmartin Jr., *Galleons and Galleys* (London: Cassell, 2002), p. 130.

8. Lynch, *Spain*, pp. 96, 97; António Henrique de Oliveira Marques, *History of Portugal* (New York: Columbia University Press, 1972), p. 215; Tracy, *Emperor Charles V*, p. 166; Williams, "Mediterranean Conflict," p. 48; Guilmartin, *Galleons and Galleys*, p. 132.

9. Tracy, *Emperor Charles V*, pp. 170, 171; Inalcik, *Ottoman Empire*, pp. 36, 37.

10. Tracy, *Emperor Charles V*, pp. 180, 181.

11. Salvador de Madariaga, *Hernán Cortés* (Westport, CT: Greenwood Press, 1979), pp. 475–77.

12. Tracy, *Emperor Charles V*, pp. 171, 176, 179; Madariaga, *Hernán Cortés*, p. 477.

13. Madariaga, *Hernán Cortés*, pp. 480–84; Irving A. Leonard, "Cortés's Remains—and a Document," *Hispanic American Historical Review* 29, no. 1 (1949): 55–59.

14. Lynch, *Spain*, pp. 65, 78.

15. Tracy, *Emperor Charles V*, pp. 290–96.

16. Lynch, *Spain*, pp. 90–92; Tracy, *Emperor Charles V*, pp. 184, 188, 190.

17. Tracy, *Emperor Charles V*, pp. 191–94.

18. Ibid., pp. 187, 196–202, 226.

19. Ibid., pp. 202, 203.

20. Jonathan I. Israel, *The Dutch Republic* (Oxford: Clarendon Press, 1995), pp. 130–33; Tracy, *Emperor Charles V*, pp. 254–56, 271–73.

21. Tracy, *Emperor Charles V*, pp. 91, 92, 103, 258, 271–73.

22. Ibid., pp. 266–68.

23. Ehrenberg, *Capital and Finance*, pp. 42–45; Tracy, *Emperor Charles V*, p. 167; Williams, "Mediterranean Conflict," p. 49; Van Der Wee, *Antwerp Market*, pp. 110, 343–48, 352, 355; Jervis Wegg, *Antwerp* (London: Methuen, 1916), p. 292.

24. Tracy, *Emperor Charles V*, p. 6; Lynch, *Spain*, pp. 99, 101.

25. Tracy, *Emperor Charles V*, pp. 206, 207.

26. Ibid., pp. 214, 215, 221–25.

27. Ibid., pp. 225–29, 248.

28. Robert C. West, "Early Silver Mining in New Spain," in *In Quest of Mineral Wealth*, ed. Alan K. Craig and Robert C. West (Baton Rouge: Geoscience Publications, Louisiana State University, 1994), pp. 120–22; Carlos Prieto, *Mining in the New World* (New York: McGraw-Hill, 1973), pp. 131, 132.

29. James M. Lockhart, *The Men of Cajamarca* (Austin: Institute of Latin American Studies, University of Texas Press, 1972), pp. 182–88.

30. Laura Escobar, "Potosí," in Pedro Querezaza and Elizabeth Ferrer, *Potosí* (New York: Americas Society Art Gallery in Association with Fundación BHN, La Paz, 1997), p. 14; Prieto, *Mining in the New World*, pp. 30, 31; Lewis Hanke, *The Imperial City of Potosí* (The Hague: Nijhoff, 1956), p. 1; Herbert S. Klein, *Bolivia* (New York: Oxford University Press, 1982), p. 34; Peter J. Bakewell, *Silver and Entrepreneurship in Seventeenth Century Potosí* (Albuquerque: University of New Mexico Press, 1988), pp. 18, 19.

31. West, "Early Silver Mining," p. 132; Peter J. Bakewell, "Registered Silver Production in the Potosí District, 1550–1735," *Jahrbuch für Geschichte von Staat, Wirtschaft und Gesellschaft Lateinamerikas* 12 (1975): 92, 93; Prieto, *Mining in the New World*, pp. 31, 33; Adam Szaszdi, "Preliminary Estimate of Gold and Silver Production in America," in *Precious Metals in the Later Medieval and Early Modern Worlds*, ed. J. F. Richards (Durham, NC: Carolina Academic Press, 1983), p. 216; Santiago Ramirez, *Noticia Histórica de la Riqueza Mineria de México* (Mexico: Oficina Tipográfica de la Secretaría de Fomento, 1884), pp. 100–102; Alan K. Craig, "Spanish Colonial Silver Beneficiation at Potosí," in Craig and West, *In Quest of Mineral Wealth*, pp. 274, 281.

32. Lockhart, *Men of Cajamarca*, pp. 16, 182–88.

33. Ibid., pp. 182–88.

34. Rafael Varon Gabar, *Francisco Pizarro and His Brothers* (Norman: University of Oklahoma Press, 1997), pp. 285, 288, 294, 295; Hayward Keniston, *Francisco de los Cobos* (Pittsburgh: University of Pittsburgh Press, 1960), p. 308.

35. West, "Early Silver Mining," p. 132; Bakewell, "Registered Silver," pp. 92, 93; Prieto, *Mining in the New World*, pp. 31, 33; Szaszdi, "Preliminary Estimate," p. 216; Ramirez, *Noticia Histórica*, pp. 100–102.

36. Harry E. Cross, "South American Bullion Production and Export," in Richards, *Precious Metals*, pp. 401, 410; Szaszdi, "Preliminary Estimate," pp. 175, 176; Tracy, *Emperor Charles V*, pp. 104, 106.

37. Ehrenberg, *Capital and Finance*, pp. 97, 100, 140, 144, 222–23; Wegg, *Antwerp*, p. 298; K. S. Mathew, *Indo-Portuguese Trade and the Fuggers of Germany* (New Delhi: Manohar, 1997), p. 19.

38. Ehrenberg, *Capital and Finance*, p. 245.

39. Van Der Wee, *Antwerp Market*, pp. 165, 362, 363.

40. Ibid., pp. 156–59.

41. Wegg, *Antwerp*, pp. 291, 292, 323–25; Van Der Wee, *Antwerp Market*, pp. 121, 187.

42. Wegg, *Antwerp*, pp. 98, 197, 284, 305, 328–31; Van Der Wee, *Antwerp Market*, pp. 163, 179–82, 184, 186.

43. Israel, *Dutch Republic*, pp. 14, 15, 19, 29–34; Tracy, *Emperor Charles V*, pp. 11, 69–71.

44. Israel, *Dutch Republic*, pp. 55–57, 62, 64, 113, 116, 130–33.

45. Wegg, *Antwerp*, p. 295; Ehrenberg, *Capital and Finance*, pp. 101, 102; Victor von Klarwill, ed., *The Fugger News-Letters* (New York: G. P. Putnam, 1924), p. xxv.

46. Ehrenberg, *Capital and Finance*, pp. 103, 104.

47. Ibid., pp. 104, 106; Pierre Vilar, *A History of Gold and Money, 1450–1920* (Atlantic Highlands, NJ: Humanities Press, 1976), pp. 145, 146, 161.

48. Vilar, *Gold and Money*, pp. 145, 146.

49. Ibid., p. 148; Tracy, *Emperor Charles V*, pp. 242, 243, 300, 301.

50. Wegg, *Antwerp*, pp. 204, 206, 229, 230, 260; Hermann Kellenbenz, *The Rise of the European Economy* (London: Weidenfeld & Nicolson, 1976), p. 186.

51. Geoffrey Parker, *The Military Revolution* (Cambridge University Press, 1996), pp. 17–24, 45, 61.

52. Ibid., pp. 9–11, 17–24; Tracy, *Emperor Charles V*, pp. 30, 31.

53. Parker, *Military Revolution*, p. 12.

54. Tracy, *Emperor Charles V*, pp. 32, 33, 111–13.

55. Ibid., pp. 232–34, 268; Lynch, *Spain*, pp. 102–104; Inalcik, *Ottoman Empire*, p. 37.

56. Lynch, *Spain*, pp. 102–104; Tracy, *Emperor Charles V*, pp. 234–38.

57. Ehrenberg, *Capital and Finance*, pp. 107–109.

58. Tracy, *Emperor Charles V*, pp. 239–42; Lynch, *Spain*, pp. 102–104.

59. Tracy, *Emperor Charles V*, pp. 109, 111.

60. Ibid., pp. 241–46, 302.

61. Ibid., pp. 245, 246; Lynch, *Spain*, p. 81.

62. Lynch, *Spain*, p. 81; Tracy, *Emperor Charles V*, pp. 246, 251.

63. Ehrenberg, *Capital and Finance*, pp. 105, 112, 180, 255; Wegg, *Antwerp*, pp. 281, 282, 309; Mathew, *Indo-Portuguese Trade*, p. 133; Lynch, *Spain*, pp. 83, 106.

64. Parker, *Military Revolution*, p. 63; Vilar, *Gold and Money*, p. 148.

65. Tracy, *Emperor Charles V*, pp. 269, 70.

66. Ehrenberg, *Capital and Finance*, pp. 113, 114; Wegg, *Antwerp*, p. 332.

67. Ehrenberg, *Capital and Finance*, pp. 114, 115, 280; Wegg, *Antwerp*, p. 332.

68. Parker, *Military Revolution*, p. 63; Van Der Wee, *Antwerp Market*, pp. 213–15, 221; Tracy, *Emperor Charles V*, p. 101.

69. Ehrenberg, *Capital and Finance*, pp. 111, 116, 149, 226.

70. Ehrenberg, *Capital and Finance*, pp. 117, 121, 128; Klarwill, *Fugger News-Letters*, pp. xxv, xxvi, xxvii.

71. Lynch, *Spain*, pp. 177, 178.


## CHAPTER SIX: MINING REVOLUTION IN SPANISH AMERICA

1. Harry E. Cross, "South American Bullion Production and Export," in *Precious Metals in the Later Medieval and Early Modern Worlds*, ed. J. F. Richards (Durham, NC: Carolina Academic Press, 1983), pp. 401, 410; John U. Nef, "Silver Production in Central Europe," *Journal of Political Economy* 49, no. 4 (1941): 576, 578; Pierre Vilar, *A History of Gold and Money, 1450–1920* (Atlantic Highlands, NJ: Humanities Press, 1976), pp. 29, 30.

2. Jaime Vicens Vives, *Approaches to the History of Spain* (Berkeley: University of California Press, 1967), p. 317; Lawrence Anderson, *The Art of the Silversmith in Mexico, 1519–1936* (New York: Oxford University Press, 1941), p. 124; David C. Goodman, *Power and Penury* (Cambridge: Cambridge University Press, 1988), pp. 164–69.

3. Robert C. West, "Early Silver Mining in New Spain," in *In Quest of Mineral Wealth*, ed. Alan K. Craig and Robert C. West (Baton Rouge: Geoscience Publications, Louisiana State University, 1994), p. 126; Carlos Prieto, *Mining in the New World* (New York: McGraw-Hill, 1973), p. 24.

4. West, "Early Silver Mining," pp. 126, 127, 133; Prieto, *Mining in the New World*, p. 24; Ramon Sanchez Flores, "Technology of Mining in Colonial Mexico," in Craig and West, *In Quest of Mineral Wealth*, p. 139.

5. Otis E. Young Jr., "Black Legends and Silver Mountains," in Craig and West, *In Quest of Mineral Wealth*, pp. 111–14; West, "Early Silver Mining," p. 127; Flores, "Technology of Mining," pp. 140, 144, 145.

6. D. M. Dunlop, "Sources of Gold and Silver in Islam According to al-Hamdani," *Studia Islamica* 8 (1957): 47, 48.

7. Young, "Black Legends," p. 114; West, "Early Silver Mining," pp. 127, 128; Alan Probert, "Bartolomé de Medina: The Patio Process and the Sixteenth Century Silver Crisis," *Journal of the West* 8:95, 96; Goodman, *Power and Penury*, p. 175.

8. Pierre Vilar, *A History of Gold and Money, 1450–1920* (Atlantic Highlands, NJ: Humanities Press, 1976), p. 116; West, "Early Silver Mining," pp. 127, 128.

9. West, "Early Silver Mining," pp. 131, 132.

10. Peter J. Bakewell, *Silver Mining and Society in Colonial Mexico* (Cambridge: Cambridge University Press, 1971), pp. 1, 4, 6, 9; Samuel Salinas Alvarez, *Historia de los Caminos de Mexico* (Mexico: Banco Nacional de Obras y Servicios Publicos, 1994), p. 220.

11. Bakewell, *Silver Mining*, pp. 13–15; West, "Early Silver Mining," p. 132.

12. Anderson, *Art of the Silversmith*, pp. 122, 123, 277, 279.

13. Ian Blanchard, "England and the International Bullion Crisis," in *Precious Metals in the Age of Expansion*, ed. Hermann Kellenbenz (Stuttgart: Klett-Cotta, 1981), pp. 91, 108, 109.

14. West, "Early Silver Mining," pp. 131, 134.

15. Translation of Georgius Agricola, *De Re Metallica*, trans. Herbert C. Hoover and Lou Henry Hoover (New York: Dover Publications, 1950), pp. 297–300, 607–15.

16. John Percy, *Metallurgy, Silver and Gold* (London: John Murray, 1880), pp. 559–62.

17. Adam Szaszdi, "Preliminary Estimate of Gold and Silver Production in America," in Richards, *Precious Metals*, pp. 189, 194, 195; Percy, *Metallurgy*, p. 562; Blanchard, "England," pp. 89, 105–10.

18. Ruth Pike, *Aristocrats and Traders* (Ithaca, NY: Cornell University Press, 1972), pp. 6, 8, 11; James D. Tracy, *Emperor Charles V, Impresario of War* (Cambridge: Cambridge University Press, 2002), pp. 96, 97, 100.

19. Probert, "Bartolomé de Medina," p. 94.

20. Ibid., p. 95.

21. Ibid., pp. 96, 99.

22. Ibid., pp. 92, 99, 100.

23. Ibid., pp. 100, 101.

24. Ibid., p. 102.

25. Ibid., pp. 102–104; Joceyln Houseman and David A. Johnson, "Spanish American Silver," *Interdisciplinary Science Reviews* 16, no. 3 (1991): 246; David A. Johnson and Karl Whittle, "The Chemistry of the Hispanic-American Amalgamation Process," *Journal of the Royal Society of Chemistry, Dalton Trans.* (1999): 4239–42.

26. Probert, "Bartolomé de Medina," pp. 104, 105; Goodman, *Power and Penury*, p. 175.

27. Bakewell, *Silver Mining*, pp. 134–40.

28. Prieto, *Mining in the New World*, pp. 78, 80; Goodman, *Power and Penury*, pp. 175, 76; Probert, "Bartolomé de Medina," p. 102; Robert C. West, "The Mining Community in Northern New Spain," *Ibero-Americana* 30 (1949): 32, 33; Houseman and Johnson, "Spanish American Silver," p. 246; Szaszdi, "Preliminary Estimate," p. 189.

29. Goodman, *Power and Penury*, pp. 195, 196, 202.

30. Probert, "Bartolomé de Medina," pp. 91, 107, 108.

31. Ibid., pp. 108–10, 114.

32. Ibid., pp. 113, 114, 117, 120, 121.

33. Bakewell, *Silver Mining*, pp. 130, 131, 137–43.

34. Philip Wayne Powell, *Soldiers, Indians and Silver* (Berkeley: University of California Press, 1952), p. 14; West, "Mining Community," pp. 47, 51; Bakewell, *Silver Mining*, p. 15.

35. Bakewell, *Silver Mining*, pp. 41, 42, 142.

36. Ibid., pp. 44, 50, 51.

37. Ibid., pp. 9–12.

38. Ibid., p. 241, Prieto, *Mining in the New World*, p. 160.

39. Goodman, *Power and Penury*, pp. 196–98.

40. Salinas, *Historia*, pp. 218, 222; Powell, *Soldiers*, pp. 18, 22; Ross Hassig, *Trade, Tribute and Transportation* (Norman: University of Oklahoma Press, 1985), pp. 175, 176.

41. Hassig, *Trade*, pp. 161, 164–66, 193, 196; David G. Ringrose, "Carting in the Hispanic World," *Hispanic American Historical Review* 50, no. 1 (1970): 34–38; Max L. Moorhead, "Spanish Transportation in the Southwest," *New Mexico Historical Review* 32 (1957): 109–11; Bakewell, *Silver Mining*, p. 22; West, "Mining Community," pp. 77, 87.

42. Bakewell, *Silver Mining*, p. 241; Prieto, *Mining in the New World*, p. 160; Szaszdi, "Preliminary Estimate," pp. 162, 188; Bakewell, "Registered Silver," pp. 132–35.

43. Vives, *Economic History*, p. 322; West, "Mining Community," pp. 25, 47, 53, 57, 66, 72; Woodrow W. Borah, "New Spain's Century of Depression," *Ibero-Americana* 35 (1951): 19, 21, 25, 32, 36, 37.

44. Robert C. West, *Colonial Placer Mining in Colombia* (Baton Rouge: Louisiana State University Press, 1952), pp. 7–11, 22.

45. Prieto, *Mining in the New World*, pp. 29, 30, 60; Vilar, *Gold and Silver*, p. 113; Szaszdi, "Preliminary Estimate," pp. 178, 185.

46. Bakewell, *Silver and Entrepreneurship*, pp. 18, 19; Szaszdi, "Preliminary Estimate," pp. 213–15.

47. Alan K. Craig, "Spanish Colonial Silver Beneficiation at Potosí," in Craig and West, *In Quest of Mineral Wealth*, pp. 274–81; Bakewell, *Silver and Entrepreneurship*, p. 20.

48. West, "Mining Community," pp. 17, 23; D. A. Brading and Harry E. Cross, "Colonial Silver Mining: Mexico and Peru," *Hispanic American Historical Review* 52, no. 4 (1972): 549; Houseman and Johnson, "Spanish American Silver," p. 245.

49. Gwendolin B. Cobb, "Supply and Transportation for the Potosí Mines," *Hispanic American Historical Review* 29, no. 1 (1949): 25, 26, 34, 35.

50. Pedro Querezaza and Elizabeth Ferrer, *Potosí* (New York: Americas Society Art Gallery in Association with Fundación BHN, La Paz, 1997), p. 8; Prieto, *Mining in the New World*, pp. 29, 30; Herbert S. Klein, *Bolivia* (New York: Oxford University Press, 1982), pp. 35, 36.

51. Cobb, "Supply and Transportation," pp. 29, 34–36.

52. Ibid., pp. 26, 27, 29.

53. Prieto, *Mining in the New World*, pp. 79, 81; Szaszdi, "Production Estimate," pp. 195–97.

54. Szaszdi, "Production Estimate," pp. 195–97; Prieto, *Mining in the New World*, pp. 79, 81, 158.

55. Goodman, *Power and Penury*, p. 176.

56. Peter J. Bakewell, "Registered Silver," pp. 92, 93.

57. Szaszdi, "Production Estimate," pp. 197, 198, 215, 217; Carlos Sempat Assadourian, "The Colonial Economy," *Journal of Latin American Studies* 24, Supplement (1992): 58–60; Houseman and Johnson, "Spanish American Silver," p. 245; Percy, *Metallurgy*, p. 650; Vilar, *Gold and Silver*, pp. 119–22; Craig, "Spanish Colonial Silver," pp. 275, 282.

58. Peter J. Bakewell, "Technological Change in Potosí: The Silver Boom of the 1570s," *Jahrbuch für Geschichte von Staat, Wirtschaft und Gesellschaft Lateinamerikas* 14 (1977): 59–61.

59. Bakewell, "Technological Change," pp. 62, 63; Assadourian, "The Colonial Economy," pp. 58–60; Houseman and Johnson, "Spanish American Silver," p. 245.

60. Bakewell, "Technological Change," p. 64; National Geographic Society, *The Inca* (Washington: May 2002), map.

61. Goodman, *Power and Penury*, pp. 183, 184.

62. Szaszdi, "Production Estimate," pp. 199, 201; Klein, *Bolivia*, pp. 41, 42.

63. Craig, "Spanish Colonial Silver," pp. 275–79, 281; Bakewell, *Silver and Entrepreneurship*, p. 20; Szaszdi, "Production Estimate," pp. 199, 200; Goodman, *Power and Penury*, pp. 176, 177; Leonard A. Curchin, *Roman Spain* (New York: Routledge, 1991), p. 137.

64. Bakewell, "Technological Change," pp. 67–76.

65. Ibid., pp. 67, 75, 92–95.

## CHAPTER SEVEN: CASH FLOW SQUEEZE
## IN THE SPANISH NETHERLANDS

1. John Lynch, *Spain, 1516–1598* (Oxford: Blackwell, 1991), pp. 197–99, 202.

2. David C. Goodman, *Power and Penury* (Cambridge: Cambridge University Press, 1988), pp. 151, 152, 163.

3. Ibid., pp. 152–61.

4. Hermann Kellenbenz, "Final Remarks," in *Precious Metals in the Age of Expansion*, ed. Hermann Kellenbenz (Stuttgart: Klett-Cotta, 1981), pp. 321, 322; Goodman, *Power and Penury*, pp. 164–69, 172.

5. Lynch, *Spain*, pp. 183, 207, 255, 256, 269–72.

6. David R. Ringrose, *Madrid and the Spanish Economy* (Berkeley: University of California Press, 1983), pp. 7, 312, 313, 331.

7. Lynch, *Spain*, pp. 143, 144.

8. Jaime Vicens Vives, *Approaches to the History of Spain* (Berkeley: University of California Press, 1967), pp. 97, 98, 100, 103; Lynch, *Spain*, pp. 116, 117, 123–25.

9. Lynch, *Spain*, pp. 110–14.

10. Ruth Pike, *Enterprise and Adventure* (Ithaca, NY: Cornell University Press, 1966), pp. 48, 168.

11. Ibid., pp. 65–68, 79, 80.

12. Ibid., pp. 84–85; John Elliott, "Ottoman-Habsburg Rivalry," in *Suleyman the Second and His Time*, Halil Inalcik and Cemal Kafadar (Istanbul: Isis Press, 1993), pp. 157, 158.

13. Ann Williams, "Mediterranean Conflict," in *Suleyman the Magnificent and His Age*, ed. Metin Kunt and Christine Woodhead (London; New York: Longman, 1995), pp. 49, 51.

14. Lynch, *Spain*, pp. 236–38; Williams, "Mediterranean Conflict," pp. 51–53; Goodman, *Power and Penury*, pp. 154, 155; John F. Guilmartin Jr., *Galleons and Galleys* (London: Cassell, 2002), p. 215.

15. Rhoads Murphey, "Silver Production in Rumelia According to an Official Ottoman Report Circa 1600," *Sudost-Forschungen* 39 (1980): 78, 79, 86; Salih Ozbaran, "Ottoman Naval Policy in the South," in Kunt and Woodhead, *Suleyman the Magnificent*, p. 68; Frederic C. Lane, "Venetian Shipping during the Commercial Revolution," in *Crisis and Change in the Venetian Economy in the Sixteenth and Seventeenth Centuries*, ed. Brian Pullan (London: Methuen, 1968), pp. 34, 36.

16. Williams, "Mediterranean Conflict," pp. 51–53; Guilmartin, *Galleons and Galleys*, p. 133; Lynch, *Spain*, pp. 225, 226, 234; Halil Sahillioglu, "The Role of International Monetary Movements in Ottoman Monetary History, 1300–1750," in *Precious Metals in the Later Medieval and Early Modern Worlds*, ed. J. F. Richards (Durham, NC: Carolina Academic Press, 1983), p. 280; Ozbaran, "Ottoman Naval Policy," pp.

217–18; Sevket Pamuk, *A Monetary History of the Ottoman Empire* (Cambridge: Cambridge University Press, 2000), p. 133; Lane, "Venetian Shipping," pp. 50–54; K. S. Mathew, *Indo-Portuguese Trade and the Fuggers of Germany* (New Delhi: Manohar, 1997), p. 230.

17. Lynch, *Spain*, pp. 219, 227–32, 237, 38.

18. Carlos Sempat Assadourian, "The Colonial Economy," *Journal of Latin American Studies* 24, Supplement (1992): pp. 56, 57, 65; Lynch, *Spain*, pp. 240, 241; Halil Inalcik, *The Ottoman Empire*, trans. Norman Itzkowitz and Colin Imber (New Rochelle, NY: Orpheus Publishing, 1973), p. 137.

19. Lynch, *Spain*, pp. 241, 242; Guilmartin, *Galleons and Galleys*, p. 136.

20. Geoffrey Parker, *The Military Revolution* (Cambridge; New York: Cambridge University Press, 1996), pp. 88, 89; Goodman, *Power and Penury*, p. 96; Lynch, *Spain*, pp. 243–45; Guilmartin, *Galleons and Galleys*, pp. 141, 147, 148; Harold Bloom, "Introduction," in Miguel de Cervantes, *Don Quixote*, trans. Edith Grossman (New York: Ecco, 2003), p. xxv.

21. Lynch, *Spain*, pp. 245, 246; Andrew C. Hess, "The Battle of Lepanto and Its Place in Mediterranean History," *Past and Present* 57 (1972): 53, 54, 61; Guilmartin, *Galleons and Galleys*, p. 149.

22. Hess, "The Battle of Lepanto," pp. 63–66; Inalcik, *Ottoman Empire*, p. 41.

23. Jonathan I. Israel, *The Dutch Republic* (Oxford: Clarendon Press, 1995), pp. 136–38.

24. Lynch, *Spain*, pp. 288–90.

25. Ibid., pp. 290, 291.

26. Israel, *Dutch Republic*, pp. 41, 44, 48, 52, 74.

27. Geert Mak, *Amsterdam* (London: Harvill Press, 1999), pp. 59, 61–64, 67, 69; Herman Van Der Wee, *The Growth of the Antwerp Market and the European Economy* (Louvain: Biblioteque de l'Universite, 1963), pp. 152, 153; Lynch, *Spain*, pp. 91, 94, 104, 293–95; Israel, *Dutch Republic*, pp. 79, 82–84, 93, 100.

28. James D. Tracy, *Emperor Charles V, Impresario of War* (Cambridge: Cambridge University Press, 2002), pp. 308, 309; Guido Marnef, *Antwerp in the Age of Reformation*, trans. J. C. Grayson (Baltimore: Johns Hopkins University Press, 1996), p. 5.

29. Van Der Wee, *Antwerp Market*, pp. 222–27, 363–68; Jonathan I.

Israel, *Dutch Primacy in World Trade* (Oxford: Clarendon Press, 1989), p. 29.

30. Van Der Wee, *Antwerp Market*, pp. 233, 234, 236; Israel, *Dutch Republic*, pp. 145, 149, 151, 153.

31. Israel, *Dutch Republic*, pp. 139, 155–57, 161; Lynch, *Spain*, pp. 296–98; Geoffrey Parker, *The Army of Flanders and the Spanish Road, 1567–1659* (Cambridge: Cambridge University Press, 1972), pp. 106–108, 131, 132.

32. Israel, *Dutch Republic*, pp. 162–66, 181; Victor von Klarwill, ed., *The Fugger News-Letters* (New York: G. P. Putnam, 1924), p. 3; Lynch, *Spain*, pp. 298, 299; Parker, *Army of Flanders*, pp. 139, 140.

33. Parker, *Army of Flanders*, p. 140.

34. Israel, *Dutch Republic*, pp. 170–75, 178–79, 181–84.

35. Parker, *Army of Flanders*, pp. x, 3–5, 13, 25, 27.

36. Van Der Wee, *Antwerp Market*, p. 241; Lynch, *Spain*, pp. 139, 140, 302.

37. Van Der Wee, *Antwerp Market*, pp. 228, 237, 238; Lynch, *Spain*, pp. 152–54.

38. Clarence Henry Haring, *Trade and Navigation between Spain and the Indies in the Time of the Hapsburgs* (Gloucester, MA: P. Smith, 1964), pp. 69–77; Lynch, *Spain*, pp. 165–70.

39. Haring, *Trade and Navigation*, p. 116; J. H. Parry, *Age of Reconnaissance* (New York: New American Library, 1963), pp. 183, 184; Guilmartin, *Galleons and Galleys*, pp. 102, 103.

40. Lynch, *Spain*, pp. 311, 312, 315, 318.

41. Vives, *Economic History*, pp. 325–28; D. A. Brading and Harry E. Cross, "Colonial Silver Mining: Mexico and Peru," *Hispanic American Historical Review* 52, no. 4 (1972): 574.

42. Ruth Pike, *Aristocrats and Traders* (Ithaca, NY: Cornell University Press, 1972), pp. 17–19; Earl J. Hamilton, *American Treasure and the Price Revolution in Spain, 1501–1650* (Cambridge, MA: Harvard University Press, 1934), pp. 23, 24, 64.

43. Lynch, *Spain*, pp. 150–52, 155.

44. Parker, *Army of Flanders*, pp. 10, 14, 16, 134.

45. Harry E. Cross, "South American Bullion Production and Export," in Richards, *Precious Metals*, pp. 404, 405; James Conklin, "The Theory of Sovereign Debt and Spain Under Philip II," *Journal of Political Economy* 106, no. 3 (1998): 508; Parker, *Army of Flanders*, pp. 135, 146.

46. Parker, *Army of Flanders*, pp. 232–34.

47. Ibid., pp. 26, 27, 136, 141.

48. Ibid., p. 137.

49. Ibid., pp. 145, 148; Conklin, "Theory of Sovereign Debt," pp. 489, 490.

50. Parker, *Army of Flanders*, pp. 152, 153; Conklin, "Theory of Sovereign Debt," pp. 491, 500–502.

51. Parker, *Army of Flanders*, pp. 232–34; Lynch, *Spain*, pp. 300–302; Tracy, *Emperor Charles V*, p. 303; Parker, *Military Revolution*, p. 59.

52. A. W. Lovett, "The Castilian Bankruptcy of 1575," *Historical Journal* 23, no. 4 (1980): 901–906.

53. Conklin, "Theory of Sovereign Debt," pp. 508, 509; Lovett, "Castilian Bankruptcy," pp. 907–909; Albert Lovett, "The General Settlement of 1577," *Historical Journal* 25, no. 1 (1982): 21.

54. Conklin, "Theory of Sovereign Debt," p. 509.

55. Ibid., p. 492; Lovett, "General Settlement," p. 3.

56. Conklin, "Theory of Sovereign Debt," pp. 485–92, 504, 505; Lovett, "General Settlement," p. 3; James C. Boyagian, *Portuguese Bankers at the Court of Spain* (New Brunswick, NJ: Rutgers University Press, 1983), pp. 1–3; Parker, *Army of Flanders*, pp. 148–50.

57. Conklin, "Theory of Sovereign Debt," pp. 484–85, 510; Lovett, "Castilian Bankruptcy," p. 905; Lovett, "General Settlement," p. 5.

58. Lovett, "General Settlement," pp. 4, 5, 7, 12–14.

59. Conklin, "Theory of Sovereign Debt," pp. 504–505.

60. Parker, *Army of Flanders*, pp. 234, 235.

61. Harry E. Cross, "South American Bullion Production and Export," in Richards, *Precious Metals*, pp. 404, 405; Conklin, "Theory of Sovereign Debt," p. 508, Parker, *Army of Flanders*, pp. 135, 146.

62. Parker, *Army of Flanders*, pp. 150, 151, 236, 237; Van Der Wee, *Antwerp Market*, pp. 240, 254; Parker, *Military Revolution*, p. 59; Lynch, *Spain*, p. 303.

63. Lynch, *Spain*, p. 304.

64. Parker, *Army of Flanders*, pp. 150, 151, 236, 237; Lovett, "General Settlement," pp. 8–11, 13, 14; Lynch, *Spain*, p. 304.

65. Parker, *Army of Flanders*, pp. 238–40; Lovett, "General Settlement," pp. 17–19; Conklin, "Theory of Sovereign Debt," p. 510.

66. Conklin, "Theory of Sovereign Debt," pp. 496, 497; Parker, *Army of Flanders*, p. 238; Lovett, "General Settlement," p. 15; Parker, *Military Revolution*, p. 76

67. Conklin, "Theory of Sovereign Debt," pp. 497–500; Boyagian, *Portuguese Bankers*, p. 5.

68. Mathew, *Portuguese Trade*, pp. 114–18, 173, 174.

69. Wegg, *Antwerp*, pp. 301, 332; Hermann Kellenbenz, *The Rise of the European Economy* (London: Weidenfeld & Nicolson, 1976), pp. 177–79; Klarwill, *Fugger News-Letters*, p. xxvi.

70. Lynch, *Spain*, pp. 170–72; Pike, *Enterprise and Adventure*, pp. 94–97.

71. Lynch, *Spain*, pp. 305–307.

72. Ibid., pp. 186–88, 190.

73. António Henrique de Oliveira Marques, *History of Portugal* (New York: Columbia University Press, 1972), pp. 311, 312.

74. Ibid., pp. 307, 308, 312, 313; Hess, "Battle of Lepanto," pp. 67, 68; Ozbaran, "Ottoman Naval Policy," p. 68; Lane, "Venetian Shipping," p. 38; Christine Woodhead, "Introduction," in Kunt and Woodhead, *Suleyman the Magnificent*, p. 119; Christine Woodhead, "Perspectives on Suleyman," in Kunt and Woodhead, *Suleyman the Magnificent*, p. 170; Lynch, *Spain*, p. 249; J. D. Fage, *An Introduction to the History of West Africa* (Cambridge: Cambridge University Press, 1955), pp. 28–31.

75. Bernard H. M. Vlekke, *Nusantara* (The Hague: W. van Hoeve, 1959), pp. 81, 85, 88.

76. Marques, *History of Portugal*, pp. 285–87, 292, 310, 318, 321.

77. Ibid., pp. 314, 315; Lynch, *Spain*, p. 327.

78. Israel, *Dutch Republic*, pp. 183–87, 192–93, 197, 201; Lynch, *Spain*, pp. 304, 305.

## CHAPTER EIGHT: CONVERGENCE IN THE FAR EAST

1. K. S. Mathew, *Indo-Portuguese Trade and the Fuggers of Germany* (New Delhi: Manohar, 1997), pp. 113, 124–28, 149; Pierre Vilar, *A History of Gold and Money, 1450–1920* (Atlantic Highlands, NJ: Humanities Press, 1976), pp. 100, 101; A. R. Disney, *Twilight of the Pepper Empire* (Cambridge, MA: Harvard University Press, 1978), p. 36; Salik

Ozbaran, "Expansion in the Southern Seas," in *Suleyman the Second and His Time*, Halil Inalcik and Cemal Kafadar (Istanbul: Isis Press, 1993), pp. 214, 215; António Henrique de Oliveira Marques, *History of Portugal* (New York: Columbia University Press, 1972), p. 261.

2. Mathew, *Portuguese Trade*, p. xii; James C. Boyagian, *Portuguese Trade in Asia Under the Habsburgs, 1580–1640* (Baltimore: Johns Hopkins University Press, 1993), pp. 8–10, 13–14; George D. Winius, "Two Lusitanian Variations on a Dutch Theme," in *Companies and Trade*, ed. Leonard Blusse and Femme Gaastra (Leiden: Leiden University Press, 1981), p. 120; Niels Steensgaard, *The Asian Trade Revolution of the Seventeenth Century* (Chicago: University of Chicago Press, 1974), pp. 193, 198.

3. J. H. Parry, *Age of Reconnaissance* (New York: New American Library, 1963), pp. 115–22; Geoffrey Parker, *The Military Revolution* (Cambridge; New York: Cambridge University Press, 1996), pp. 89, 90; Richard W. Unger, *The Ship in the Medieval Economy* (London: Croom Helm; Montreal: McGill-Queen's University Press, 1980), pp. 231, 234–36; Marques, *History of Portugal*, pp. 177, 189, 262; John F. Guilmartin Jr., *Galleons and Galleys* (London: Cassell, 2002), pp. 82, 97.

4. Disney, *Pepper Empire*, pp. 16–23; M. C. Richlefs, *A History of Modern Indonesia* (Bloomington: Indiana University Press, 1981), p. 23.

5. Disney, *Pepper Empire*, pp. 23, 24; M. N. Pearson, *Coastal Western India* (New Delhi: Concept, 1981), pp. 31, 33, 43–46, 77, 80, 84; Marques, *History of Portugal*, pp. 215, 250.

6. Rhoads Murphey, "Silver Production in Rumelia According to an Official Ottoman Report Circa 1600," *Sudost-Forschungen* 39 (1980): 235–37.

7. Pearson, *Coastal Western India*, pp. 50, 51, 68, 76, 81, 86, 106, 108; Disney, *Pepper Empire*, pp. 2, 4, 18–20, 27, 28; Boyagian, *Portuguese Trade*, pp. 74, 75; Parry, *Age of Reconnaissance*, p. 246; C. R. Boxer, *The Portuguese Seaborne Empire* (New York: Knopf, 1969), p. 59; Holden Furber, *Rival Empires of Trade in the Orient* (Minneapolis: University of Minnesota Press, 1976), pp. 10, 11, 27.

8. Jan Huyghen van Linschoten, "Travels in India," in *Early Travels in India*, ed. J. Talboys Wheeler (Delhi: Deep Publications, 1974), pp. 169–71, 184.

9. Richlefs, *Modern Indonesia*, pp. 30, 31; M. A. P. Meilink-Roelofsky, *Asian Trade and European Influence in the Indonesian Arche-*

*pelago, 1500–1630* (The Hague: Nijhoff, 1962), pp. 133–45, 149, 164–70; Parker, *Military Revolution*, pp. 105–106, 112.

10. Meilink-Roelofsky, *Asian Trade*, pp. 146, 151, 153; C. R. De Silva, "Trade in Ceylon Cinnamon in the Sixteenth Century," *CJHSS* 3, no. 2 (1973): 110, 123, 125, 126; Linschoten, "Travels," p. 223.

11. Meilink-Roelofsky, *Asian Trade*, pp. 170–72; Disney, *Pepper Empire*, p. 26.

12. Meilink-Roelofsky, *Asian Trade*, pp. 159–62.

13. Marques, *History of Portugal*, p. 343; K. S. Mathew, *Indo-Portuguese Trade and the Fuggers of Germany* (New Delhi: Manohar, 1997), pp. 14, 169, 170.

14. K. S. Mathew, *Indo-Portuguese Trade and the Fuggers of Germany* (New Delhi: Manohar, 1997), pp. 171–73.

15. Victor von Klarwill, ed., *The Fugger News-Letters* (New York: G. P. Putnam, 1924), pp. 38, 45.

16. Ernest S. Dodge, *Islands and Empires* (Minneapolis: University of Minnesota Press, 1976), pp. 212–16; Parry, *Age of Reconnaissance*, pp. 5–7.

17. Steve Shipp, *Macau China* (Jefferson, NC: McFarland & Co., 1997), p. 3; Clive Willis, ed., *China and Macau* (Aldershot, UK: Ashgate, 2002), pp. xiiii, xiv, xv; Edward Farmer et al., *Ming History: An Introductory Guide to Research* (Minneapolis: History Department, University of Minnesota, 1994), p. 20.

18. Shipp, *Macau*, pp. 13, 14; Willis, *China and Macau*, pp. xvi–xviii.

19. Shipp, *Macau*, p. 14; Willis, *China and Macau*, pp. xix, xx; Farmer, *Ming History*, p. 20.

20. Shipp, *Macau*, pp. 14, 15; Boxer, *Portuguese Seaborne Empire*, pp. 46, 48; Parry, *Age of Reconnaissance*, pp. 144, 145; Dodge, *Islands and Empires*, pp. 225–27; Richlefs, *Modern Indonesia*, p. 21; Leonard Y. Andaya, *The World of Maluku* (Honolulu: University of Hawaii Press, 1993), p. 119; Leonard Blussé, "Divesting a Myth: Seventeenth-Century Dutch-Portuguese Rivalry in the Far East," in *Vasco da Gama and the Linking of Europe and Asia*, ed. Antony Disney and Emily Booth (Oxford: Oxford University Press, 2000), pp. 387–91.

21. Kozo Yamamura and Tetsuo Kamiki, "Silver Mines and Sung Coins—A Monetary History of Medieval and Modern Japan in International Perspective," in *Precious Metals in the Later Medieval and Early Modern Worlds*, ed. J. F. Richards (Durham, NC: Carolina Academic

Press, 1983), pp. 336, 337; Mikiso Hane, *Premodern Japan* (Boulder, CO: Westview Press, 1991), p. 111.

22. Delmer M. Brown, *Money Economy in Medieval Japan* (New Haven, CT: Institute for Far Eastern Languages, Yale University, 1951), pp. 1–5; William Gowland, "Metals and Metal Working in Old Japan," *Transactions and Proceedings of the Japan Society* 13, 24th Session, London (1914–15): 29, 30, 42, 43, 67.

23. Delmer M. Brown, "The Importation of Gold into Japan by the Portuguese in the Sixteenth Century," *Pacific Historical Review* 26, no. 2 (1947): 126; Brown, *Money Economy*, pp. 1–5, 10, 13, 36, 67, 68; Gowland, "Metals and Metal Working," p. 30.

24. Brown, *Money Economy*, p. 33.

25. William S. Atwell, "International Bullion Flows and the Chinese Economy Circa 1530–1650," *Past and Present* 95 (May 1982): 79; Richard Von Glahn, *Fountain of Fortune: Money and Monetary Policy in China* (Berkeley: University of California Press, 1996), p. 114.

26. Yamamura and Kamiki, "Silver Mines," pp. 337, 338; Nagahara Keiji and Kozo Yamamura, "Shaping the Process of Unification: Technological Progress in 16th and 17th Century Japan," *Journal of Japanese Studies* 14, no. 1 (1988): 77–81; Brown, *Money Economy*, pp. 56–58.

27. Brown, *Money Economy*, pp. 33–35, 56; Keiji and Yamamura, "Shaping the Process," pp. 81, 82; Brown, "Importation," p. 130; Atwell, "International Bullion," pp. 69, 70, 76.

28. Olof G. Lidin, *Tanegashima: The Arrival of Europe in Japan* (Copenhagen: Nordic Institute of Asian Studies, 2002), pp. 1–6, 14, 15, 19; Atwell, "International Bullion," pp. 69, 70, 76.

29. Marques, *History of Portugal*, pp. 347, 350–53.

30. Meilink-Roelofsky, *Asian Trade*, pp. 130, 131; Boyagian, *Portuguese Trade*, pp. 29, 30.

31. Lidin, *Tanegashima*, pp. 30, 164–69.

32. Ibid., pp. 172–78.

33. A. Kobata, "The Production and Uses of Gold and Silver in 16th and 17th Century Japan," *Economic History Review* 18, Second Series, no. 2 (1965): 245, 249, 250, 253; C. R. Boxer, *The Great Ship from Amacon* (Lisbon: Centro de Estudos Historicos Ultramarinos, 1959), p. 2; Iwao Seiichi, "Japanese Foreign Trade in the 16th and 17th Centuries," *Acta Asiatica* 30 (1976): 1, 2.

34. Shipp, *Macau*, pp. 18, 23, 24; Lidin, *Tanegashima*, pp. 178, 179.

35. Shipp, *Macau*, pp. 15–17; Geoffrey C. Gunn, *Encountering Macau* (Boulder, CO: Westview Press, 1996), p. 17; Willis, *China and Macau*, p. xxiii.

36. Brown, *Money Economy*, p. 30; Shipp, *Macau*, pp. 17–19; Willis, *China and Macau*, p. xxiii.

37. Shelagh Vainker, *Chinese Silk* (New Brunswick, NJ: British Museum Press in Association with Rutgers University Press, 2004), pp. 6–9, 12, 58–60, 76.

38. Vainker, *Chinese Silk*, pp. 142, 144–46, 169.

39. Shipp, *Macau*, p. 39; Gunn, *Encountering Macau*, pp. 20, 21; Willis, *China and Macau*, p. xviii; Boxer, *Amacon*, pp. 7, 8.

40. Boxer, *Amacon*, p. 2; Seiichi, "Japanese Foreign Trade," pp. 3, 5, 7; George Bryan Souza, *The Survival of Empire* (Cambridge: Cambridge University Press, 1986), p. 17.

41. Hane, *Premodern Japan*, pp. 123, 124.

42. Keiji and Yamamura, "Shaping the Process," pp. 82, 104; Brown, *Money Economy*, pp. 56–58; Eiichi Kato, "Unification and Adaption," in *Companies and Trade*, ed. Leonard Blusse and Femme Gaastra (Leiden: Leiden University Press, 1981), p. 208.

43. Brown, *Money Economy*, pp. 58–61, 67, 68; Brown, "Importation," p. 130.

44. Hane, *Premodern Japan*, pp. 112, 114.

45. William Henry Scott, *Prehistoric Source Material for the Study of Philippine History* (Quezon City: New Day Publishers, 1985), p. 137; William Henry Scott, *Cracks in the Parchment Curtain* (Quezon City: New Day Publishers, 1985), pp. 21, 22, 86, 87; Stanley Karnow, *In Our Image: America's Empire in the Philippines* (New York: Random House, 1989), pp. 39, 40.

46. Scott, *Prehistoric Source Material*, p. 137; Scott, *Cracks*, pp. 21, 22, 86, 87; Karnow, *In Our Image*, pp. 39, 40.

47. William L. Schurz, *The Manila Galleon* (New York: E. P. Dutton, 1959), pp. 20, 21; Karnow, *In Our Image*, p. 43.

48. Karnow, *In Our Image*, pp. 43, 44; John M. Headley, "Spain's Asian Presence, 1565–1590," *Hispanic American Historical Review* 75, no. 4 (1995): 634.

49. Schurz, *Manila Galleon*, pp. 22, 23.

50. Karnow, *In Our Image*, p. 45; Scott, *Cracks*, p. 46; Headley, "Spain's Asian Presence," p. 628; Eufronio M. Alip, *Political and Cultural History of the Philippines* (Manila: Alip & Brion Publications, 1951), pp. 235, 236; Parry, *Age of Reconnaissance*, p. 195.

51. Karnow, *In Our Image*, p. 46.

52. Schurz, *Manila Galleon*, pp. 24, 26, 27; Karnow, *In Our Image*, pp. 29, 47, 67; Jose S. Arcilla, *An Introduction to Philippine History* (Manila: Ateneo Publications, 1971), p. 16; Scott, *Cracks*, pp. 47, 91; Alip, *Political and Cultural History*, pp. 271, 336, 298, 299.

53. Karnow, *In Our Image*, pp. 55, 56.

54. Alip, *Political and Cultural History*, p. 232; Karnow, *In Our Image*, pp. 55, 56.

55. Headley, "Spain's Asian Presence," pp. 634, 635; John J. Tepaske, "New World Silver, Castile and the Philippines, 1590–1800," in Richards, *Precious Metals*, pp. 434–38; Boxer, *Amacon*, p. 3; Gunn, *Encountering Macau*, pp. 21, 22; Alip, *Political and Cultural History*, pp. 283, 285; Von Glahn, *Fortune*, p. 119; Pierre Chaunu, *Les Philippines et le Pacifique des Iberiques*, vol. 11 (Paris: S.E.V.P.E.N, 1960), pp. 148–53; Boyagian, *Portuguese Trade*, p. 276; Karnow, *In Our Image*, pp. 49, 50; Scott, *Cracks*, p. 22; Arcilla, *Philippine History*, pp. 22, 23, 43.

56. Adam Szaszdi, "Preliminary Estimate of Gold and Silver Production in America," in Richards, *Precious Metals*, p. 168; Atwell, "International Bullion," pp. 72–74, 78; Yamamura and Kamiki, "Silver Mines," pp. 338, 339, 341; Furber, *Rival Empires*, p. 22; Hane, *Premodern Japan*, p. 122; Von Glahn, *Fortune*, pp. 124, 125, 138; Souza, *Survival of Empire*, pp. 67, 68, 81, 82; Chaunu, *Les Philippines*, pp. 152, 153.

57. Headley, "Spain's Asian Presence," pp. 634, 635, 637, 639; Karnow, *In Our Image*, p. 51; Pedro A. Gagelonia, *The Filipino Historian* (Manila: Feucci, 1970), pp. 53, 56, 61.

58. Tsing Yuan, "The Silver Trade between America and China," in *Precious Metals in the Age of Expansion*, ed. Hermann Kellenbenz (Stuttgart: Klett-Cotta, 1981), p. 267; Alip, *Political and Cultural History*, pp. 237, 240, 256, 258; Dennis O. Flynn and Arturo Giraldez, "Cycles of Silver," *Journal of World History* 13, no. 2 (2002): 397–404; Parry, *Age of Reconnaissance*, p. 195; Atwell, "International Bullion," pp. 80–86; Von Glahn, *Fortune*, p. 135.

59. Marques, *History of Portugal*, pp. 315–17; Seiichi, "Japanese Foreign Trade," p. 2.

60. Marques, *History of Portugal*, p. 331; Meilink-Roelofsky, *Asian Trade*, pp. 126–30.

61. Marques, *History of Portugal*, pp. 271–74; Boyagian, *Portuguese Trade*, p. 12.

62. David Birmingham, *The Portuguese Conquest of Angola* (New York: Oxford University Press, 1965), pp. 16–19; David Birmingham, "The Date and Significance of the Imbangala Invasion of Angola," *Journal of African History* 6, no. 2 (1965): 20–24, 146–48; Lawrence W. Henderson, *Angola* (Ithaca, NY: Cornell University Press, 1979), pp. 83, 84; M. D. D. Newitt, "The Portuguese on the Zambezi," *Journal of African History* 10, no. 1 (1969): 67–77; Malyn Newitt, *A History of Mozambique* (London: Hurst & Co., 1995), pp. 53–60; Marques, *History of Portugal*, p. 340; Souza, *Survival of Empire*, p. 77; Boyagian, *Portuguese Trade*, pp. 32, 63.

63. Marques, *History of Portugal*, pp. 277–81, 298, 346, 347, 353.

64. Ibid., pp. 241–43; D. W. Davies, *A Primer of Dutch Seventeenth Century Overseas Trade* (The Hague: M. Nijoff, 1961), pp. 95, 96; Halil Inalcik, *The Ottoman Empire*, trans. Norman Itzkowitz and Colin Imber (New Rochelle, NY: Orpheus Publishing, 1973), p. 45; Murphey, "Silver Production," pp. 75–79, 88, 94, 96.

65. Mathew, *Indo-Portuguese Trade*, pp. 114–18, 171–78, 181–87; Boyagian, *Portuguese Trade*, pp. 21, 22.

66. Mathew, *Indo-Portuguese Trade*, pp. 181–83; Boyagian, *Portuguese Trade*, p. 22.

67. Marques, *History of Portugal*, p. 278; Vilar, *Gold and Money*, pp. 93–95.

68. James C. Boyagian, *Portuguese Bankers at the Court of Spain* (New Brunswick, NJ: Rutgers University Press, 1983), p. 6; Boyagian, *Portuguese Trade*, pp. 30, 31, 38, 267, 278, 309.

69. Boyagian, *Portuguese Trade*, pp. 41–46, 52, 82, 87, 139, 143, 282, 316.

70. Ibid., pp. 45–50, 65, 137–41, 320.

71. Boyagian, *Portuguese Trade*, pp. 76–82.

72. Shipp, *Macau*, pp. 32–37, 41; Gunn, *Encountering Macau*, pp. 15, 16, 18, 23.

73. Shipp, *Macau*, pp. 26–28, 35; Willis, *China and Macau*, pp. xxiv, xxvi; Souza, *Survival of Empire*, pp. 25–28, 31, 32; Boyagian, *Portuguese Trade*, p. 84.

74. Boxer, *Amacon*, pp. 12, 15, 18; Michael Cooper, "The Mechanics of the Macao-Nagasaki Silk Trade," *Monumenta Nipponica* 27, no. 4 (1972): 426, 427.

75. Boxer, *Amacon*, pp. 5–7, 17; Shipp, *Macau*, pp. 37–40; Souza, *Survival of Empire*, pp. 46, 47, 52.

76. Boxer, *Amacon*, pp. 8–11, 14–17; Souza, *Survival of Empire*, pp. 55, 56.

77. Hane, *Premodern Japan*, pp. 114, 115.

78. Kobata, "Production and Uses," pp. 252–54; Brown, "Importation," pp. 129, 133; Boxer, *Amacon*, p. 2.

79. Brown, "Importation," pp. 61–64, 91, 92; Boxer, *Amacon*, p. 7; Harry E. Cross, "South American Bullion Production and Export," in Richards, *Precious Metals*, p. 403; Von Glahn, *Fortune*, p. 134.

80. William B. Hauser, "Osaka Castle and Tokugawa Authority in Western Japan," in *The Bakufu in Japanese History*, ed. Jeffrey P. Mass and William B. Hauser (Stanford, CA: Stanford University Press, 1985), pp. 155, 157, 160; Brown, *Money Economy*, pp. 58–61, 72–76, 82, 83, 91, 92; Brown, "Importation," p. 133.

81. Hane, *Premodern Japan*, pp. 15, 116, 117.

82. Parker, *Military Revolution*, pp. 108–109; Guilmartin, *Galleons and Galleys*, pp. 184–88.

83. Brown, "Importation," pp. 129, 133; Boxer, *Amacon*, p. 2; Brown, *Money Economy*, pp. 69–72; Hamish Todd, "The British Library's Sado Mining Scrolls," *British Library Journal* 24, no. 1 (1998): 130; Tessa Morris-Suzuki, *The Technological Transformation of Japan* (Cambridge: Cambridge University Press, 1994), p. 44.

84. Hane, *Premodern Japan*, pp. 15, 116, 117; Parker, *Military Revolution*, pp. 109, 111; Guilmartin, *Galleons and Galleys*, pp. 188, 189.

85. Leonard Blussé, "Divesting a Myth: Seventeenth-Century Dutch-Portuguese Rivalry in the Far East," in Disney and Booth, *Vasco da Gama*, pp. 387–91; Furber, *Rival Empires*, p. 23; Marques, *History of Portugal*, pp. 337, 350, 351; Hane, *Premodern Japan*, pp. 124–26; Seiichi, "Japanese Foreign Trade," p. 7.

# CHAPTER NINE:
# UNIMAGINABLE WEALTH AND SQUANDER

1. Peter J. Bakewell, "Registered Silver Production in the Potosí District, 1550–1735," *Jahrbuch für Geschichte von Staat, Wirtschaft und Gesellschaft Lateinamerikas* 12 (1975): 67, 75.

2. Ramon Sanchez Flores, "Technology of Mining in Colonial Mexico," in *In Quest of Mineral Wealth*, ed. Alan K. Craig and Robert C. West (Baton Rouge: Geoscience Publications, Louisiana State University, 1994), pp. 142, 145, 146; Adam Szaszdi, "Preliminary Estimate of Gold and Silver Production in America," in *Precious Metals in the Later Medieval and Early Modern Worlds*, ed. J. F. Richards (Durham, NC: Carolina Academic Press, 1983), pp. 196, 200; Carlos Prieto, *Mining in the New World* (New York: McGraw-Hill, 1973), pp. 78–80; Alan K. Craig, "Spanish Colonial Silver Beneficiation at Potosí," in Craig and West, *In Quest of Mineral Wealth*, pp. 278, 279; Peter J. Bakewell, *Silver and Entrepreneurship in Seventeenth Century Potosí* (Albuquerque: University of New Mexico Press, 1988), p. 31; David C. Goodman, *Power and Penury* (Cambridge: Cambridge University Press, 1988), pp. 193, 194.

3. Herbert S. Klein, *Bolivia* (New York: Oxford University Press, 1982), pp. 36–39, 287; Carlos Sempat Assadourian, "The Colonial Economy," *Journal of Latin American Studies* 24, Supplement (1992): 59; Peter J. Bakewell, "Technological Change in Potosí: The Silver Boom of the 1570s," *Jahrbuch für Geschichte von Staat, Wirtschaft und Gesellschaft Lateinamerikas* 14 (1977): 60.

4. Laura Escobar, "Potosí," in Pedro Querezaza and Elizabeth Ferrer, *Potosí* (New York: Americas Society Art Gallery in Association with Fundación BHN, La Paz, 1997), pp. 18, 20; Pierre Vilar, *A History of Gold and Money, 1450–1920* (Atlantic Highlands, NJ: Humanities Press, 1976), p. 123; Klein, *Bolivia*, p. 42; Joceyln Houseman and David A. Johnson, "Spanish American Silver," *Interdisciplinary Science Reviews* 16, no. 3 (1991): 245, 246.

5. Goodman, *Power and Penury*, pp. 180, 183, 187–89; Pierre Chaunu, *Seville et L'Atlantique, 1504–1650*, vol. 8, *La Conjoncture* (Paris: Institute des Hautes Etudes de L'Amerique Latine, 1959), pp. 1959–61.

6. Escobar, "Potosí," pp. 16, 17; Craig, "Spanish Colonial Silver," pp. 275–79, 282; Vilar, *Gold and Money*, p. 125.

7. Goodman, *Power and Penury*, pp. 190, 192; Houseman and Johnson, "Spanish American Silver," p. 246.

8. Goodman, *Power and Penury*, pp. 192, 206; John Percy, *Metallurgy, Silver and Gold* (London: John Murray, 1880), pp. 650, 651.

9. Bakewell, "Registered Silver," p. 85; C. B. Kroeber, "The Mobilization of Philip II's Revenue in Peru, 1590–96," *Economic History Review* 10, no. 3 (1958): 40, 41; M. F. Lang, "New Spain's Mining Depression and the Supply of Quicksilver from Peru, 1600–1700," *Hispanic American Historical Review* 48, no. 4 (1968): 632–41; Szaszdi, "Preliminary Estimate," pp. 201, 202.

10. Escobar, "Potosí," pp. 14, 16; Vilar, *Gold and Money*, p. 131; Lewis Hanke, *The Imperial City of Potosí* (The Hague: Nijhoff, 1956), pp. 2, 3, 18, 19; Bakewell, *Silver and Entrepreneurship*, p. 16.

11. Prieto, *Mining in the New World*, pp. 32, 33; Vilar, *Gold and Money*, pp. 91, 132; Bakewell, *Silver and Entrepreneurship*, p. 30.

12. Assadourian, "The Colonial Economy," pp. 61–65; Klein, *Bolivia*, pp. 43–50; Gwendolin B. Cobb, "Supply and Transportation for the Potosí Mines," *Hispanic American Historical Review* 29, no. 1 (1949): 33.

13. Klein, *Bolivia*, pp. 51–56.

14. J. H. Parry, *Age of Reconnaissance* (New York: New American Library, 1963), p. 179.

15. Cobb, "Supply and Transportation," pp. 30–40, 44; Bakewell, *Silver and Entrepreneurship*, p. 24.

16. Parry, *Age of Reconnaissance*, p. 182; Clarence Henry Haring, *Trade and Navigation between Spain and the Indies in the Time of the Hapsburgs* (Gloucester, MA: P. Smith, 1964), p. 117; James C. Boyagian, *Portuguese Trade in Asia Under the Habsburgs, 1580–1640* (Baltimore: Johns Hopkins University Press, 1993), p. 239.

17. Parry, *Age of Reconnaissance*, pp. 178, 179; Escobar, "Potosí," p. 28; Vilar, *Gold and Money*, p. 106; Szaszdi, "Preliminary Estimate," p. 218; Boyagian, *Portuguese Trade*, p. 292.

18. Klein, *Bolivia*, pp. 57, 58; Richard L. Garner, "Long Term Silver Mining Trends in Spanish America," *American Historical Review* 93, no. 4 (1988): 907–10; Percy, *Metallurgy*, pp. 654, 655; Bakewell, "Registered Silver," pp. 86, 87, 90, 101; Szaszdi, "Preliminary Estimate," pp. 202, 207, 213.

19. Robert C. West, *Colonial Placer Mining in Colombia* (Baton Rouge: Louisiana State University Press, 1952), pp. 10–17, 23.

20. Ibid., pp. 22–26, 44, 48.

21. Ibid., pp. 31–33.

22. Estadisticas Históricas de México, Instituto Nacionale de Estadística, Geografía e Informática, vol. 1, *Aguascalientes*, Cuarta Edición (1999), p. 469; Harry E. Cross, "South American Bullion Production and Export," in Richards, *Precious Metals*, p. 410.

23. Parry, *Age of Reconnaissance*, p. 185.

24. Ibid., p. 196, Ernest S. Dodge, *Islands and Empires* (Minneapolis: University of Minnesota Press, 1976), p. 240; John Lynch, *Spain, 1516–1598* (Oxford: Blackwell, 1991), pp. 318, 319; John Maynard Keynes, *A Treatise on Money*, vol. 2 (London: Macmillan & Co., 1950), pp. 156, 157.

25. Murdo J. Macleod, "Spain and America: The Atlantic Trade, 1492–1720," in *The Cambridge History of Latin America*, vol. 1, ed, Leslie Bethell (Cambridge: Cambridge University Press, 1984), pp. 359, 360; Adam Szaszdi, "Preliminary Estimate of Gold and Silver Production in America," in Kellenbenz, *Precious Metals*, pp. 376, 377; Parry, *Age of Reconnaissance*, p. 185; Lynch, *Spain*, pp. 319–21.

26. Van Der Wee, *Antwerp Market*, pp. 255–57.

27. Ibid.

28. Geoffrey Parker, *The Army of Flanders and the Spanish Road, 1567–1659* (Cambridge: Cambridge University Press, 1972), pp. 153–55.

29. Jonathan I. Israel, *The Dutch Republic* (Oxford: Clarendon Press, 1995), pp. 208, 212, 216; Lynch, *Spain*, p. 331.

30. Lynch, *Spain*, p. 307.

31. Victor von Klarwill, ed., *The Fugger News-Letters* (New York: G. P. Putnam, 1924), p. 75; Parker, *Army of Flanders*, pp. 240–41; Van Der Wee, *Antwerp Market*, pp. 258–265; Israel, *Dutch Republic*, pp. 218, 219.

32. Parker, *Army of Flanders*, pp. 241, 242; Klarwill, *Fugger News-Letters*, p. 53.

33. Herman Van Der Wee, *The Growth of the Antwerp Market and the European Economy* (Louvain: Biblioteque de l'Universite, 1963), pp. 258–65; Israel, *Dutch Republic*, pp. 218, 219.

34. Parker, *Army of Flanders*, pp. 239, 242; Goodman, *Power and Penury*, pp. 88, 89.

35. Parker, *Army of Flanders*, p. 244.

36. Szaszdi, "Preliminary Estimate," pp. 194, 195; Lynch, *Spain*, pp. 172–75; Geoffrey Parker, *The Military Revolution* (Cambridge; New York: Cambridge University Press, 1996), p. 104.

37. Parker, *Army of Flanders*, p. 243; Lynch, *Spain*, pp. 337, 338; I. A. A. Thompson, "The Appointment of the Duke of Medina Sidonia to the Command of the Spanish Armada," *Historical Journal* 12, no. 2 (1969): 202.

38. Parker, *Army of Flanders*, p. 243; Israel, *Dutch Republic*, pp. 220, 223, 225, 233, 239; Lynch, *Spain*, p. 332; J. C. A. Schokkenbroek, "The Dutch and the Spanish Armada 1588," in *God's Obvious Design*, ed. P. Gallagher and D. W. Cruickshank (London: Tamesis Books Limited, 1990), p. 102.

39. Lynch, *Spain*, pp. 332, 333, 339.

40. Ibid., p. 111; Schokkenbroek, "The Dutch and the Spanish Armada," p. 103.

41. Haring, *Trade and Navigation*, pp. 49–53; John F. Guilmartin Jr., *Galleons and Galleys* (London: Cassell, 2002), pp. 154, 155.

42. Lynch, *Spain*, pp. 333–35; Thompson, "Duke of Medina Sidonia," p. 197.

43. Lynch, *Spain*, p. 336.

44. Ibid., pp. 117, 118, 337; James D. Tracy, *Emperor Charles V, Impresario of War* (Cambridge: Cambridge University Press, 2002), p. 313; Thompson, "Duke of Medina Sidonia," p. 202.

45. Lynch, *Spain*, pp. 337–40.

46. Thompson, "Duke of Medina Sidonia," pp. 202, 205.

47. Lynch, *Spain*, pp. 340, 341; Thompson, "Duke of Medina Sidonia," pp. 204–206, 213.

48. Thompson, "Duke of Medina Sidonia," pp. 209–13; Boyagian, *Portuguese Trade*, p. 23.

49. Unger, *The Ship in the Medieval Economy*, pp. 256–58; Carla Rahn Phillips, "The Caravel and the Galleon," in *Cogs, Caravels and Galleons*, ed. Robert J. Gardiner (London: Conway Maritime Press, 1994), pp. 100–102; John F. Guilmartin Jr., "Guns and Gunnery," in Gardiner, *Cogs, Caravels and Galleons*, pp. 147, 148; Guilmartin, *Galleons and Galleys*, pp. 158, 159.

50. Unger, *The Ship in the Medieval Economy*, pp. 256–58; Ian Friel, "The Carrack," in Gardiner, *Cogs, Caravels and Galleons*, pp. 83, 87; Phillips, "The Caravel and the Galleon," pp. 100–103; Martin Elbe, "The

Caravel and the Galleon," in Gardiner, *Cogs, Caravels and Galleons*, p. 98; Parker, *Military Revolution*, pp. 90, 91; Guilmartin, "Guns and Gunnery," pp. 144–49.

51. Goodman, *Power and Penury*, pp. 89, 103–108; Jose Luis Casada Soto, "Atlantic Shipping in Sixteenth Century Spain and the 1588 Armada," in *England, Spain and the Gran Armada*, ed. M. J. Rodriguez-Salado and Simon Adams (Edinburgh: John Donald, 1991), pp. 111, 113; Phillips, "The Caravel and the Galleon," p. 104; Unger, *The Ship in the Medieval Economy*, pp. 259, 260.

52. Parker, *Military Revolution*, pp. 92, 93; Phillips, "The Caravel and the Galleon," pp. 105, 106; Soto, "Atlantic Shipping," pp. 114, 116, 117, 123; Armando da Silva Saturnino Monteiro, "The Decline and Fall of Portuguese Seapower, 1583–1663," *Journal of Military History 65*, no. 1 (2001): 11, 12.

53. Hugo O'Donnell y Duque de Estrada, "The Army of Flanders and the Invasion of England," in Rodriguez-Salado and Adams, *England, Spain*, pp. 218, 220, 225, 231, 232; Guilmartin, *Galleons and Galleys*, pp. 170, 178.

54. Schokkenbroek, "The Dutch and the Spanish Armada," pp. 103, 105, 106; Parker, *Army of Flanders*, p. 244.

55. Schokkenbroek, "The Dutch and the Spanish Armada," pp. 106–10; Guilmartin, *Galleons and Galleys*, p. 178.

56. Lynch, *Spain*, pp. 342–44; Guilmartin, *Galleons and Galleys*, p. 170.

57. Soto, "Atlantic Shipping," pp. 119, 120; O'Donnell, "The Army of Flanders and the Invasion of England," p. 233; Guilmartin, *Galleons and Galleys*, p. 179.

58. Lynch, *Spain*, pp. 344, 345; Phillips, "The Caravel and the Galleon," p. 106; Soto, "Atlantic Shipping," p. 120.

59. Parker, *Military Revolution*, pp. 93, 94; Guilmartin, *Galleons and Galleys*, p. 174.

60. Klarwill, *Fugger News-Letters*, pp. 89, 105, 126, 127, 129.

61. Kroeber, "The Mobilization," p. 439.

62. Lynch, *Spain*, pp. 346, 347; Soto, "Atlantic Shipping," p. 113.

63. Jonathan I. Israel, *Conflicts of Empires* (London: Hambledon Press, 1997), pp. 26–29; Lynch, *Spain*, pp. 350, 362.

64. Parker, *Army of Flanders*, pp. 244, 245; Israel, *Conflicts*, pp. 26–29.

65. Lynch, *Spain*, pp. 351, 352.

66. Parker, *Army of Flanders*, pp. 245, 246.

67. Lynch, *Spain*, pp. 348, 349, 353, 354; Parker, *Army of Flanders*, p. 247.

68. Lynch, *Spain*, pp. 136, 137; Parry, *Age of Reconnaissance*, pp. 245–48.

69. John J. Tepaske, "New World Silver, Castile and the Philippines, 1590–1800," in Richards, *Precious Metals*, pp. 431, 32; Lynch, *Spain*, pp. 137, 138.

70. Tracy, *Emperor Charles V*, p. 303.

71. Kroeber, "The Mobilization," pp. 439, 40.

72. Ibid., pp. 442–45.

73. Ibid., p. 446; Goodman, *Power and Penury*, pp. 177–82, 207.

74. Kroeber, "The Mobilization," pp. 447–49; Bakewell, *Silver and Entrepreneurship*, p. 16.

75. Lynch, *Spain*, pp. 140, 141, 348, 349.

76. Ibid., p. 142.

77. Parker, *Army of Flanders*, p. 247; Lynch, *Spain*, pp. 354, 356.

78. Lynch, *Spain*, pp. 355, 356, 365.

## CHAPTER 10: THE DUTCH ADVANCE

1. Geoffrey Parker, *The Army of Flanders and the Spanish Road, 1567–1659* (Cambridge: Cambridge University Press, 1972), pp. 249–50.

2. Ibid., pp. 247–49; Jonathan I. Israel, *The Dutch Republic* (Oxford: Clarendon Press, 1995), pp. 242, 263, 267, 270, 272.

3. James C. Boyagian, *Portuguese Bankers at the Court of Spain* (New Brunswick, NJ: Rutgers University Press, 1983), pp. 6, 7; James C. Boyagian, *Portuguese Trade in Asia Under the Habsburgs, 1580–1640* (Baltimore: Johns Hopkins University Press, 1993), pp. 87–91.

4. Jonathan I. Israel, *Conflicts of Empires* (London: Hambledon Press, 1997), pp. 1–5; Ramon Carande, *Carlos V y Sus Banqueros* (Barcelona: Critica, 1977), p. 16.

5. Israel, *Dutch Republic*, pp. 254, 255, 258–62; Geoffrey Parker, *The Military Revolution* (Cambridge; New York: Cambridge University Press, 1996), pp. 13, 39; Parker, *Army of Flanders*, pp. 249–51; John F. Guilmartin Jr., *Galleons and Galleys* (London: Cassell, 2002), p. 192.

6. Israel, *Dutch Republic*, pp. v–vii, 2.

7. Ibid., pp. 108, 111, 113, 114.

8. Ibid., pp. 308, 309.

9. Ibid., pp. 116–18; Jonathan I. Israel, *Dutch Primacy in World Trade* (Oxford: Clarendon Press, 1989), pp. 23, 24; D. W. Davies, *A Primer of Dutch Seventeenth Century Overseas Trade* (The Hague: M. Nijoff, 1961), pp. 1, 8.

10. Israel, *Dutch Republic*, p. 118; Israel, *Dutch Primacy*, pp. 18, 22, 23; Richard W. Unger, *The Ship in the Medieval Economy* (London: Croom Helm; Montreal: McGill-Queen's University Press, 1980), p. 223.

11. Violet Barbour, *Capitalism in Amsterdam in the Seventeenth Century* (Ann Arbor: University of Michigan Press, 1963), pp. 13–15; Geert Mak, *Amsterdam* (London: Harvill Press, 1999), p. 7.

12. Davies, *A Primer*, pp. 11–13.

13. Israel, *Dutch Republic*, pp. 311, 312; Israel, *Dutch Primacy*, p. 34; C. R. Boxer, *The Dutch Seaborne Empire* (New York: Knopf, 1965), p. 19.

14. Mak, *Amsterdam*, pp. 9–12, 16–20.

15. Ibid., pp. 16, 21–24.

16. Ibid., pp. 31, 35, 122.

17. Ibid., pp. 36, 42, 46, 54, 56, 58.

18. Ibid., pp. 55, 71, 75, 80–85.

19. Ibid., pp. 54, 55, p. 88–91.

20. Ibid., pp. 91, 99.

21. Unger, *The Ship in the Medieval Economy*, pp. 223–26, 252; Richard W. Unger, "The Fluit," in *Cogs, Caravels and Galleons*, ed. Robert J. Gardiner (London: Conway Maritime Press, 1994), pp. 115, 116.

22. Unger, *The Ship in the Medieval Economy*, pp. 262, 263; Unger, "The Fluit," p. 121.

23. Unger, "The Fluit," pp. 122, 124, 126, 127.

24. Davies, *A Primer*, pp. 38, 42.

25. Ibid., pp. 38, 39.

26. Unger, *The Ship in the Medieval Economy*, pp. 263–65; Parker, *Military Revolution*, p. 99; Davies, *A Primer*, pp. 8, 10, 11.

27. Davies, *A Primer*, p. 16.

28. Ibid., pp. 10, 11, 16–18, 23.

29. Israel, *Dutch Primacy*, pp. 50, 51, 53, 55.

30. J. H. Parry, *Age of Reconnaissance* (New York: New American Library, 1963), p. 186; Eric Williams, *From Columbus to Castro* (New York: Harper & Row, 1971), p. 29; António Henrique de Oliveira Marques, *History of Portugal* (New York: Columbia University Press, 1972), pp. 361, 369; Boyagian, *Portuguese Bankers*, pp. 9–12.

31. Davies, *A Primer*, pp. 24–26, 39.

32. Ibid., pp. 46, 47, Ernest S. Dodge, *Islands and Empires* (Minneapolis: University of Minnesota Press, 1976), p. 238.

33. K. S. Mathew, *Indo-Portuguese Trade and the Fuggers of Germany* (New Delhi: Manohar, 1997), pp. 90, 92.

34. Davies, *A Primer*, p. 33.

35. Israel, *Dutch Primacy*, pp. 60–62; Marques, *History of Portugal*, pp. 374, 375.

36. Davies, *A Primer*, pp. 113, 114; Israel, *Dutch Primacy*, pp. 62, 64, 66.

37. Michael Edwardes, *Ralph Fitch: Elizabethan in the Indies* (London: Faber, 1972), pp. 7, 15, 21; Ralph Fitch, "The Voyage of Mr. Ralph Fitch, Merchant of London," in *A General Collection of the Best and Most Interesting Voyages and Travels in All Parts of the World*, vol. 9, ed. John Pinkerton (Philadelphia: Kimber & Conrad, 1810–12), pp. 407, 409.

38. Fitch, "The Voyage," pp. 422, 423.

39. Parry, *Age of Reconnaissance*, pp. 193–97; Dodge, *Islands and Empires*, pp. 240, 241, 245; Edwardes, *Ralph Fitch*, pp. 7, 155.

40. Parry, *Age of Reconnaissance*, pp. 193, 197; Dodge, *Islands and Empires*, pp. 240, 241.

41. Om Prakash, *The Dutch East India Company and the Economy of Bengal, 1630–1720* (Princeton, NJ: Princeton University Press, 1985), p. 9; Dodge, *Islands and Empires*, p. 238; Israel, *Dutch Republic*, p. 320; Israel, *Dutch Primacy*, pp. 67, 68; M. C. Richlefs, *A History of Modern Indonesia* (Bloomington: Indiana University Press, 1981), p. 24.

42. Davies, *A Primer*, pp. 49, 50; M. A. P. Meilink-Roelofsky, *Asian Trade and European Influence in the Indonesian Archepelago, 1500–1630* (The Hague: Nijhoff, 1962), p. 173.

43. Prakash, *Dutch East India Company*, p. 9; Parry, *Age of Reconnaissance*, pp. 197–99; C. R. Boxer, *The Portuguese Seaborne Empire*

(New York: Knopf, 1969), pp. 22, 23; Israel, *Dutch Primacy*, p. 68; Davies, *A Primer*, p. 50.

44. Victor von Klarwill, ed., *The Fugger News-Letters* (New York: G. P. Putnam, 1924), pp. 222–24, 229, 257.

45. Dodge, *Islands and Empires*, pp. 241, 242; Davies, *A Primer*, p. 71; Alexander H. De Groot, "The Organization of Western European Trade in the Levant," in *Companies and Trade*, ed. Leonard Blusse and Femme Gaastra (Leiden: Leiden University Press, 1981), p. 233.

46. Dodge, *Islands and Empires*, p. 241; Niels Steensgaard, "The Companies as a Specific Institution in the History of European Expansion," in Blusse and Gaastra, *Companies and Trade*, pp. 248–50; Meilink-Roelofsky, *Asian Trade*, pp. 192–94; Boyagian, *Portuguese Trade*, pp. 107, 108.

47. Dodge, *Islands and Empires*, pp. 241, 242.

48. Israel, *Dutch Primacy*, pp. 68–71.

49. Dodge, *Islands and Empires*, pp. 243, 244; Israel, *Dutch Republic*, pp. 321–22; Boxer, *Dutch Seaborne*, pp. 23, 24; Mak, *Amsterdam*, p. 116; Femme Gaastra, "The Shifting Balance of Trade of the Dutch East India Company," in Blusse and Gaastra, *Companies and Trade*, p. 50.

50. Israel, *Dutch Primacy*, pp. 70, 71; Richlefs, *Modern Indonesia*, p. 25; Mak, *Amsterdam*, p. 117.

51. Parry, *Age of Reconnaissance*, p. 198; Marques, *History of Portugal*, p. 335; Davies, *A Primer*, p. 53.

52. Parker, *Military Revolution*, pp. 106, 107; Meilink-Roelofsky, *Asian Trade*, pp. 174, 175; Boyagian, *Portuguese Trade*, 151; Armando da Silva Saturnino Monteiro, "The Decline and Fall of Portuguese Seapower, 1583–1663," *Journal of Military History* 65, no. 1 (2001): 19; Mensun Bound et al., "The Dutch East Indiaman, Nassau, Lost at the Battle of Cape Rachado, Straits of Malacca, 1606," in *Excavating Ships of War*, ed. Mensun Bound, International Maritime Archaeology Series, vol. 2 (Oswestry: A. Nelson, 1998), pp. 89, 101.

53. Monteiro, "Portuguese Seapower," pp. 14–19; Phillips, "The Caravel and the Galleon," pp. 102, 113, 114; Guilmartin, "Guns and Gunnery," pp. 149, 150; Guilmartin, *Galleons and Galleys*, p. 163; Bound et al., "The Dutch East Indiaman," pp. 89, 101.

54. Meilink-Roelofsky, *Asian Trade*, pp. 178–80, 182.

55. Parry, *Age of Reconnaissance*, pp. 253, 254; Israel, *Dutch Republic*, pp. 322, 323; Meilink-Roelofsky, *Asian Trade*, pp. 181, 182.

56. Sushil Chaudhury and Michel Morineau, "Introduction," in *Merchants, Companies and Trade*, ed. Sushil Chaudhury and Michel Morineau (Cambridge: Cambridge University Press, 1999), pp. 2, 4, 8, 9.

57. Richlefs, *Modern Indonesia*, pp. 33, 34, 37, 39; Meilink-Roelofsky, *Asian Trade*, pp. 182, 183.

58. Dodge, *Islands and Empires*, pp. 238–44; Boxer, *Portuguese Seaborne Empire*, p. 62; Richlefs, *Modern Indonesia*, pp. 25, 26; Parry, *Age of Reconnaissance*, p. 250.

59. Bound et al., "The Dutch East Indiaman," pp. 84–89.

60. Disney, *Pepper Empire*, pp. 62, 63.

61. Boyagian, *Portuguese Bankers*, pp. 14–16; Boyagian, *Portuguese Trade*, pp. 89–96.

62. Meilink-Roelofsky, *Asian Trade*, pp. 184, 185; George Bryan Souza, *The Survival of Empire* (Cambridge: Cambridge University Press, 1986), p. 33; Monteiro, "Portuguese Seapower," p. 17; Guilmartin, *Galleons and Galleys*, pp. 191–94.

63. Steensgaard, "The Companies," pp. 194–202, 206, 207.

64. Disney, *Pepper Empire*, p. 64.

65. Davies, *A Primer*, p. 72.

66. Ibid., pp. 70, 71; Eiichi Kato, "The Japanese-Dutch Trade in the Formative Period of the Seclusion Policy," *Acta Asiatica* 30 (1976): 34; Eiichi Kato, "Unification and Adaption," in Blusse and Gaastra, *Companies and Trade*, p. 215.

67. Kato, "Japanese-Dutch Trade," pp. 35, 36; Kato, "Unification," pp. 215, 216.

68. Mikiso Hane, *Premodern Japan* (Boulder, CO: Westview Press, 1991), pp. 131, 133.

69. Ibid., p. 133; William B. Hauser, "Osaka Castle and Tokugawa Authority in Western Japan," in *The Bakufu in Japanese History*, ed. Jeffrey P. Mass and William B. Hauser (Stanford, CA: Stanford University Press, 1985), pp. 154, 160, 161, 170, 171.

70. Delmer M. Brown, *Money Economy in Medieval Japan* (New Haven, CT: Institute for Far Eastern Languages, Yale University, 1951), pp. 58–61; Harry E. Cross, "South American Bullion Production and Export," in Richards, *Precious Metals*, p. 403; Richard Von Glahn, *Fountain of Fortune: Money and Monetary Policy in China* (Berkeley: University of California Press, 1996), p. 136; Souza, *Survival of Empire*, p. 55.

71. Souza, *Survival of Empire*, pp. 71, 72.

72. Delmer M. Brown, "The Importation of Gold into Japan by the Portuguese in the Sixteenth Century," *Pacific Historical Review* 26, no. 2 (1947): 129, 133; C. R. Boxer, *The Great Ship from Amacon* (Lisbon: Centro de Estudos Historicos Ultramarinos, 1959), pp. 2, 5–7; Brown, *Money Economy*, pp. 69–72; Tessa Morris-Suzuki, *The Technological Transformation of Japan* (Cambridge: Cambridge University Press, 1994), pp. 43, 45; Hamish Todd, "The British Library's Sado Mining Scrolls," *British Library Journal* 24, no. 1 (1998): 139–43; William S. Atwell, "International Bullion Flows and the Chinese Economy Circa 1530–1650," *Past and Present 95* (May 1982): 70–72.

73. Kato, "Japanese-Dutch Trade," pp. 46–48; Von Glahn, *Fortune*, pp. 121, 136; Souza, *Survival of Empire*, pp. 50, 54; Charles David Sheldon, *The Rise of the Merchant Class in Tokugawa Japan* (Locust Valley, NY: Association for Asian Studies, J. J. Augustin, 1958), p. 36; Boyagian, *Portuguese Bankers*, p. 220.

74. C. R. Boxer, "Plata es Sangre," *Philippines Studies* 18, no. 3 (1970): 458–61; Boyagian, *Portuguese Bankers*, p. 220; Von Glahn, *Fortune*, pp. 138, 140; Edward Farmer et al., *Ming History: An Introductory Guide to Research* (Minneapolis: History Department, University of Minnesota, 1994), p. 20.

75. Kato, "Japanese-Dutch Trade," p. 36; Kato, "Unification," pp. 217–20.

76. Leonard Blussé, "Divesting a Myth: Seventeenth-Century Dutch-Portuguese Rivalry in the Far East," in *Vasco da Gama and the Linking of Europe and Asia*, ed. Antony Disney and Emily Booth Booth (Oxford: Oxford University Press, 2000), pp. 387–91; Holden Furber, *Rival Empires of Trade in the Orient* (Minneapolis: University of Minnesota Press, 1976), p. 23; Marques, *History of Portugal*, pp. 337, 350, 351; Davies, *A Primer*, p. 73.

77. Kato, "Japanese-Dutch Trade," pp. 42–44, 55; Von Glahn, *Fortune*, p. 283; Souza, *Survival of Empire*, pp. 51, 57–61, 161, 162; Boyagian, *Portuguese Trade*, pp. 190, 204, 207.

78. Eric Axelson, *Portuguese in South-East Africa* (Johannesburg: Witwatersrand University Press, 1960), p. 2; Dodge, *Islands and Empires*, p. 239; Souza, *Survival of Empire*, pp. 19, 20; Boyagian, *Portuguese Trade*, p. 263.

79. Parry, *Age of Reconnaissance*, pp. 186–88; Clarence Henry Haring, *Trade and Navigation between Spain and the Indies in the Time of the Hapsburgs* (Gloucester, MA: P. Smith, 1964), p, 119; Israel, *Dutch Republic*, p. 325.

80. Davies, *A Primer*, p. 114.

81. Ibid., pp. 114, 115.

## CHAPTER ELEVEN: CAPITAL OF THE WORLD

1. Brian Pullan, "Editor's Introduction," in *Crisis and Change in the Venetian Economy in the Sixteenth and Seventeenth Centuries*, ed. Brian Pullan (London: Methuen, 1968), pp. 6, 7; Frederic C. Lane, "Venetian Shipping during the Commercial Revolution," in Pullan, *Crisis and Change*, p. 41.

2. Jan de Vries, *European Urbanization* (Cambridge, MA: Harvard University Press, 1984), p. 36.

3. Noble David Cook, *Demographic Collapse, Indian Peru, 1520–1620* (Cambridge: Cambridge University Press, 1981), pp. 54, 70, 114.

4. Nicolas Sanchez-Albornoz, "The Population of Colonial Spanish America," in *The Cambridge History of Latin America*, vol. 2, ed. Leslie Bethell (Cambridge: Cambridge University Press, 1984), p. 4; Simon Collier et al., eds., *Cambridge Encyclopedia of Latin America* (Cambridge: Cambridge University Press, 1992), pp. 131, 132; W. Borah, "New Spain's Century of Depression," *Ibero-Americana* 35 (1951): 3; Murdo J. Macleod, "Spain and America: The Atlantic Trade, 1492–1720," in Bethell, *Cambridge History of Latin America*, vol. 1, pp. 356, 357.

5. Cook, *Demographic Collapse*, pp. 94, 151, 215.

6. J. D. Fage, *An Introduction to the History of West Africa* (Cambridge University Press, 1955), pp. 58, 59.

7. António Henrique de Oliveira Marques, *History of Portugal* (New York: Columbia University Press, 1972), pp. 374–76, 378; D. W. Davies, *A Primer of Dutch Seventeenth Century Overseas Trade* (The Hague: M. Nijoff, 1961), pp. 33, 34; P. C. Emmer, "The West Indies Company," in *Companies and Trade*, ed. Leonard Blusse and Femme Gaastra (Leiden: Leiden University Press, 1981), p. 85.

8. Emmer, "The West Indies Company," p. 77; Davies, *A Primer*, p.

123; Carlos Prieto, *Mining in the New World* (New York: McGraw-Hill, 1973), pp. 155–57; Marques, *History of Portugal*, p. 232; George D. Winius, "Two Lusitanian Variations on a Dutch Theme," in Blusse and Gaastra, *Companies and Trade*, pp. 128–32.

9. Violet Barbour, *Capitalism in Amsterdam in the Seventeenth Century* (Ann Arbor: University of Michigan Press, 1963), pp. 17, 18, 28, 29; Geert Mak, *Amsterdam* (London: Harvill Press, 1999), pp. 94, 118, 119; Richard W. Unger, "The Fluit," in *Cogs, Caravels and Galleons*, ed. Robert J. Gardiner (London: Conway Maritime Press, 1994), p. 117.

10. Mak, *Amsterdam*, pp. 99, 103, 108, 110, 126, 131, 133, 177; Barbour, *Capitalism in Amsterdam*, pp. 60, 62, 64, 71.

11. Barbour, *Capitalism in Amsterdam*, pp. 15–18, 24, 25; James C. Boyagian, *Portuguese Trade in Asia Under the Habsburgs, 1580–1640* (Baltimore: Johns Hopkins University Press, 1993), pp. 176–83, 213, 216, 312.

12. Barbour, *Capitalism in Amsterdam*, pp. 31–33, 35–41; John F. Guilmartin Jr., "Guns and Gunnery," in Gardiner, *Cogs, Caravels and Galleons*, pp. 149, 150; John F. Guilmartin Jr., *Galleons and Galleys* (London: Cassell, 2002), p. 163; James P. Delgado, ed., *Encyclopedia of Underwater and Maritime Archaeology* (New Haven, CT: Yale University Press, 1998), p. 454; Davies, *A Primer*, pp. 13, 14.

13. C. R. Boxer, *The Portuguese Seaborne Empire* (New York: Knopf, 1969), pp. 26, 27; Barbour, *Capitalism in Amsterdam*, pp. 22, 27, 31; Richard W. Unger, *The Ship in the Medieval Economy* (London: Croom Helm; Montreal: McGill-Queen's University Press, 1980), p. 245; John Lynch, *Spain, 1516–1598* (Oxford: Blackwell, 1991), p. 121.

14. Om Prakash, *The Dutch East India Company and the Economy of Bengal, 1630–1720* (Princeton, NJ: Princeton University Press, 1985), p. 11; F. S. Gaastra, "The Exports of Precious Metal from Europe to Asia by the Dutch East India Company, 1602–1795," in *Precious Metals in the Later Medieval and Early Modern Worlds*, ed. J. F. Richards (Durham, NC: Carolina Academic Press, 1983), pp. 451, 452, 470, 475; Boxer, *Dutch Seaborne*, p. 21; Barbour, *Capitalism in Amsterdam*, pp. 52–56; Mensun Bound et al., "The Dutch East Indiaman, Nassau, Lost at the Battle of Cape Rachado, Straits of Malacca, 1606," in *Excavating Ships of War*, ed. Mensun Bound, International Maritime Archaeology Series, vol. 2 (Oswestry: A. Nelson, 1998), pp. 104, 105.

15. Barbour, *Capitalism in Amsterdam*, pp. 49, 50; Mak, *Amsterdam*, pp. 99, 100; Pierre Vilar, *A History of Gold and Money, 1450–1920* (Atlantic Highlands, NJ: Humanities Press, 1976), p. 196; Emmer, "The West Indies Company," p. 76.

16. Barbour, *Capitalism in Amsterdam*, pp. 24, 29, 33, 34, 52–56.

17. Jonathan I. Israel, *Dutch Primacy in World Trade* (Oxford: Clarendon Press, 1989), pp. 75–77; Barbour, *Capitalism in Amsterdam*, pp. 74–76, 80, 81.

18. Geoffrey Parker, *The Military Revolution* (Cambridge; New York: Cambridge University Press, 1996), pp. 106–107; Boyagian, *Portuguese Trade*, pp. 148–51, 205, 291; Gaastra, "The Exports of Precious Metal," pp. 59–62.

19. Mak, *Amsterdam*, pp. 111, 115–17.

20. Israel, *Dutch Primacy*, pp. 45–48; Davies, *A Primer*, pp. 11–13, 28–32; Mak, *Amsterdam*, p. 117.

21. J. H. Parry, *Age of Reconnaissance* (New York: New American Library, 1963), pp. 203, 204; Prieto, *Mining in the New World*, p. 154; Marques, *History of Portugal*, p. 228; Pierre H. Boulle, "French Mercantilsm," in Blusse and Gaastra, *Companies and Trade*, p. 103.

22. Philip Viereck, *The New Land* (New York: John Day, 1967), pp. 121, 124.

23. Ibid., pp. 122–25, 132.

24. Ibid., pp. 126–29.

25. Ibid., pp. 129–36.

26. Davies, *A Primer*, pp. 136, 137, 144; Emmer, "The West Indies Company," p. 76.

# INDEX